T0298703

Introduction to Photocatalysis

Explore the intriguing world of photocatalysis with *Introduction to Photocatalysis: Fundamentals and Applications*. This book explores the complexities of photocatalytic processes, investigating the contributing elements, nano-photocatalyst manufacturing methodologies, and their wide applications in the energy and environmental sectors.

Additionally, sophisticated modification approaches that may be used to improve the efficiency of visible light–driven processes (such as doping and plasmonics photocatalysis) are discussed. Key features include novel methodologies of photocatalysts, providing an insight into fundamentals and methodology, and examples of efficient applications of photocatalysis such as wastewater treatment, hydrogen production, and CO_2 reduction. Later chapters discuss the commercial aspects of photocatalysis to help guide future entrepreneurs.

The book is useful for advanced undergraduates and graduate students in a range of subjects such as physics, biotechnology, and biochemistry. This book will also prove invaluable for researchers and scientists in photocatalysis, and chemical engineers and chemists in industry R&D working on wastewater treatment and renewable sources of energy. It stands out as a modernised version of current literature that bridges the gap between scholars and students.

Dr Tahir Iqbal Awan has over 20 years of teaching and research experience, and works as Associate Professor in Physics at University of Gujrat (UoG), Punjab, Pakistan. He has significantly contributed to research in the field of Nanoscience and has published more than 200 research articles with an impact factor of over 500 in well-reputed international research journals. He has also received international awards including 'Outstanding Reviewer' Award winner 2019 for *Journal of Physics D: Applied Physics*.

Dr Sumera Afsheen is working as Associate Professor (as an entomologist) at University of Gujrat (UoG), Punjab, Pakistan. She successfully completed three research projects as Principal Investigator (PI)/Co-PI. She is a member of various research/academic statutory bodies both within and outside the university and served as Chairperson-Zoology, Director-UoG City campus, and Warden (Girls Hostel) at University of Gujrat.

Iqra Maryam is one of the most prominent emerging researchers in the field of Nanoscience (photocatalysis, material science) and is presently working as Research Assistant in the NRPU research project at the Department of Physics, University of Gujrat (UoG), Punjab, Pakistan. She significantly contributed to developing a photocatalytic reactor to be used for wastewater treatment and has published various research articles in well-reputed international journals in this area.

Introduction to Photocatalysis
Fundamentals and Applications

Dr Tahir Iqbal Awan, Dr Sumera Afsheen, and
Iqra Maryam

CRC Press
Taylor & Francis Group
Boca Raton London New York

CRC Press is an imprint of the
Taylor & Francis Group, an **informa** business

Designed cover image: New Africa/Shutterstock.com

First edition published 2024
by CRC Press
2385 NW Executive Center Drive, Suite 320, Boca Raton FL 33431

and by CRC Press
4 Park Square, Milton Park, Abingdon, Oxon, OX14 4RN

CRC Press is an imprint of Taylor & Francis Group, LLC

© 2024 Taylor & Francis Group, LLC

ISBN: 978-1-032-51651-6 (hbk)
ISBN: 978-1-032-51670-7 (pbk)
ISBN: 978-1-003-40335-7 (ebk)

DOI: 10.1201/9781003403357

Typeset in Sabon
by Apex CoVantage, LLC

Contents

Acknowledgement

Dr Tahir Iqbal (Principal Investigator) greatly appreciates the assistance of the Higher Education Commission (HEC) through NRPU Project no. 20-14755 for this work.

Part I

Fundamentals

Chapter 1

Introduction to Photocatalysis

ABSTRACT

Photocatalysis is a green process which offers a sustainable solution to many problems. It can be employed for degradation of organic pollutants from contaminated water, hydrogen evolution to solve the energy crisis, CO_2 reduction, antimicrobial activity etc. Nano-photocatalysts have promising characteristics such as high surface area, tunable bandgap, and high absorption of irradiated light. In this chapter, we discuss the history, background theory, and working mechanism of photocatalysis followed by the factors affecting the photocatalytic efficiency and their limitations. A brief literature review about semiconductor metal oxides such as TiO_2, ZnO, CuO, and WO_3 is also given to highlight the latest innovations in the field of photocatalysis for various applications.

1.1 INTRODUCTION

Photocatalysis is a combination of two terms: *photo*, which originated from the word *photon* meaning "light", and *catalyst*, a material which can boost the rate of reaction. A substance that works like a catalyst by absorbing light in a chemical reaction is known as a photocatalyst. Photocatalysts can adjust the rate of reactions in the presence of light. During this phenomenon, when a semiconductor material is exposed to light, a chemical reaction takes place and as a consequence an electron-hole pair is created. These reactions can be divided into two groups according to the characterizing state of the reactants i.e. homogeneous and heterogeneous photocatalysis. In the homogenous photocatalysis, the reactants are in the same phase, whereas for heterogeneous photocatalysis, they are not in the same phase.[1]

In the 1960s, Akira Fujishima and his team began their observations on the photoelectrolysis of water by utilizing *n*-type semiconductors which is TiO_2 electrodes. In 1972, a breakthrough occurred in the field of photocatalysis

DOI: 10.1201/9781003403357-2

3

when Akira Fujishima and Kenichi Honda reported the successful electro-chemical photolysis of water under ultraviolet (UV) light using TiO_2 and platinum electrodes to produce hydrogen gas.[2] It was the first clean and cost-effective method of hydrogen production. Afterwards, in 1977, Nozik reported that photoactivity of electrochemical photolysis could be enhanced by the incorporation of noble metals.[3] Schematic illustrations of the Fujishima-Honda photoelectrochemical (PEC) cell is shown in Figure 1.1. The anode and cathode materials are connected directly without an electric

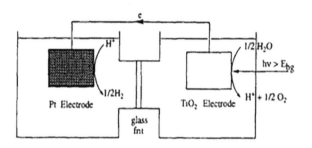

Fujishima-Honda Photoelectrochemical cell

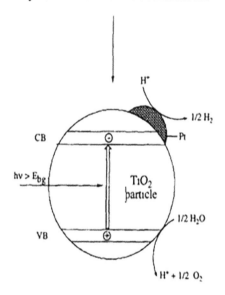

Micro-photoelectrochemical cell

Figure 1.1 The Fujishima-Honda photoelectrochemical cell (top), and a micro version of the cell, for the dissociation of water into H_2 and O_2.

Source: Used with permission of Royal Society of Chemistry from Mills, A., R.H. Davies, and D. Worsley, *Water purification by semiconductor photocatalysis*. Polish Journal of Environmental Studies, 1993. **22**(6): pp. 417–425.

circuit in a water-splitting PEC cell, and a water-splitting reaction on the material is expected because it is accessible, especially in powder form. This composite (metal/semiconductor) structure appears appealing for macro-sized PEC cells. The metal component is known as the "co-catalyst" in this respect, while the semiconductor is referred to as the "photocatalyst".[4]

Photocatalysis research and development, particularly in electrochemical water splitting, is still ongoing, although the method has not yet been commercialized. Researchers are hunting for a stable, cost-effective, active, and energy-efficient material for hydrogen production from water.

Photocatalysis has also been attracting increasing attention to remove pollutants from water. Researchers discovered many types of photocatalysts that could address removal of various contaminants such as insecticides, dyes, fatal inorganic salts, household and commercial wastewater, and heavy metals. Until now, various semiconductor materials have been discovered for photocatalytic water splitting, although many of these materials do not absorb sufficient visible light.[4]

1.2 DIFFERENT TYPES OF MATERIALS AS PHOTOCATALYST

Many factors are involved in material selection for photocatalytic applications. In photocatalysis reactions, both reduction and oxidation take place. Therefore, one must choose a material in which both oxidation and a reduction reaction can occur simultaneously. Based on electronic properties of materials, there are three categories: insulators, conductors, and semiconductors. The insulators have a high bandgap, and they require a high amount of energy for the oxidation and reduction reactions. The higher the bandgap, the more energy will be required for the oxidation and reduction reactions and vice versa. So, by using insulator as a catalyst, one cannot split water molecules as high energy is necessary to initiate the reaction. Therefore, insulators are not suitable as photocatalysts. Figure 1.2 represents the different types of photocatalysts and improvements for photocatalytic activity and CO_2 reduction.

The class of conductors cannot be a suitable photocatalyst due to the fact that their valence band and conduction band overlap. Only free electrons are available, which means an oxidation reaction can occur only if the conductor is used as a catalyst. The necessary condition for a photocatalyst is the simultaneous occurrence of oxidation as well as reduction reaction. Therefore, conductors are also not suitable for photocatalyst. Semiconductors as a photocatalyst are very suitable as they have moderate bandgaps (1.5–3.5 eV), have the properties to do both oxidation and reduction reactions, and their absorption wavelengths fall into the UV-visible region (350–700 nm). Most common semiconductor materials are metal oxides, sulphides, dichalcogenide etc. These materials have high photostability, high light absorption, and tunable bandgap which makes them useful as photocatalyst.

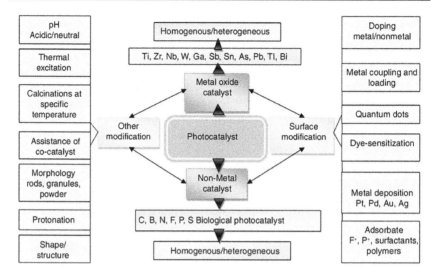

Figure 1.2 Photocatalyst types and suggested adjustments to boost photocatalyst activity and CO$_2$ photoreduction.

Source: Used with permission of Walter de Gruyter and Company from Meryem, S.S., et al., *An overview of the reaction conditions for an efficient photoconversion of CO$_2$*. Reviews in Chemical Engineering, 2018. **34**(3): pp. 409–425.

1.3 BASIC PRINCIPLES AND MECHANISM OF PHOTOCATALYSIS

A photocatalytic reaction principally relies on the light energy of incident radiation and the properties of catalyst used. Mostly, as a photocatalyst, semiconductor materials (including transition metal oxides, sulphides, dichalcogenides etc.) are utilized because their bandgaps help them harness the UV-visible radiation of the solar spectrum. Irradiation of light stimulates the redox (oxidation and reduction) reaction in semiconductors due to which they behave as sensitizers. The working principle is the same for natural and artificial photocatalysis. In natural photocatalysis, solar radiation is used as a primary source of photons, whereas in artificial photocatalysis, UV-visible lamps provide the incident photons. Generally, the photocatalytic activity for most materials is superior under UV-visible light as compared to natural sunlight which attributes to the fact that the solar spectrum only has about 5% UV light and approximately 50% visible light. This is one of the motives for the extensive research on different materials for production of active photocatalysts in the visible region.[5-7] Strong oxidizing agents are required for deterioration of organic contaminants (dyes for example) into less harmful form. In the process of photocatalysis, the free electron pair produced by the absorption of an incident photon reacts with oxygen and water to produce

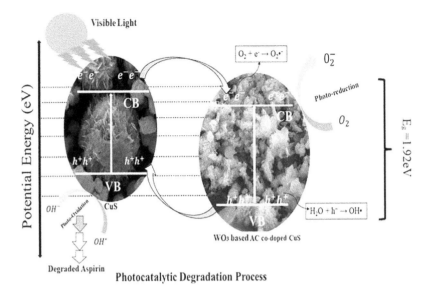

Figure 1.3 The basic principal mechanism of photocatalysis for degradation of aspirin (an organic contaminant) using activated carbon co-doped CuS nanoparticles.

powerful hydro-oxides.[8,9] The basic principal mechanism of photocatalytic degradation of aspirin by activated carbon-based CuS nanoparticles is shown in Figure 1.3.

The basic steps involved in semiconductor photocatalysis are as follows:

i. When the surface of the semiconductor is exposed to light energy i.e. photons, the valence electrons in the semiconductor excite and move towards the conduction band. For this agitation, the incoming photon energy should be similar to the semiconductor's bandgap energy.

ii. Excited valence electrons leave behind holes in the valence band. These holes contribute to oxidizing the donor molecules. Furthermore, hydroxyl ions are formed as a result of the interaction between water molecules with holes. Because of their great oxidizing ability, these hydroxyl ions are responsible for the degradation.

iii. Superoxide ions are also produced due to the reaction of conduction band electrons with the oxygen. These ions are responsible for the oxidation reactions.[8–11]

1.3.1 Oxidation Mechanism

To explain the oxidation mechanism of photocatalysis, let us consider TiO_2 as a photocatalyst. When TiO_2 is exposed to UV light, electron-hole pairs are generated. Due to this exposure of light on photocatalyst, electrons become

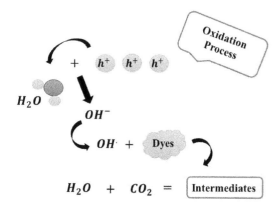

Figure 1.4 Representation of oxidation mechanism in photocatalysis.

energized and move towards the conduction band from the valence band of TiO_2. This happens at the photoexcitation stage of the semiconductor.[8] Also, absorbed water present on the surface of the TiO_2 photocatalyst is oxidized by the positive holes of TiO_2 created in the valence band when negative electrons move towards the conduction band and because of this reaction hydroxyl radicals (*OH*) are produced. These hydroxyl ions further react with organic matter present in dyes and degrade the organic pollutants. Radical chain reactions are initiated due to intermediate radicals in organic compounds if oxygen remains present during the whole process. These radical chains consume oxygen and experience decomposition of organic matter into CO_2 and H_2O,[9,11,12] as shown in Figure 1.4. This results in oxidative decomposition.[1]

1.3.2 Reduction Mechanism

In the reduction mechanism of photocatalysis, oxygen dissolved in air molecules is reduced due to the pairing reactions. These pairing reactions occur as a substitute of hydrogen production because oxygen can be easily reduced. After the excitation of electrons in the conduction band in the presence of photocatalyst TiO_2, these electrons react with the dissolved oxygen present in air to produce superoxide anions. Anions and intermediate products produced in oxidative mechanism become attach and convert into hydrogen peroxide (H_2O_2) and then into water (H_2O). This reduction is preferable in organic matter but not that much in water, and that is why more positive holes are produced for a large number of organic compounds that leads to the formation of reduction in carrier recombination. This results in increasing photocatalytic activity. The reduction mechanism of photocatalysis is shown in Figure 1.5.[9,11]

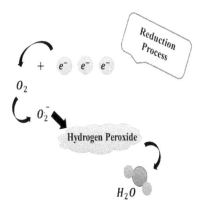

Figure 1.5 Representation of reduction mechanism in photocatalysis.

1.3.3 Factors Affecting Photocatalytic Activity

Photocatalysis is a sophisticated reaction which is governed by several parameters such as the concentration/amount of photocatalyst, pH of the solution, temperature, intensity of light, irradiation time, concentration of dye etc. Concentration of both the catalyst and pollutant affects the degradation/removal percentage of pollutant.

It is quite challenging to evaluate the effects of pH on the photodegradation procedure since there are three different reactions that might be contributing towards the degradation process of contaminants. First is the hydroxyl radical assault, second is the direct oxidation in the presence of positive holes, and third is direction reeducation of contaminant with the conduction band electron. Type of catalyst and pH determine the overall contribution/role of each process in the reaction. Separation of photo-generated charge carriers is affected by the pH of the solution along with the sorption-desorption process on the surface of the photocatalyst. TiO_2 for example, an excellent photocatalyst, is amphoteric in nature. Therefore, it can have either positive or negative charge on the surface. Therefore, change in pH can immensely affect the adsorption of pollutants onto the surface of TiO_2. The rate of photocatalytic degradation of methylene blue (MB) was found to accelerate with rising pH, according to Bubacz et al.[13] Negative TiO_2 electrostatic interactions with the MB cation at basic pH result in substantial adsorption and a correspondingly rapid rate of dye removal. The higher proton concentration in acidic media (pH < 5) inhibits the photodegradation of the dye, resulting in a lower degradation efficiency. The presence of hydroxyl ions, on the other hand, neutralizes the acidic end-products produced by the photodegradation process in alkaline conditions (pH > 10). When the initial pH of the reaction medium was held at 11, a sharp decline in degradation was noticed. Another very important aspect is the nature of dye (one of the most extensively used model pollutants in research) whether it is anionic or cationic. This also has a tremendous effect on the choice of pH for the reaction. Figure 1.6(a, b) depicts the effect of dye concentration and effect of pH for TiO$_2$/alum sludge.[14]

Various photocatalysts have different lattice defects which influences their photocatalytic activity. Additionally, the active surface area and surface

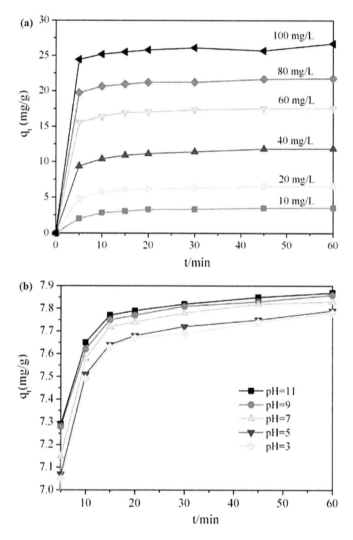

Figure 1.6 Depicting the effect of dye concentration (a) and pH (b) on photocatalytic degradation of TiO$_2$/alum sludge.

Source: Used with permission of Royal Society of Chemistry from Geng, Y., et al., *Study on adsorption of methylene blue by a novel composite material of TiO$_2$ and alum sludge*. RSC Advances, 2018. **8**(57): pp. 32799–32807.

impurities of the catalyst have an impact on the pollutant's adsorption behaviour as well as the lifespan and electron-hole pair recombination. In some photodegradation processes, a high surface area may be crucial because a lot of adsorbed organic molecules speed up the reaction. Intensity of irradiated light is a major factor affecting the catalytic rate constant. It was observed that at lower intensities of light (below 20 mW/cm^2), the photocatalytic degradation rate increases linearly with intensity i.e. first order, whereas for midrange values (about 25 mW/cm^2), the dependence of the rate constant becomes half order. At higher light intensities, the degradation rate becomes independent since the number of sites to absorb photon are not increasing, therefore after saturation, a further increase in intensity does not affect the rate constant.

1.4 LIMITATIONS/SHORTCOMINGS OF PHOTOCATALYSIS

Photocatalysis is a green process. It has many applications such as wastewater treatment for removal of organic pollutants,[15] CO$_2$ reduction,[16] hydrogen production for energy applications,[17] antimicrobial applications[18] etc. There are many difficulties for the industrial application of this technology. For instance, if we discuss water treatment, photocatalysis cannot be employed on a large scale due to the following reasons. First, it is very difficult to separate nano-photocatalysts from water after the treatment is completed. We can use alternative methods such as using beads, electrodes, or any deposit nanolayer of photocatalysts on another bulk surface, but the approach has its own drawbacks. With the reduced surface area and limitation to the availability of proper light, the overall efficiency does not improve. Second, most of the photocatalysts are active in the UV-visible region which hinders the full potential of the material. Therefore, the efficiency of most photocatalysts in their pure state is not very impressive. Third, the high photoelectron/hole recombination rate is another factor for low efficiency. Overcoming the low efficiency problem is a viable solution because if the efficiency of the nano-photocatalyst is improved tremendously, the large-scale application of photocatalysis will become possible. Doping the photocatalysts with metal or non-metals, co-doping, and coupling with other semiconductor materials are some of the methods adopted to cover these deficiencies. The goal of incorporation is the shifting of the bandgap of the photocatalyst towards the visible region so that the energy of the solar spectrum can be used more efficiently. In most cases, it is observed that doping or composition reduces the electron-hole recombination rate of many photocatalysts which is a positive improvement for photocatalysis. Moreover, sometimes, the size of nano-photocatalyst is also reduced along with the enhanced morphological properties such as production of new defects which provides active sites for the photocatalytic reaction.

1.5 LITERATURE REVIEW

1.5.1 Titanium Dioxide

TiO_2 is among the most researched photocatalysts for the last few decades because of its wide bandgap, eco-friendly nature, and high stability. It is used for treatment of wastewater to remove harmful organic dyes.[19] The versatile properties of TiO_2 make it a promising material for different applications like treatment of wastewater, CO_2 reduction, hydrogen production, shelf-life applications etc. TiO_2 poses different structures, wide bandgap, good texture, large amount of binding energy, low cost, and high-speed surface activation.[20] With alteration in size at the nano level, different types of defects can be produced in the semiconductor materials. Therefore, the desirable, distinct, and novel properties can be achieved for different applications.

The ability of TiO_2 to harvest solar light is better as contrasted to many other semiconductor materials. Tunable bandgap and high photo and chemical stability make it a suitable photocatalyst.[21] The bandgap, size, shape, phase purity, method of preparation, and surface area of the semiconductor play a crucial role in photocatalytic activity. Therefore, researchers are investigating efficient methods for optimizing the photocatalytic activity of TiO_2.[22] Recent advancements in photocatalytic activity of TiO_2 are mentioned later.

Improvements in the material by doping with metals/non-metals have significant impact in changing its bandgap towards desired values. Metals/non-metals doping improves the electronic structure of the material, enlarges the visible solar spectrum absorption range, and increases the stability and charge mobility of the material. It also reduces the nanoparticles size, increases the surface area, and creates defects which enhance the photocatalytic activity.[23] Matiullah Khan et al. reported Ag and V co-doped TiO_2 photocatalyst using a hydrothermal method. There was no calcination process after the hydrothermal treatment. The aim of this research was to study the enhanced photocatalytic properties of TiO_2. Thus, different characterization techniques such as X-ray diffraction (XRD), scanning electron microscopy (SEM), transmission electron microscopy (TEM), UV-visible spectroscopy, and X-ray photoelectron spectroscopy (XPS) were used. The activity of the photocatalyst was examined. The co-doping of V and Ag in TiO_2 decreased the crystal's size, providing more specific area for reaction, and enhanced the light absorption rate of the photocatalyst, improving the tendency to separate the photo-generated charge carriers. Due to all these factors, the photocatalytic degradation of MB dye was enhanced. Moreover, the reusability experiment revealed the improved stability of Ag-doped TiO_2.[24]

Zhang et al. reported the nanorods of Au-doped TiO_2 by seed-mediated method and used it for the degradation of rhodamine B (RhB) dye. The visible absorption spectrum of the Au-doped TiO_2 samples increased due to modified bandgap which is a good sign for photocatalytic activity. The movement of electrons from valance to conduction band and production of oxygen radicals was also enhanced for Au-doped TiO_2. The recombination rate of holes and electrons is reduced with doping which is due to the production of Schottky barriers.[25] Wu et al. published their work about the nanowires of Cu-doped TiO_2 using the solvothermal method to degrade the methyl orange (MO) dye. The ionic radius of copper is comparable to titanium ion; therefore, it easily substitutes into the crystal structure of TiO_2 to change its optical properties. It was noted that the surface area of nanomaterial is increased after doping which has shown to be an effective contribution for the dye's degradation. Moreover, the absorption in the visible region was increased and the recombination rate was reduced as compared to pure TiO_2 for better photocatalytic efficiency.[26,36]

Many researchers have shown that co-doping different materials is a competent way to increase photocatalytic efficiency. M. Khairy et al. in 2022 published their work about the nanoparticles of Ag and Au co-doped TiO_2 photocatalyst embedded in rGO using the precipitation decomposition method.[69] The nanomaterial was utilized to degrade the MB dye when the solution was irradiated by visible light. The results of XRD, energy-dispersive X-ray spectroscopy (EDX), SEM, and TEM proved the successful addition of Au and Ag in TiO_2. Many unique properties were observed which played a significant role in enhancing the photocatalytic activity of the samples. In the Au and Ag co-doped TiO_2 photocatalyst, the Ag inhibits the recombination rate of electrons and holes produced on the surface of semiconductor by light. The electrons were trapped by Ag. The hydroxyl radicals were formed by the reaction of a trapped electron with water, and the degradation of dye took place. The Au increases the charge mobility and also produced the barriers to enlarge the charge separation. The stability of the photocatalyst also improved because of bimetallic combination. In the comparison with these two doped materials i.e. Ag-doped TiO_2 and Au-doped TiO_2 photocatalysts, the Au and Ag co-doped TiO_2 photocatalyst posed the greatest efficiency for remediation of MB dye.[27]

Khalid et al. has reported the nanoparticles of Ag-loaded N-TiO_2 using a sol-gel method.[28] The samples were utilized to investigate the CO_2 reduction into hydrocarbons under UV-visible light at 288°K. XRD, XPS, and UV-visible analyses confirmed the preparation of N-Ag co-doped TiO_2 photocatalysts. Brunauer-Emmett-Teller (BET) results revealed the enhancement in surface area of the modified nanomaterial. The value for the bandgap of the N-Ag co-doped TiO_2 was reduced, enhancing its visible light absorption. The study of photocatalytic activity showed that the efficiency of N, Ag-doped TiO_2 is higher as compared to undoped TiO_2. Shaban et al. published their work regarding the Ni/Cr co-doped TiO_2 prepared by hydrothermal method. The nanoparticles were employed for the degradation of MB dye under visible light. Various characterization techniques such as XRD, Raman spectroscopy, dynamic light scattering (DLS), SEM, and BET were utilized to investigate the optical, morphological, and structural properties. The bandgap of co-doped samples was reported to be 2.45 eV. In 1.5 hours, 96% degradation of MB was reported.[29]

The next way to modify the structure as well as properties of TiO_2 is the coupling of TiO_2 with other semiconductors. It is a promising approach to polish the different properties of the materials and to overcome the flaws. Different researchers have proved that the modification of TiO_2 with other semiconductors results in the enhancement of photocatalytic performance.[30] Dong et al. and Jin et al., in the year 2015, fabricated the nanoparticles of Ag_3PO_4-TiO_2 composite using in situ precipitation technique and ball milling technique, respectively.[70,71] To check the photocatalytic efficiency/performance of the prepared samples, RhB under visible light was used. The edge of the conduction band and the valance band edge have mostly negative and a positive value of TiO_2 and Ag_3PO_4, respectively, because electrons travel from TiO_2 to Ag_3PO_4 and generate more electrons, and photocorrosion is also restricted. These factors play a crucial role in photocatalytic performance. The Ag_3PO_4-TiO_2 composite harvests the greater portion of the solar spectrum than pure TiO_2 and Ag_3PO_4 due to the large number of electron-hole pairs generated which further produces the hydroxyl radical with the reaction of water, and these hydroxyl radicals react with RhB dye and degrade them. In the Ag_3PO_4-TiO_2 composite, the charge mobility increased, and the surface area of the samples also increased. It was proved that Ag_3PO_4-TiO_2 has better performance than pure TiO_2.[31]

Pant et al., in 2013 and M. Pérez-González et al. in 2019, prepared the Ag-loaded zinc oxide (ZnO)/RGO and Ag doped ZnO/TiO_2 flower-shaped nanoparticles by hydrothermal and solgel method to test the photocatalytic activity.[72,73] The photoluminescence (PL) result showed that Ag-doped ZnO-TiO_2 composites have minimum intensity which means that in the composite the recombination rate reduced and there was excellent charge mobility. The stability of the photocatalyst also improved due to Ag loading on the ZnO-TiO_2. It was noted that the Ag-loaded ZnO-TiO_2 posed better results for dyes degradation than pure ZnO.[32] Ganesh et al. reported the flower-shaped nanoparticles of Cadmium Sulfide (CdS-TiO_2) composites for RhB, the solution was irradiated with the light of a halogen lamp.[74] The XRD, EDX, TEM, and SEM were the characterization techniques used for the confirmation of doping, purity, size, and shapes of the samples. The results demonstrate that the CdS doping into TiO_2 increases the separation/distance between charges and also increases the charge mobility. A number of defects, hydroxyl radicals, and superoxide ions were produced which plays a vital role in photocatalytic performance. The photocatalytic activity showed that CdS-TiO_2 composites are promising and a potential candidate for the dye's degradation.[33]

Paušová et al. in 2019 reported activated carbon–based TiO_2 composite materials using a commercial active carbon. The hydrothermal method was used to prepare the composites. BET analysis revealed that the prepared composites have high surface area for better adsorption of azo-dye.[34] Another study of activated carbon–based TiO_2 by Gu et al.[35] reported the removal of ibuprofen at room temperature under 15W UV light for 4 hours. The effect of the initial concentration of contaminant and pH was also investigated. A removal of 92% was observed.

1.5.2 Tungsten Trioxide

Shahmoradi and Byrappa synthesize WO_3-doped TiO_2 by using the hydrothermal method of preparation of nanoparticles. They prepared a composite of WO_3 with TiO_2.[75] The degradation rate was checked for Amaranth and Brilliant Blue for colouring food (FCF) dyes. The photocatalytic results have shown that tungsten-doped TiO_2 has higher photocatalytic efficiency than pure TiO_2 i.e., WO_3-doped TiO_2 showed 93.25% and 74.39% photocatalytic efficiency for Amaranth and Brilliant Blue FCF dyes, respectively, in sunlight irradiation. Also, under UV light irradiation, the performance was 70.14% and 42.90%, respectively.[36]

Li et al. reported the preparation method for WO_3/TiO_2 photocatalyst.[76] The catalyst was prepared by supercritical pretreatment. The photocatalyst structure was analyzed by XRD, SEM, diffuse reflectance spectroscopy (DRS), and also by PL. The results of these prepared photocatalysts showed that WO_3 and Cr doping in well-known photocatalyst TiO_2 results in the increase of crystal surface area. This doping results in the decrease of bandgap. Due to the decrease in the bandgap of TiO_2, the optical response shifted from UV to visible region. Photocatalytic efficiency was tested on RhB dye.[37]

Seung Yong Chai et al. synthesized WO_3 composite with TiO_2.[77] They made a modification onto the surface of the TiO_2. This modification activates the WO_3/TiO_2 nano-photocatalyst. This modification ultimately results in the increase of photocatalytic activity/degradation of toxic dye i.e., gaseous 2-propanol. Crystal structure analyses were made by XRD, SEM, TEM, Raman spectra and also by UV-visible. It has been reported that WO_3 with molar ratio 10%, calcinated for 700°C gives maximum activity performance. And 10 mol% of WO_3 shows 20 times enhanced photocatalytic activity as compared to pure TiO_2.[38]

Fan et al. synthesized TiO_2 nanocomposite with Cr.[78] This photocatalyst was prepared through self-evaporation. To characterize the prepared, the photocatalyst XRD, XPS, TEM, and UV-visible techniques were applied. The Cr^{3+} doping concentration of ions was varied from 0.1 to 1 mol%. Characterization results tell us that the size of the prepared photocatalyst is about 8 nm. By comparing the pure and Cr-doped TiO_2, it has been revealed that doping increases the photo absorption rate especially under $\lambda > 460$ nm and towards the visible region.[39] Pan and Wu prepared a co-doped composite of TiO_2 with Cr and N.[79] For this purpose, the sol-gel route was adopted. This particular doping creates an accumulation of Cr and N ions in titania (TiO_2) by XRD technique. To check the photocatalytic activity of the co-doped photocatalyst, MB was employed. Due to extra imperfections of TiO_2 during preparation, the efficiency of the contamination removal was slowed.[40]

It is reported that the H_2 can be generated, and MO can be degraded from the one-dimensional (1D) WO_3 nanotubes of dimensions 85 nm in diameter and 250 nm in length using the water-splitting process called the PEC process. In this method, the sample is solar illuminated. Production of H_2 depends on the exposure time i.e. it increases by increasing the exposure time. Due to the well-optimized functional and stable physiochemical properties of WO_3 film with bandgap (2.4–2.8 eV), it has a highly active surface area that is useful for electrochromic layers and applications for the production of H_2.[41–43] Moreover, hydrogen can also be produced by doped tungsten trioxide (WO_3) with graphene oxide. This is done by the hydrothermal method in which synthesized WO_3/rGO composite (surface area 62.03 m²/g) is used for the sonocatalytic degradation of Congo red and production of H_2. XPS and Raman analysis confirm the formation of the required composite. XRD is used for characterization of the prepared samples. In this method, degradation efficiency is 56% to 94%, and hydrogen can be produced from 424.5 up to 825.8 μ mol/h.g.[44]

For the evolution of H_2, carbon nanotube (CN) and WO_3 composite has been utilized. To do photocatalysis for the production of H_2, a dispersion method is used for the preparation of composites of mesoporous CN/WO_3. This technique is more efficient because of twofold behaviour due to an increase in the surface area and the induction of solid-state Z-scheme (for charge separation).[45,46] Doped WO_3 with copper is highly efficient for the photocatalytic activity. First, the hydrothermal method is used to prepare the hybrid nanostructures of the $Cu_3V_2O_8$-WO_3. As a precursor, $Na_2WO_4.2H_2O$ and copper acetate are used. For efficient evolution of hydrogen, the quantities of hybrid $Cu_3V_2O_8$ are 0.2%, 0.5%, 1.0%, 2.0%, and 3.0% in WO_3 (10–15 nm or 70–195 nm is required). This technique is well oriented for charge separation and chemical energy applications due to enhanced evolution of hydrogen.[47] Tahir et al.[48] published their work on hydrothermal synthesis of activated carbon-based WO_3 for removal of RhB dye. Various techniques were employed to study the optical, structural, and morphological properties. The best photocatalytic removal of RhB dye was observed by 2% AC-WO_3 under UV-visible light irradiation.

CO_2 is one of the fundamental compounds contributing to climate change. Due to the exponential increase in CO_2 concentration caused by fossil fuel usage, it becomes a major environmental problem. Semiconductor photocatalysts have an ability to convert CO_2 into hydrocarbons by utilizing water like an electron donor.[49]

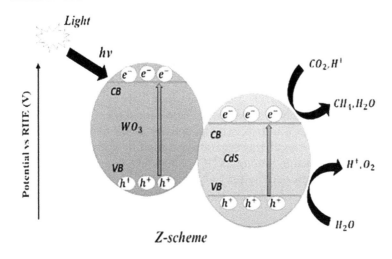

Figure 1.7 Schematic of the charge transfer and separation in the CdS-WO₃ Z-scheme photocatalytic system under visible light.

The majority of the catalysts used for photocatalytic CO_2 reduction have larger bandgap semiconductors. They can also produce sufficient negative electrons for CO_2 reduction but still work in the presence of UV light, such as TiO_2, $SrTiO_3$, ZnS, and others. For photocatalytic CO_2 reduction to solar fuels, we use the hydrothermal method in which mesoporous WO_3 was prepared that was used as a catalytic activity. Through the solid-liquid phase arc discharge technique, we synthesized extremely thin, single-crystal WO_3 nanosheets in an aqueous solution which shows improvement in photocatalytic CO_2 reduction.

CdS nanoparticles, having small granules, cover a portion of the surface of WO_3 nanosheets in a CdS-WO_3 heterostructure having appropriate CdS content, forming a CdS-WO_3 Z-scheme photocatalytic system (Figure 1.7). Photo-induced electron-hole pairs are formed in WO_3 and CdS when exposed to visible light.

Photo-induced holes are expected to stay in the valence band of WO_3, while electrons are transported to the valence band of CdS by ohmic contact. The enhanced electrons on the surface of CdS are expected to facilitate CO_2 reduction and the production of CH_4. Figure 1.7 indicates that Z-scheme heterostructure enhances photocatalytic CO_2 reduction to form CH_4. CdS-WO_3 heterostructure samples had a greater CO_2 adsorption capacity than pure WO_3 hollow spherical samples because of their larger specific surface areas. By increasing efficient interaction between CO_2 and CdS, such a form of adsorption may deliver high CO_2 amounts to CdS nanoparticles. A concentrated CO_2 molecule on a surface of the catalyst can accelerate a reaction that generates electrons, thus increasing photocatalytic CO_2 reduction.[50] Two-dimensional (2D) WO_3 nanosheets were successfully synthesized and utilized for photocatalytic CO_2 reduction which have a great amount of 2D $LaTiO_2N$ photocatalyst. As a zeolite semiconductor, $LaTiO_2N$ has a wide bandgap (2.14 eV) and so exhibits activity when exposed with visible light.[51] The increased charge separation caused by the development of a Z-scheme heterojunction among $LaTiO_2N$ and WO_3 may account for the enhanced photocatalytic activity for CO_2 reduction.

1.5.3 Zinc Oxide

Fu and Fu reported ZnO nanoparticles synthesized using biosynthesis and a hydrothermal approach.[52] A comparative study has been done for photocatalytic application against MB dyes. Nanoparticles have sizes between 50 nm and 180 nm. They have rodlike shape, and bandgap is about 3.07 eV. Photocatalytic activity of synthesized samples was investigated against MB dye for 180 minutes under UV-visible lamp. Biosynthesis sample has shown maximum photocatalytic activity 92.45%, whereas dye removal for the hydrothermal sample was 67.47%. Babitha et al. (2019)[53] reported microwave heating and sol-gel method for synthesizing ZnO nanoparticles and nanosheets, respectively. Photocatalytic application of both samples was investigated. An organic contaminant, MB dye was taken as the model pollutant. UV-visible analysis revealed the bandgaps of 3.42 eV and 3.23 eV for ZnO nanoparticles and nanosheets, respectively. Optical, physical, and chemical properties were examined from XRD, SEM, EDX, DRS, and PL techniques. ZnO nanoparticles have maximum photocatalytic degradation ability as compare to ZnO nanosheets due to small size and have high surface-to-volume ratio.

Khalaf et al., in 2019, published a biosynthesis method of ZnO nanoparticles using microalgae *Chlorella* extract. Synthesized material has shown an absorbance peak at 362 nm in UV-visible spectrometer and average 20 nm size, highly crystalline nature, and a hexagonal wurtzite structure was confirmed by SEM and XRD characterization techniques. Synthesized nanoparticles were used to treat dibenzothiophene (DBT) organic pollutant under natural sunlight. Removal of 97% was observed which was reduced to 93% after five cycles exhibiting high stability of the material.[54]

Sahoo et al. (2019) reported a physical method by mechanical grinding for the synthesis of ZnO nanoparticles. In this work, properties and catalytic activity of synthesized ZnO were compared with an ungrinded and standard sample of ZnO. In the presence of UV light, degradation of MB dye was investigated. Synthesized nanoparticles have average size and diameter of

300 nm and 59.34 nm, respectively, with the rate constant of 0.18 min^{-1} and 0.06 min^{-1}. ZnO nanoparticles synthesized in the lab have extremely low cost and are suitable for photocatalytic degradation of dye from polluted water.[55] Another research group, Kumar et al., reported a chemical method for ZnO synthesis. They examined the influence of ZnO in bulk and in the form of nanoparticles for water treatment against organic pollutant dyes. These chemically synthesized nanoparticles have sizes in the range of 36 to 63 nm, and bandgap is about 3.20 eV. After 5 hours, ZnO nanoparticles degraded 90.7% of pollutant dye, whereas the removal percentage was 79.35% for bulk ZnO under natural light irradiation. Kinetic study revealed the rate constant of 0.018 min^{-1} for ZnO nanoparticles. The increased difference between the rate constants of ZnO nanoparticles reported in this work as compared to the work of Sahoo et al. (2019) can be attributed to the different light sources used which revealed that degradation efficiency or removal percentage for ZnO nanoparticles is superior under UV light as compared to natural sunlight.[56]

Bomila et al., in 2019, published a comparative study using pure and dual doped (Ce-La, La-Gd, and Gd-Ce) ZnO nanoparticles. They synthesized ZnO nanoparticles by a wet chemical method. The fabricated material had a hexagonal wurtzite crystal structure with size 40 nm, 14 nm, 16 nm, and 28 nm, respectively, for pure ZnO, Ce-La, La-Gd, and Gd-Ce dual doped nanoparticles. The lowest bandgap of 2.81 eV was observed for Ce-La doped ZnO under natural light irradiation.[57] Singh et al. (2019) reported their work on the synthesis of pure and Mn-doped ZnO nanoparticles. Pure ZnO nanoparticles' bandgap was found to be 3.31 eV. They have observed that variation in the concentration of Mn (1%, 2%, 3%, and 4%) has increased the bandgap from 3.41 to 3.51 eV. A strong green emission peak at 545 nm was observed for pure ZnO which was shifted towards the weak green region at 565 nm for 4% Mn-ZnO sample. Morphology and size were also affected by doping. Pure ZnO nanoparticles had pyramid structures with average size from 30 to 45 nm, while doped ZnO nanoparticles are spherical structures of 15 to 30 nm. These samples were used for degradation of different dyes such as MB, MO, and Congo red. Maximum degradation efficiency of 88%, 93.5%, and 93% for MB, MO, and Congo red, respectively, was observed for 4% Mn-doped sample due to its wide bandgap.[58]

1.5.4 Copper Oxide

CuO is a p-type semiconductor. It has a narrow bandgap of 1.2 to 1.51 eV. It is considered to be a good photocatalyst because of its narrow bandgap, good stability, low cost, easy manufacturing process, high absorption, good conducting properties, non-toxic nature and so on. From the narrow bandgap of material, we expected a high conversion efficiency. However, some groups suggested that CuO shows very little photocatalytic properties under visible light, but with the help of a small amount of H_2O_2, the efficiency of catalytic activity can be greatly enhanced.[59] Degradation of brominated flame retardants using CuO with H_2O_2, reported by Vinothkumar et al., has shown an increased activity of the material.[60] Interaction of CuO with hydrogen peroxide produces electron spin resonance spectrum just like Cu^{+2} which may attribute to the changed electronic configuration of CuO; therefore, H_2O_2 might behave as a co-catalyst for activating CuO for photocatalysis. Photocatalytic activity depends on the shape and size of the nanostructure. This explains the different photocatalytic activity for different nanostructures. Umadevi et al. reported that the solution combustion method produces nanoflowers of CuO using glycine as fuel. The prepared material has a monoclinic crystalline structure. The bandgap of the synthesized material was reported to be 3 eV, and its photocatalytic activity was measured to be 0.016/min.[61]

Scuderi et al. reported nanowires of CuO and Cu_2O prepared by hydrothermal process. SEM images show a mean length of approximately 80 nm length. Photocatalytic activity of MB was studied with the synthesized material. Results showed that the catalytic activity of CuO nanowires is much higher than Cu_2O nanowires. Variation in catalytic activity is due to thermal defects.[62] Iqbal et al. in 2022 reported the work on pure and La-doped CuO nanoparticles via a green approach using leaf extract of *Citrus medica* Linn. for degradation of MB dye in the presence of solar light.[64] XRD, EDX, and SEM were used to study the structural and morphological properties of samples, whereas UV-vis and PL spectrometry were utilized to investigate the optical properties. The bandgap observed for pure CuO was 3.03 eV which reduced to 2.71 eV for 2% La-doped sample. Particle size showed a decreasing trend with an increase in doping concentration (40.82–31.89 nm). The best photocatalytic activity of 84% was observed for 2% doped sample due to smaller size, tuned bandgap, and suppressed electron-hole recombination.[15] Konar et al. reported an aqueous-based chemical precipitation method to prepare different morphologies of CuO nanoparticles. Flower-shaped, rodlike, spherical, star-shaped nanostructures were synthesized. The catalytic activity of all shapes was compared. Synthesized nanoparticles are used to reduce 4-nitro phenol into 4-amino phenol. The activity of star-shaped nanoparticles was observed to be the highest.[63]

A chemical method was used to prepare CuO and Mn-doped CuO in different concentrations. The bandgap emission peak was observed at 424 nm. The bandgap of CuO nanowires was calculated using UV-vis spectrometer, and it is about 2.6 eV which is higher than the bandgap of bulk copper oxide 1.85 eV. Cd-doped CuO was prepared using a hydrothermal method.[65] Field emission scanning electron microscopy (FESEM) and EDX were used to study the morphology of the fabricated material. It showed that Cd does not have much influence on the surface morphology of CuO. Catalytic degradation of MB was observed. An optical bandgap of 1.56 eV was reported.

Using Cd-doped CuO, 92.33% MB was degraded. The catalytic efficiency of CuO is enhanced due to lattice expansion which increases charge mobility. La-doped nanosheets of CuO were prepared using the electrodeposition method.[66] FESEM, XRD, and EDX were used to study structural and morphological properties of prepared nanomaterial. EDX showed that La substituted the Cu atom without affecting the geometry of the structure. SEM showed a petal-like CuO nanostructure. The optical bandgap was between 1.1 and 1.3 eV. As compared to undoped CuO, 11.5% better photocatalytic degradation of La-doped CuO was reported. Sellaiyan et al. reported a solution combustion method to prepare pure and Ta-doped CuO nanoparticles.[67] The effect of annealing temperature on optical and morphological properties was studied. XPS analysis shows that the valence state of Cu, O, and Tb exists on the surface of particles. Photocatalytic degradation of MO and MB was carried out. After 2 hours, pure CuO showed degradation of 85%, and the doped material annealed at 600°C showed degradation of 92%.[68] Chen et al. reported Ce-doped CuO nanoparticles synthesized using a hydrothermal method. XRD and EDX showed that a very small amount of Ce ions has substituted the Cu ions. UV-visible spectroscopy showed that absorption of Ce-doped CuO is much higher than CuO nanoparticles. For different concentrations, different bandgaps like 2.07 eV, 1.93 eV, 2.03 eV, and 2.04 eV were observed. Photocatalytic degradation of MB shows that Ce-doped CuO has much better degradation efficiency as compared to pure CuO. The efficiency of doped material has reached up to 90.04%.[69]

1.6 SUMMARY

Photocatalysis is a redox reaction which can initiate reactions with high activation energy when exposed to light of suitable frequency and intensity. Major applications of photocatalysis include wastewater treatment for dye degradation, CO_2 reduction for air purification, hydrogen production for dealing with the energy crisis, and antimicrobial activity for medical applications. Basic parameters which immensely affect the capabilities of photocatalyst are pH, temperature, amount of catalyst and dye, morphology, and bandgap. Generally, pH of 6 to 9 is found suitable for most of the photocatalytic processes. UV-visible light is suitable for most semiconductor metal oxide photocatalysts. Researchers are working on the precise tuning of bandgap to harness the solar spectrum more efficiently. The intensity of light also plays a significant role in catalytic efficiency. The factors hindering the industrial-scale application of photocatalysis (specifically hydrogen production and water treatment) include the difficulty of removing nanoparticles from wastewater after treatment, low efficiency of reactors, along with high electron-hole pair recombination rate and poor absorption of visible light for most semiconductor photocatalysts.

REFERENCES

1. Saravanan, R., F. Gracia, and A. Stephen, *Basic principles, mechanism, and challenges of photocatalysis.* In *Nanocomposites for Visible Light-Induced Photocatalysis* (pp. 19–40). Springer, 2017.
2. Nakata, K., and A. Fujishima, *TiO_2 photocatalysis: Design and applications.* Journal of Photochemistry and Photobiology C: Photochemistry Reviews, 2012. **13**(3): pp. 169–189.
3. Nozik, A.J., *Photochemical diodes.* Applied Physics Letters, 1977. **30**(11): pp. 567–569.
4. Maeda, K., *Photocatalytic water splitting using semiconductor particles: History and recent developments.* Journal of Photochemistry and Photobiology C: Photochemistry Reviews, 2011. **12**(4): pp. 237–268.
5. Mandade, P., *Introduction, basic principles, mechanism, and challenges of photocatalysis.* In *Handbook of Nanomaterials for Wastewater Treatment* (pp. 137–154). Elsevier, 2021.
6. Saravanan, R., et al., *ZnO/Ag/CdO nanocomposite for visible light-induced photocatalytic degradation of industrial textile effluents.* Journal of Colloid and Interface Science, 2015. **452**: pp. 126–133.
7. Gust, D., T.A. Moore, and A.L. Moore, *Realizing artificial photosynthesis.* Faraday Discussions, 2012. **155**: pp. 9–26.
8. Hagen, J., *Industrial Catalysis: A Practical Approach.* John Wiley & Sons, 2015.
9. Khan, M.M., S.F. Adil, and A. Al-Mayouf, *Metal Oxides as Photocatalysts* (pp. 462–464). Elsevier, 2015.
10. Rehman, S., et al., *Strategies of making TiO_2 and ZnO visible light active.* Journal of Hazardous Materials, 2009. **170**(2–3): pp. 560–569.
11. Rajeshwar, K., et al., *Heterogeneous photocatalytic treatment of organic dyes in air and aqueous media.* Journal of Photochemistry and Photobiology C: Photochemistry Reviews, 2008. **9**(4): pp. 171–192.
12. Qu, X., P.J. Alvarez, and Q. Li, *Applications of nanotechnology in water and wastewater treatment.* Water Research, 2013. **47**(12): pp. 3931–3946.
13. Bubacz, K., et al., *Methylene blue and phenol photocatalytic degradation on nanoparticles of anatase TiO_2.* Polish Journal of Environmental Studies, 2010. **19**(4): pp. 685–691.
14. Geng, Y., et al., *Study on adsorption of methylene blue by a novel composite material of TiO_2 and alum sludge.* RSC Advances, 2018. **8**(57): pp. 32799–32807.
15. Iqbal, T., et al., *Green synthesis of novel lanthanum doped copper oxide nanoparticles for photocatalytic application: Correlation between experiment and COMSOL simulation.* Ceramics International, 2022. **48**(10): pp. 13420–13430.
16. Fu, J., et al., *Product selectivity of photocatalytic CO_2 reduction reactions.* Materials Today, 2020. **32**: pp. 222–243.
17. Kumaravel, V., et al., *Photocatalytic hydrogen production using metal doped TiO_2: A review of recent advances.* Applied Catalysis B: Environmental, 2019. **244**: pp. 1021–1064.
18. Chauhan, A., et al., *Photocatalytic dye degradation and antimicrobial activities of pure and Ag-doped ZnO using cannabis sativa leaf extract.* Scientific Reports, 2020. **10**(1): pp. 1–16.

19. Martins, A.S., L. Nuñez, and M.R. de Vasconcelos Lanza, *Enhanced photoelectrocatalytic performance of TiO_2 nanotube array modified with WO_3 applied to the degradation of the endocrine disruptor propyl paraben.* Journal of Electroanalytical Chemistry, 2017. **802**: pp. 33–39.

20. Ramos-Delgado, N., et al., *Solar photocatalytic activity of TiO_2 modified with WO_3 on the degradation of an organophosphorus pesticide.* Journal of Hazardous Materials, 2013. **263**: pp. 36–44.

21. Bhanvase, B., T. Shende, and S.H. Sonawane, *A review on graphene–TiO_2 and doped graphene–TiO_2 nanocomposite photocatalyst for water and wastewater treatment.* Environmental Technology Reviews, 2017. **6**(1): pp. 1–14.

22. Zhang, Y., et al., *WO3 modification of MnO_x/TiO_2 catalysts for low temperature selective catalytic reduction of NO with ammonia.* Chinese Journal of Catalysis, 2012. **33**(9–10): pp. 1523–1531.

23. Peiris, S., et al., *Recent development and future prospects of TiO_2 photocatalysis.* Journal of the Chinese Chemical Society, 2021. **68**(5): pp. 738–769.

24. Khan, M., et al., *Visible-light-active silver-, vanadium-codoped TiO_2 with improved photocatalytic activity.* Journal of Materials Science, 2017. **52**(10): pp. 5634–5640.

25. Zhang, Y., et al., *Non-uniform doping outperforms uniform doping for enhancing the photocatalytic efficiency of Au-doped TiO_2 nanotubes in organic dye degradation.* Ceramics International, 2017. **43**(12): pp. 9053–9059.

26. Wang, M., et al., *Inorganic-modified semiconductor TiO_2 nanotube arrays for photocatalysis.* Energy & Environmental Science, 2014. **7**(7): pp. 2182–2202.

27. Yang, L., et al., *Photocatalytic reduction of Cr (VI) on WO_3 doped long TiO_2 nanotube arrays in the presence of citric acid.* Applied Catalysis B: Environmental, 2010. **94**(1–2): pp. 142–149.

28. Khalid, N., et al., *Highly visible light responsive metal loaded N/TiO_2 nanoparticles for photocatalytic conversion of CO_2 into methane.* Ceramics International, 2017. **43**(9): pp. 6771–6777.

29. Shaban, M., et al., *Ni-doped and Ni/Cr co-doped TiO_2 nanotubes for enhancement of photocatalytic degradation of methylene blue.* Journal of Colloid and Interface Science, 2019. **555**: pp. 31–41.

30. Klimczak, M., et al., *High-throughput study of the effects of inorganic additives and poisons on NH3-SCR catalysts—part I: V_2O_5–WO_3/TiO_2 catalysts.* Applied Catalysis B: Environmental, 2010. **95**(1–2): pp. 39–47.

31. Chen, M., et al., *Effect of Ce doping into V_2O_5-WO_3/TiO_2 catalysts on the selective catalytic reduction of NO_x by NH_3.* Journal of Rare Earths, 2017. **35**(12): pp. 1206–1215.

32. Mu, S., et al., *Surface modification of TiO_2 nanoparticles with a C60 derivative and enhanced photocatalytic activity for the reduction of aqueous Cr (VI) ions.* Catalysis Communications, 2010. **11**(8): pp. 741–744.

33. Dinari, M., and A. Haghighi, *Surface modification of TiO_2 nanoparticle by three-dimensional silane coupling agent and preparation of polyamide/ modified-TiO_2 nanocomposites for removal of Cr (VI) from aqueous solutions.* Progress in Organic Coatings, 2017. **110**: pp. 24–34.

34. Paušová, Š., et al., *Composite materials based on active carbon/TiO_2 for photocatalytic water purification.* Catalysis Today, 2019. **328**: pp. 178–182.

35. Gu, Y., et al., *Adsorption and photocatalytic removal of Ibuprofen by activated carbon impregnated with TiO_2 by UV-Vis monitoring.* Chemosphere, 2019. **217**: pp. 724–731.

36. Wu, M., et al., *Surface modification of TiO_2 nanotube arrays with metal copper particle for high efficient photocatalytic reduction of Cr (VI).* Desalination and Water Treatment, 2016. **57**(23): pp. 10790–10801.

37. Acharya, R., B. Naik, and K. Parida, *Cr (VI) remediation from aqueous environment through modified-TiO_2-mediated photocatalytic reduction.* Beilstein Journal of Nanotechnology, 2018. **9**(1): pp. 1448–1470.

38. Wang, N., et al., *Visible light photocatalytic reduction of Cr (VI) on TiO_2 in situ modified with small molecular weight organic acids.* Applied Catalysis B: Environmental, 2010. **95**(3–4): pp. 400–407.

39. Al-Awadi, A.S., et al., *Role of TiO_2 nanoparticle modification of Cr/MCM41 catalyst to enhance Cr-support interaction for oxidative dehydrogenation of ethane with carbon dioxide.* Applied Catalysis A: General, 2019. **584**: p. 117114.

40. Nasirian, M., and M. Mehrvar, *Modification of TiO_2 to enhance photocatalytic degradation of organics in aqueous solutions.* Journal of Environmental Chemical Engineering, 2016. **4**(4): pp. 4072–4082.

41. Lai, C.W., *Photocatalysis and photoelectrochemical properties of tungsten trioxide nanostructured films.* The Scientific World Journal, 2014. **2014**.

42. Berger, S., et al., *High photocurrent conversion efficiency in self-organized porous WO_3.* Applied Physics Letters, 2006. **88**(20): p. 203119.

43. Li, W., et al., *Visible light photoelectrochemical responsiveness of self-organized nanoporous WO_3 films.* Electrochimica Acta, 2010. **56**(1): pp. 620–625.

44. Yadav, A., Y. Hunge, and S.-W. Kang, *Porous nanoplate-like tungsten trioxide/reduced graphene oxide catalyst for sonocatalytic degradation and photocatalytic hydrogen production.* Surfaces and Interfaces, 2021. **24**: p. 101075.

45. Kailasam, K., et al., *Mesoporous carbon nitride-tungsten oxide composites for enhanced photocatalytic hydrogen evolution.* ChemSusChem, 2015. **8**(8): pp. 1404–1410.

46. Liao, G., et al., *Emerging polymeric carbon nitride Z-scheme systems for photocatalysis.* Cell Reports Physical Science, 2021. **2**(3): p. 100355.

47. Tahir, M., et al., *Tuning the photocatalytic performance of tungsten oxide by incorporating $Cu_3V_2O_8$ nanoparticles for H_2 evolution under visible light irradiation.* Journal of Electrochemical Energy Conversion and Storage, 2020. **17**(1): p. 011002.

48. Tahir, M.B., et al., *Activated carbon doped WO_3 for photocatalytic degradation of rhodamine-B.* Applied Nanoscience, 2020. **10**(3): pp. 869–877.

49. Wang, H., et al., *Enhanced photocatalytic CO_2 reduction to methane over $WO_3 \cdot 0.33\,H_2O$ via Mo doping.* Applied Catalysis B: Environmental, 2019. **243**: pp. 771–779.

50. Jin, J., et al., *A hierarchical Z-scheme CdS-WO_3 photocatalyst with enhanced CO_2 reduction activity.* Small, 2015. **11**(39): pp. 5262–5271.

51. Lin, N., et al., *Construction of a 2D/2D WO_3/$LaTiO_2N$ direct Z-scheme photocatalyst for enhanced CO_2 reduction performance under visible light.* ACS Sustainable Chemistry & Engineering, 2021. **9**(40): pp. 13686–13694.

52. Fu, L., and Z. Fu, *Plectranthus amboinicus leaf extract–assisted biosynthesis of ZnO nanoparticles and their photocatalytic activity.* Ceramics International, 2015. **41**(2): pp. 2492–2496.

53. Babitha, N., et al., *Enhanced antibacterial activity and photo-catalytic properties of ZnO nanoparticles:* Pedalium murex *plant extract-assisted synthesis.* Journal of Nanoscience and Nanotechnology, 2019. **19**(5): pp. 2888–2894.

54. Khalafi, T., F. Buazar, and K. Ghanemi, *phycosynthesis and enhanced photocatalytic activity of zinc oxide nanoparticles toward organosulfur pollutants.* Scientific Reports, 2019. **9**(1): p. 6866.

55. Sahoo, R., A. Mundamajhi, and S.K. Das, *Growth of ZnO nanoparticles prepared from cost effective laboratory grade ZnO powder and their application in UV photocatalytic dye decomposition.* Journal of Materials Science: Materials in Electronics, 2019. **30**(5): pp. 4541–4547.

56. Karnan, T., and S.A.S. Selvakumar, *Biosynthesis of ZnO nanoparticles using rambutan (*Nephelium lappaceum *L.) peel extract and their photocatalytic activity on methyl orange dye.* Journal of Molecular Structure, 2016. **1125**: pp. 358–365.

57. Bomila, R., S. Suresh, and S. Srinivasan, *Synthesis, characterization and comparative studies of dual doped ZnO nanoparticles for photocatalytic applications.* Journal of Materials Science: Materials in Electronics, 2019. **30**(1): pp. 582–592.

58. Singh, J., et al., *The effect of manganese doping on structural, optical, and photocatalytic activity of zinc oxide nanoparticles.* Composites Part B: Engineering, 2019. **166**: pp. 361–370.

59. Tran, T.H., and V.T. Nguyen, *Copper oxide nanomaterials prepared by solution methods, some properties, and potential applications: A brief review.* International Scholarly Research Notices, 2014. **2014**.

60. Vinothkumar, P., et al., *Effect of reaction time on structural, morphological, optical and photocatalytic properties of copper oxide (CuO) nanostructures.* Journal of Materials Science: Materials in Electronics, 2019. **30**(6): pp. 6249–6262.

61. Umadevi, M., A.J. Christy, and B. Spectroscopy, *Synthesis, characterization and photocatalytic activity of CuO nanoflowers.* Spectrochimica Acta Part A: Molecular and Biomolecular Spectroscopy, 2013. **109**: pp. 133–137.

62. Scuderi, V., et al., *Photocatalytic activity of CuO and Cu_2O nanowires.* Materials Science in Semiconductor Processing, 2016. **42**: pp. 89–93.

63. Konar, S., et al., *Shape-dependent catalytic activity of CuO nanostructures.* Journal of Catalysis, 2016. **336**: pp. 11–22.

64. Iqbal, M., et al., *Influence of Mn-doping on the photocatalytic and solar cell efficiency of CuO nanowires.* Inorganic Chemistry Communications, 2017. **76**: pp. 71–76.

65. Wang, Y., et al., *Synthesis and enhanced photocatalytic property of feather-like Cd-doped CuO nanostructures by hydrothermal method.* Applied Surface Science, 2015. **355**: pp. 191–196.

66. Yan, B., et al., *Synthesis and enhanced photocatalytic property of La-doped CuO nanostructures by electrodeposition method.* Journal of Materials Science: Materials in Electronics, 2016. **27**(5): pp. 5389–5394.

67. Devi, L.V., et al., *Synthesis, defect characterization and photocatalytic degradation efficiency of Tb doped CuO nanoparticles.* Advanced Powder Technology, 2017. **28**(11): pp. 3026–3038.

68. Chen, Y., et al., *Effect of doping Ce ions on morphology and photocatalytic activity of CuO nanostructures.* Crystal Research and Technology, 2019. 54(9): p. 1900033.

69. Khairy, M., E.M. Kamar, and M.A. Mousa, *Photocatalytic activity of nano-sized Ag and Au metal-doped TiO$_2$ embedded in rGO under visible light irradiation.* Materials Science and Engineering: B, 2022. 286: p. 116023.

Chapter 2

Introduction to Plasmonic Photocatalysis

ABSTRACT

In recent years, plasmonic nanomaterials have emerged as a highly interesting choice for many photocatalytic processes. Their localized surface plasmon resonance brings in some special properties that resolve some of the inconveniences associated with conventional semiconductor-based photocatalysis. In this chapter, we present the latest advances made in the field of plasmonic photocatalysis which includes an introductory section to define the key types of plasmonic nanomaterials available, including the latest alternatives labeled. Following the primary catalytic application areas, a second section of the chapter deals with liquid-phase reactions for pollutant treatment and a variety of organic reactions to add value to compounds under moderate circumstances. The third portion of the chapter discusses two distinct uses of nanoplasmonic photocatalysts in gas-phase processes comprising the removal of volatile organic chemicals and the conversion of carbon dioxide into valuable energy-related chemical products. Finally, the fourth section of the chapter discusses the most recent uses of plasmonics as potential enzyme-like substitutions in biochemical reactions including cofactor molecule regulation and mimetic behaviour.

2.1 INTRODUCTION

The utilization of solar energy has become increasingly important in recent years due to its safety and effectiveness as a power source. However, traditional photocatalysis is not the most effective method of collecting light for solar energy because it only captures around 5% of ultraviolet (UV) light, which is the most effective type of light harvesting, and mostly uses semiconductors as catalysts. In the quest for novel materials that can significantly improve the performance of traditional photocatalysts, plasmonic

DOI: 10.1201/9781003403357-3

nanoparticles have newly captured the consideration of the scientific community. These materials can be used in a diversity of fields, containing sensors, biomedicine, and photocatalysis, and offer several benefits over their larger counterparts. One of the most intriguing features of these materials is their localized surface resonance band, which is in addition to their size. When the free charge carriers are in resonance with the incoming light, the localized surface plasmon resonance (LSPR) causes them to vibrate collectively and uniformly on the surface of the nanoparticle. This resonance modifies the interface of the nanostructures in the electric field, leading to an effect on the molecules or materials in close proximity. The choice of plasmonic material utilized in different phases can have a significant effect on the effectiveness of the LSPR for photocatalysis. These materials can be divided into three main groups: pure, hybrid, and alternative plasmonic materials.[1,2]

Plasmonic nanomaterials have emerged as a promising option for numerous photocatalytic processes due to their unique properties resulting from LSPR. These tools offer several advantages over traditional semiconductor-based photocatalysis. Despite being a reasonably new field, significant advancements have been made, particularly in the synthesis of novel structures capable of absorbing light across the entire solar spectrum. Nanoplasmonic materials have been employed in various processes. This chapter discusses the latest developments in plasmonic photocatalysis. It begins with an introduction that describes the core categories of plasmonic nanomaterials presented and lists recent alternatives. The second section covers liquid-phase reactions for detoxifying contaminants and adding value to various organic interactions under favourable conditions, beyond the primary catalytic application areas. The third section focuses on the treatment of volatile organic pollutants and the conversion of CO_2 into useful energy-related chemical products using nanoplasmonic photocatalysts. Finally, the fourth section summarizes recent plasmonics applications as potential enzyme-like substitutes for biological processes that require cofactor molecule regulation and their simulated behaviour.

2.1.1 Pure Plasmonic Materials

The nano-photocatalysts in this category are built entirely of unstructured plasmonic metals as well as photocatalytic insulators, allowing the metal nanomaterial to do all the chemistry. Activation of the plasmon band in this case affects the production of higher densities of charge carriers on the nanoparticle interface. The molecules that have been adsorbed on the surfaces of these carriers can then participate in reduction or oxidation reactions. The method described is frequently stated as direct transfer of charge, meanwhile the charge is transferred straight from the higher levels of energy in the metallic material to the lower, vacant orbitals of the adsorbent, so introducing the chemical reaction.[3] The mechanism of pure plasmonic materials is shown in Figure 2.1.

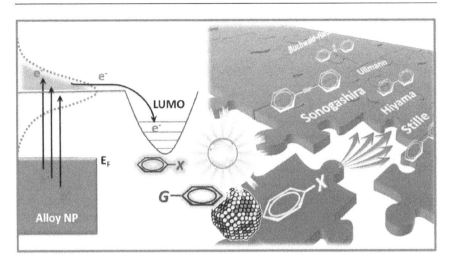

Figure 2.1 The mechanism of pure plasmonic materials.

Source: Reprinted with permission from Xiao, Q., et al., *Visible light-driven cross-coupling reactions at lower temperatures using a photocatalyst of palladium and gold alloy nanoparticles.* ACS Catalysis, 2014. 4(6): pp. 1725–1734. Copyright © 2014 American Chemical Society.

Additionally, plasmonic nanoparticles exhibit a characteristic known as local heat production. Upon excitation, the charges generated by the plasmonic nanoparticle can go through either radiative or non-radiative relaxation. Non-radiative relaxation leads to the production of heat due to e-e and e-p interactions localized on the surface of the nanoparticles. This heat can activate thermal reactions when it is transferred to the surrounding environment. The temperature that can affect the surface of the metal depends on its inherent properties. The absorption cross section of silver is approximately 10 times greater than that of gold, making it suitable for higher temperature ranges above 500°C to 700°C, which is typically the maximum range for gold nanoparticles in photocatalytic processes.[4] For instance, Hallett-Tapley et al. used gold plasmonic nanoparticles to selectively oxidize alcohols.[5]

In this work, the Au nanoparticles' plasmon band was stimulated, and the alcohol molecules around it were selectively oxidized using a green laser beam and light-emitting diode (LED). The researcher put forth two potential pathways for the process; one included direct transfer of electrons from gold (Au) to H_2O_2, resulting in radicals that start a chain reaction; the other linked peroxide oxidation and reduction to the generation of heat. Another example of effective photocatalytic activity involves the utilization of nanorattles made of an Au nanosphere and a gold/silver nanoshell.[6] The increased output of this structure is attributed to the development of electromagnetic hot zones at the junction between the two dissimilar materials. The enhanced density of reactive oxygen molecules in these hot zones causes the oxidation process. Scientists have attempted to create anisotropic nanoparticles that can absorb light in the visible and even infrared (IR) spectra. For the Suzuki coupling process, hexagonal Pb nanoplates have been developed. According to their aspect ratio, Trinh et al. plasmonic lead nanoparticles can have plasmon bands that are anywhere throughout the visible and near-infrared (NIR) spectra. The Suzuki reaction was carried out on the catalyst surface using energetic electrons produced by the LSPR. Nanomaterials made of metal nanoparticles based on photochemical insulators might also be categorized under this category. The author states that materials with large bandgap energies, such as ZrO_2, Al_2O_3, or SiO_2, cannot be excited by the visible or near-UV region of the spectra. Liu et al.[7,8] used Ag-Cu alloy nanoparticles with ZrO_2 assistance to selectively convert nitro-compounds to the azoxy variants. They demonstrated that the reaction's selectivity could be controlled solely by illuminating the target in the visible area with the alloy's composition, independent of the support.

2.1.2 Hybrid Plasmonic Materials

Photocatalytic semiconductors have been widely used since 1972, with Fujishima and Honda demonstrating the use of TiO_2 to split water using UV light.[9] Despite its advantages, TiO_2 also has several drawbacks, such as limited photon efficiency, maximum carrier recombination rate, and a large bandgap of 3.2 eV, that limit its activation to UV light. To address these issues, recent efforts have focused on developing a new class of photocatalysts that incorporate metal plasmonic nanocrystals onto semiconductor surfaces. These metal nanoparticles support and fix some of the shortcomings of semiconductors, as seen in Figure 2.2, by increasing the material's capacity to absorb light via multiple methods. Since the early 1970s, TiO_2 along with additional photocatalyst semiconductors have been the focus of significant investigation for their ability to convert light energy into chemical energy. However, their low photon efficiency, high carrier recombination level, and broad bandgap make them less useful in some applications, notably in the visible and NIR areas of the electromagnetic spectrum, which contain the majority of solar energy. To address these issues, researchers have

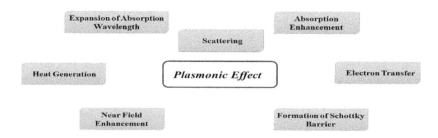

Figure 2.2 Various impacts of plasmons in hybrid plasmonic structures.

investigated the usage of hybrid materials composed of metal plasmonic nanocrystals on semiconductor interfaces.

Metal nanoparticles in the visible region of the spectrum allow the hybrid material to absorb a larger range of light, increasing the total number of photons absorbed and improving photocatalytic reaction efficiency. Furthermore, because plasmonic materials have a high absorption coefficient, they can absorb a significant portion of the light entering the hybrid, minimizing carrier recombination in the semiconductor. Furthermore, plasmonic nanoparticles scattering properties enable them to re-emit light that would not be absorbed, which may lead to additional absorption by nearby nanoparticles and an increase in the light path. Another advantage of metal-semiconductor hybrids is the establishment of a Schottky barrier at the interface between the two materials, which traps photo-generated carriers and prolongs their lifetime. This barrier can considerably reduce the rate of electron-hole recombination, thus increasing the probability of carriers participating in chemical reactions on the material's surface. Studies by Fujishima and Honda[9] and others have highlighted the limitations of traditional semiconductor photocatalysts, and recent research has shown that metal plasmonic nanocrystals on semiconductor surfaces can address some of these limitations.

Plasmonic materials show a vital role in improving the effectiveness of semiconductors for photocatalysis through various processes beyond just absorption. When light is absorbed, excited electrons proceed from the nanoparticle to the semiconductor at high energies, breaking through the Schottky barrier between the metal and semiconductor, and starting the photocatalytic activity. By manipulating the size and shape of these metal nanoparticles, their plasmon band can be shifted to the visible or even IR sections of the spectrum, making them suitable for use with nearly any wavelength. This is a critical consideration, as the excitation field can be utilized to excite materials during solar light irradiation in the remaining 95% of the overall solar spectrum. Through the production of a potent local electric field that can encourage or accelerate the creation of extra charge carriers on the photocatalyst surface, plasmonic materials can also increase the productivity of semiconductor photocatalysis. If the metal-semiconductor connection is inadequate to produce a Schottky barrier, or if another material layer is required to support the nanoparticles, plasmonic nanoparticles can generate charge carriers even when they are not in contact with the substrate. The electric field created by plasmonic materials can also facilitate the polarization of non-polar molecules adjacent to metal nanoparticles, thereby improving their ability to bind to semiconductor surfaces. Moreover, the creation of heat around the nanoparticles is another crucial point of plasmonic materials that increases the photocatalytic sensitivity of these hybrid systems. This heat can accelerate the rate of reactions and the movement of mass within the reaction system, particularly in high-energy sites. These properties make plasmonic hybrid materials a particularly intriguing choice for enhancing the environmental and economic appeal of various industrial processes.

2.1.3 Alternative Plasmonic Materials

Previously, non-metallic materials were not considered capable of carrying LSPR due to the low charge-carrier density on their surface. Recent research has nevertheless demonstrated that a number of materials may be altered to boost their free carrier concentration and hence become efficient plasmonic materials in the visible and IR spectrums.[10]

The free electron concentration of metal oxides can be increased by introducing oxygen vacancies or doping, making those suitable candidates for plasmonic materials. For instance, aluminum, zinc, cadmium, and tungsten oxides can acquire plasmonic properties in the visible or IR spectrum when doped with certain elements or when oxygen vacancies are created. Tungsten oxide, for example, can exhibit plasmonic activity in the IR range and be utilized for photocatalysis after the production of oxygen vacancies.[11,12] Additionally, chalcogenides such as CuS exhibit a unique characteristic where a decrease in the Cu percentage results in an increase in hole concentration, leading to a broad plasmon band in the IR spectrum, as depicted in Figure 2.3.[13] All of these materials can be used as plasmonic photocatalysts, employing any of the techniques discussed previously.

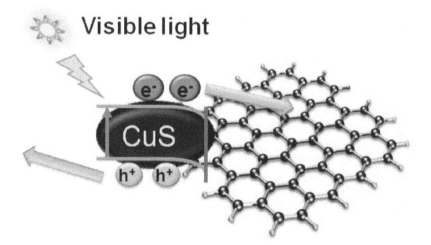

Figure 2.3 Schematic (not to scale) to show how the CuS-RGO interface's photocharge carrier dissociation and energy level alignment work.

Source: Reprinted with permission from Zhang, Y., et al., *Biomolecule-assisted, environmentally friendly, one-pot synthesis of CuS/reduced graphene oxide nanocomposites with enhanced photocatalytic performance.* Langmuir, 2012. **28**(35): pp. 12893–12900. Copyright © 2012 American Chemical Society.

2.2 NANOPLASMONIC PHOTOCATALYSIS IN LIQUID PHASE

2.2.1 Plasmonic-Based Photocatalytic Degradation of Aqueous Contaminants

Heterogeneous photocatalysis has proven to be a highly effective method for removing organic contaminants from aqueous environments.[14,15] Plasmonic nanomaterials with LSPR have been used in visible light exposure to improve the photocatalytic synthesis of chemical reactions. It is crucial to research the cleanup of aqueous phases and develop plasmonic nanostructures.[16-18] Substances such as dyes (e.g. rhodamine B, methylene blue, and methyl orange) are of significance due to the ease of monitoring their degradation. However, the impact of photosensitization on dye components speeds up the photocatalytic reactions by producing additional electrons. Thus, it is interesting to explore the potential for photoremediation of pollutants that are not coloured. Most organic and inorganic chemicals have demonstrated the value of photocatalytic removal.[19,20] Table 2.1 includes examples of organic

Table 2.1 Recent Research on Using Photocatalysts with Plasmon Assistance to Remove Aqueous Pollutants

Sr. No.	Year	Catalyst	Pollutant	Irradiation	Performance	References
1.	2016	Ag@g-C$_3$N$_4$@ BiVO$_4$	TC	300 W Xe lamp, $\lambda > 350$ nm, $\lambda > 420$ nm or $\lambda > 760$ nm	Removal of 90.8% ($\lambda > 350$ nm), 82.7% ($\lambda > 420$ nm), and 12.6% ($\lambda > 760$ nm) in 60 min	21
2.	2010	Ag$_2$SO$_3$/ Agbr GO/Ag$_1$SO$_3$/ AgBr	MO, RhB, MB	500 W Xe lamp, $\lambda > 420$ nm, 100 mW/cm^2	In 9 min, 99.9% of MO was removed, while RhB was virtually completely removed.	22
3.	2008	Ag@AgCl/ TP	RhB, X-3B, CIP, Phenol	300 W Xe lamp, $\lambda > 400$ nm	Complete removal of RhB in 8 min, X-3B in 12 min, 48% CIP elimination in 3 h, and 51% phenol removal in 3 h.	23
4.	2016	Bi/Bi$_2$WO$_6$	RhB, 4-CP	300 W Xe lamp, $\lambda > 400$ nm, 200 mW/cm^2	RhB removal was 93% after 25 min while 4-CP removal was 54.4% in 120 min.	24

(Continued)

Table 2.1 (Continued)

Sr. No.	Year	Catalyst	Pollutant	Irradiation	Performance	References
5.	2016	Ag/Ag$_2$CO$_3$-RGO	MO Phenol	350 W Xe lamp, λ > 420 nm 40 mW/cm^2	In 15 min, 93% of the MO was removed, and in 30 min, 93% of the phenol was removed.	25
6.	2015, 2004	Cu$_{2-x}$Se–g-C$_3$N$_4$	MB	500 W Xe lamp, > 420 nm	Greater than 95% removal in 2 h	26
7.	2016	Pt–BiOBr hetero-structures	PNP, PBBPA	300 W Xe lamp Simulated sunlight (320–680 nm) or visible light (400–680 nm), 150 mW/cm^2	For PNP: Complete removal in 30 min with simulated sunlight and 99% removal in 1.5 h with visible light For TBBPA: With simulated sunshine, you can get rid of it completely in 5 min.	27

Note. 4-CP, 4-chlorophenol; CIP, ciprofloxacin; MB, methylene blue; MO, methyl orange; PNP, P-nitrophenol; RhB, rhodamine B; TBBPA, tetrabromobisphenol-A; X-3B, reactive brilliant red.

chemicals, such as phenolic compounds, nitrophenols, chlorophenols, and trichloroethylene.

Hybrid plasmonic photocatalysts are commonly utilized for the elimination of pollutants, often in combination with silver and gold. However, Bi in the low range has occurred as a promising substitute to these noble metals. In addition to its direct photocatalytic activity, Bi can act as a co-catalyst to enhance the efficiency of various catalysts.

2.2.2 Organic Synthesis

The application of heterogeneous photocatalysis in organic synthesis is a more difficult challenge than organic pollutant removal.[28] Generating fine compounds of industrial significance via visible light–driven chemical processes using plasmonic nanoparticles presents an intriguing alternative approach to heterogeneous photocatalysis in organic synthesis, which is a more challenging subject.[29–31]

2.2.2.1 Oxidation Reactions

When studying plasmon-assisted photocatalyzed reactions, the selection of alcohols to be oxidized into their corresponding carbonyl part is crucial. For instance, Au/CeO$_2$ can be utilized to selectively oxidize benzyl alcohols to benzaldehydes in aqueous solutions in green LED visible light exposure.[32] Figure 2.4(A) illustrates the capacity of Au on hydrotalcite, ZnO, and benzyl alcohol to generate benzaldehyde in the vicinity of H_2O_2 and a 530 nm LED acting as a photoexcitation source (A).

A)

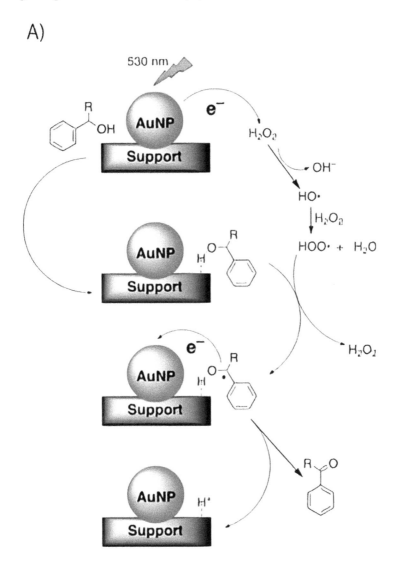

Figure 2.4(A) The hypothesis representing the plasmon-mediated oxidation of secondary phenethyl (R-CH₃) and benzyl alcohol in the vicinity of supported Au nanoparticles.

Source: Reprinted with permission from Hallett-Tapley, G.L., et al., *Supported gold nanoparticles as efficient catalysts in the solventless plasmon mediated oxidation of sec-phenethyl and benzyl alcohol.* Journal of Physical Chemistry C, 2013. 117(23): pp. 12279–12288. Copyright © 2013 American Chemical Society.

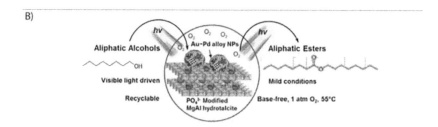

Figure 2.4(B) Direct oxidative esterification of aliphatic alcohol (using 1-octanol as an example).

Source: Reprinted with permission from Xiao, Q., et al., *Catalytic transformation of aliphatic alcohols to corresponding esters in O*$_2$ *under neutral conditions using visible-light irradiation.* Journal of the American Chemical Society, 2015. 137(5): pp. 1956–1966. Copyright © 2015 American Chemical Society.

Esterification reactions are crucial in chemical synthesis.[33] Previously, these reactions involved multistep processes that produced unwanted by-products and required extreme temperature, pressure, and pH conditions when activated acid derivatives and alcohols were combined. Figure 2.4(B) presents a one-pot approach developed by Xiao et al.,[34] which involves using Au-Pd alloy nanoparticles on a reusable photocatalyst to directly oxidative esterify aliphatic alcohols under moderate conditions (B). The reactivity is significantly influenced by the irradiation light's wavelength and intensity. According to the description, increased irradiance produces greater light-excited energy of electrons, resulting in the formation of a bigger electromagnetic field surrounding the nanoparticles (known as the LSPR field enhancement effect). Excitation of metal electrons to higher energy levels can occur when they absorb photons of a particular wavelength (e.g. 550 nm), which increases the probability of these electrons moving to anti-bonding atomic orbitals and triggering a reaction. At lower temperatures, where the thermal and photothermal effects are minimal, the impact of the wavelength becomes more pronounced, as the transfer of excited electrons plays a major role in driving photocatalytic activity.

Wang et al.[35] stated the esterification of aldehydes and alcohols at room temperature using stabilized Au nanoparticles (Au/Al$_2$O$_3$) generated by visible light. The outcomes obtained from various wavelength ranges suggest that LSPR shows a significant role in improving reaction activity in catalytic procedures, and that Au is essential for capturing visible light.

2.2.2.2 Reduction Reactions

Reduction reactions are an important aspect of fine chemical synthesis and have been extensively studied using photocatalytic methods, particularly with plasmonic materials. Using a 532 nm laser or plasmon excitation from a 530 nm LED, Au nanoparticles have been used as a catalyst in the conversion of resazurin to resorufin.[36] These Au nanoparticles have also been used to selectively reduce organic molecules when exposed to visible light or artificial sunshine. The Au/CeO_2 catalyst is particularly effective in reducing nitroaromatics to form azocompounds, hydrogenating azobenzene into hydro-azobenzene, and converting ketones to alcohols.

Moreover, the combination of Cu and graphene (Cu/graphene) has shown remarkable efficiency in the reduction of nitroaromatics to aromatic azocompounds in sunlight irradiation.[37] The temperature considerably influences the product selectivity, and it has been found that this conversion is dependent on the intensity of light. The Cu nanoparticles primarily act as a catalyst for the process by absorbing light. Notably, this conversion occurs at maximum efficiency within the visible light range of 530–600 nm, and Cu nanoparticles exhibit substantial light absorption owing to LSPR.[38]

One more example is the reduction of styrene, where hydrogen is used to produce ethylbenzene. Under visible light, this decrease is possible when Ag-Pd nanocages are present. The lead component of these photocatalysts plays an active role in hydrogenation processes, while Ag exhibits interesting plasmonic properties by converting light to heat.[33]

2.3 NANOPLASMONIC PHOTOCATALYSIS IN GAS-PHASE REACTIONS

In contrast to their traditional usage in photocatalysis in aqueous conditions, plasmonic nanoparticles have recently attracted the attention of researchers who are looking into their potential for use in gas-phase processes. With a wide range of gas-phase reactions currently under review, our focus in this segment will be on two major types: the decomposition of volatile organic compounds (VOCs) and the conversion of chemicals into energy.

2.3.1 Photodegradation of Volatile Organic Compounds

As a result of the harmful effects of prolonged exposure to air pollution on human health, it has become an increasingly pressing issue. Human-caused high concentrations of VOCs in both indoor and outdoor environments need the development of more effective solutions to this problem. VOCs are now being eliminated using a variety of approaches, including chemical procedures such as ozone treatment or UV irradiation, along with physical ones such as filtration or adsorption.[39] However, novel strategies are needed to totally remove VOCs from the environment, as present technologies, while successful in removing dangerous compounds, have severe effects on human health. Plasmonic materials provide a viable option in this context, not only for shielding against hazardous atmospheres, but also for harnessing solar energy as a safer and more cost-effective alternative to existing energy sources. While there are more scenarios involving photocatalysis for VOC removal than ever before, those that use plasmonic materials are significantly less prevalent than those that use conventional semiconductors. One of the earliest documented cases of using plasmonic materials for VOC cleanup is presented in Chen et al.'s research.[40] They employed supported Au nanoparticles on various metal oxides to achieve complete formaldehyde oxidation using visible light. To avoid interference from the visible light source, the researchers used oxide supports with high bandgaps, such as ZrO_2 or SiO_2 (with bandgaps of 5 and 9 eV, respectively), to anchor the Au nanoparticles. They were able to employ a pure plasmonic material as a result in this instance. The suggested technique makes use of two plasmonic material properties. First, the plasmon band of the Au nanoparticles can be sufficiently irradiated to raise the surrounding temperature and promote the oxidation of VOCs. Second, the intense electric field created on the surfaces of the nanoparticles can cause polar compounds like formaldehyde to totally oxidize. The successful breakdown of isopropanol using plasmonic materials by Dinh et al.[41] serves as an example of VOC degradation. The catalyst's structure plays a crucial role in its effectiveness, comprising an Au/TiO_2 thin-shell hollow nanosphere formation arranged into a three-dimensional (3D) photocatalyst. This combination enables the formation of a hybrid plasmonic structure with unique properties. According to the researchers, the numerous scattering and slow photon effects of these structures enhance the plasmonic Au nanoparticle's absorption by the catalyst's photonic shape. The operation is significantly more efficient than the standard Au/TiO_2 structure due to the increased plasmon band absorption.

Sellappan et al.[42] provide another example of VOC oxidation using plasmonic materials. In this study, Au and Ag nanoparticles were mixed with TiO_2 and various combinations to investigate the mobility of electrons after light stimulation. Under various circumstances, including a direct physical interface between the metal and semiconductor, the photocatalysis of the degradation of ethylene and methanol was determined. A Schottky barrier is formed when the nanoparticles and TiO_2 come into contact, which improves charge separation when exposed to UV radiation. The behaviour of the photocatalyst changes when the metals do not touch the aid, which is attributed to the near- and far-field effects of the plasmonic nanomaterial. Even when the metal nanoparticle is isolated from the substrate, it can still experience the strong electric field created on it.

2.3.2 Plasmonic-Driven Chemical to Energy Conversion Processes

One of the critical issues faced by the planet is the problem of global warming, which has resulted from decades of extensive use of fossil fuels.

This has disrupted the balance between CO_2 emissions and absorption, causing a severe change in the climate with large amounts of greenhouse gases present in the air and oceans. Furthermore, the depletion of fossil fuels due to increased consumption necessitates the search for alternative energy sources. Photocatalysis has emerged as a powerful technique that can address both these issues simultaneously. It can help in the production of energy-efficient products while reducing the amount of CO_2 emitted into the environment. Inoue et al. employed multiple semiconductors as photocatalysts to decrease CO_2 using light for the first time in 1979, which was an early example of this approach. It is worth noting that the first CO_2 reduction process was carried out in a liquid phase, and it was the first effort to manufacture usable products from CO_2 reduction. Several initiatives have been undertaken since then to develop photocatalysts capable of reducing CO_2 by different target chemicals. While semiconductors such as TiO_2 and ZnO have been used as photocatalysts owing to their favourable features, they also have limitations that limit their employment in a variety of contexts.[43] These drawbacks include insufficient photon absorption, high carrier recombination rates, and spectral restrictions to the UV region. To overcome these limitations, researchers have increasingly turned to plasmonic materials as a promising strategy for CO_2 reduction. In fact, Liu et al.[44] reported the first use of a plasmonic gold catalyst to produce syngas through dry reforming of CO_2 with methane. Typically, high temperatures (800°C–1000°C) are required for the conversion of two greenhouse gases into CO and H_2, which is quite intriguing since it results in energy-saving components. To take advantage of Rh's favourable properties for this synthesis and Au plasmon band to facilitate reactants, a catalyst consisting of Rh and Au supported on SBA-15 was developed. According to the researchers, the CO_2 and CH_4 particles are polarized due to the strong electric field generated at the Au nanoparticle, which facilitates their conversion to syngas. Another interesting process involving CO_2 capture is the reverse water-gas shift reaction. Upadhye et al.[45] demonstrated that by Au nanoparticles promoted on TiO_2 and CeO_2, they could increase the reaction yield by up to 1300% using visible light stimulation. They claimed that the catalyst's LSPR altered the intrinsic kinetics of the catalyst's surface reaction by altering the rate constant of either the carboxylic disintegration or the hydroxylic hydrogenation, two stages in this process. The causes of the rate increase have been proposed as hot electron production or the polarization impact of the intense electric field created on the adsorbents by the plasmonic nanoparticles. Scientists have recently explored using graphene and its derivatives as active substrates for photocatalysis, in addition to conventional semiconductor materials. Shown et al.[46] studied the CO_2 reduction behaviour of Cu nanoparticles on graphene oxide (GO) in their study (Figure 2.5). In order to evaluate photoreduction using visible light, the researchers constructed a catalyst

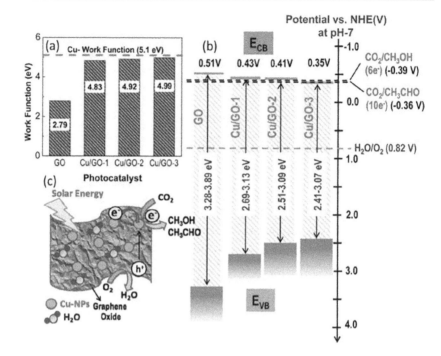

Figure 2.5 (a) Represents the work functions of GO and Cu/GO hybrids. (b) Band-edge locations of pure GO and Cu/GO hybrids in affinity to CO_2/CH_3OH and CO_2/CH_3CHO formation potential, as measured by ultraviolet photoelectron spectroscopy. (c) Diagram of the mechanism for a photocatalytic process.

Source: Reprinted with permission from Shown, I., et al., *Highly efficient visible light photocatalytic reduction of CO_2 to hydrocarbon fuels by Cu-nanoparticle decorated graphene oxide*. Nano Letters, 2014. 14(11): pp. 6097–6103. Copyright © 2014 American Chemical Society.

with a certain quantity of Cu loading. They further assessed the outcomes with GO and TiO_2 P25. The product yield of the CuGO hybrid content was 60 times larger than that of GO and 240 times greater than that of P25, with the primary products being acetaldehyde and methanol. The enhancement was attributed to the modification of the characteristic of GO by the Cu nanoparticles, which developed charge separation. Finally, several other instances of successful CO_2 hydrogenation to form methane near nanoplasmonics photocatalysts are concisely discussed later and compiled in Table 2.2. Despite its availability and widespread use as a fuel in fertilizers or an intermediary in the petrochemical industry, methane has drawn increasing attention.[47]

There are many studies that discuss the use of titanium or P25 semiconductor substrates with noble metal co-catalysts that act as active and

Table 2.2 Summary of Selected Photocatalysts for Conversion of CO_2 into CH_4

Sr. No.	Year	Catalyst	Reductant	Co-catalyst	Remarks	Reference
1.	2016	Au, In/TiO$_2$	(Au)	Gold (Au), Indium (In)	Irradiation with UV lamp (200 W; 150 mW/ cm^2) Use of monolithic reactors	48
2.	2010	Pd/TiO$_2$	Water (H$_2$O)	Lead (Pb)	Irradiation with UV-LED arrays (40 pieces) (365 nm) Fluidized bed reactors + T = 140°C	49
3.	2008	Pd/TiO$_2$	Water (H$_2$O)	Lead (Pb)	Irradiation at λ > 310 nm Formation of organic adsorbates is critical	53
4.	2015	Core-shell Pt/TiO$_2$ PtCu/TiO$_2$	H$_2$O	Cu, Pt	Light source 780 > λ > 320 nm Strong influence of co-catalysts on selectivities	54
5.	2015, 2019	Au, Cu/P25	H$_2$O	Au, Cu, Au-Cu alloys	λ= 355 and 532 nm Performance of transient absorption experiments	55

targeted catalytic centres and load reservoirs. These studies aim to extend the bandgap to the visible range by using the surface plasmon band of these materials. The selectivity of methane, which is essential for the precise synthesis of chemical intermediates, is significantly increased by the addition of Pd or Pt in comparison to TiO$_2$. However, a significant drawback is the steady deactivation of Pd when it is oxidized into PdO domains. In addition, it has been found that Au-Cu alloys, which were successfully formed, are extremely active and outperform the photocatalytic conversions of their relevant single metal equivalents. They also extend their responsiveness to the visible region, as shown by a transient absorption spectrometer.

Other investigations in this area have explored the utilization of catalytic monoliths,[48] fluidized reactors with LEDs,[49] or novel semiconductor substrates with carbon nitride or reduced graphene oxide, which offer more active co-catalysts and CO$_2$ accommodation sites.[50,51] To selectively hydrogenate CO$_2$ to CH$_4$, Zhang et al. synthesized a hybrid material consisting of Rh nanocubes sustained on Al$_2$O$_3$, as depicted in Figure 2.6. This material exhibited almost complete suppression of competitive CO generation. The plasmonic band of the metal nanoparticles was stimulated by UV and blue LED irradiation, resulting in high-energy electrons that were transported to the adsorbents.[52]

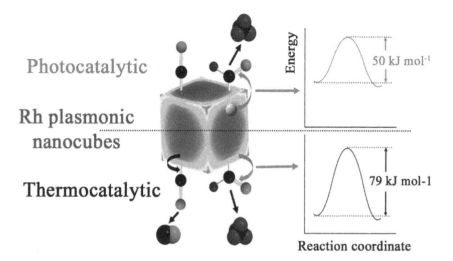

Photocatalytic

Rh plasmonic
nanocubes

Thermocatalytic

Figure 2.6 An example of recently discovered plasmonic catalysts for the hydrogenation of CO_2 into methane is the reaction mechanism on a Rh nanocube contrasting the favoured activation of CO in the thermocatalytic process with the methane route selected under photocatalytic circumstances.

Source: Adapted with permission under the terms of the CC BY-SA license 4.0, Zhang, X., et al., *Product selectivity in plasmonic photocatalysis for carbon dioxide hydrogenation.* Nature Communications, 2017. **8**(1): p. 14542. Copyright © 2017 (https://creativecommons.org/licenses/by/4.0/).

2.4 PLASMONICS IN BIOCATALYTIC PROCESSES

2.4.1 Photobiocatalysis

Photobiocatalysis is a burgeoning area of research that aims to create photocatalysts that mimic natural photosynthetic systems. While recent publications have provided a concise overview of traditional semiconductors in this field, plasmonic materials have been the focus of more recent investigations.[56] Sanchez-Iglesias et al. explored the impact of various plasmonic nanostructures shaped like Au (depicted in Figure 2.7) on the successful photoregeneration of nicotinamide adenine dinucleotide (NADH) molecules, which are essential cofactors in many enzymes in natural biochemical pathways.[57,58] Cofactor molecules show a crucial role as light harvesters in natural processes such as photosynthesis and respiration's reduction-oxidation balancing.[59] However, recent advancements in nonenzymatic pathways have only partially succeeded in reducing the drawbacks of cofactor molecules. Photocatalysts such as organic semiconductors, polymers, and dyes have historically been utilized to recycle NADH cofactors, but these compounds are inherently poor electron absorbers. To overcome this limitation, it has been suggested that an electron loop acting as a mediator be employed.

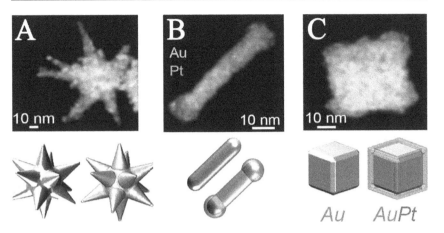

Figure 2.7 Scanning tunneling electron microscopy and energy dispersive x-ray spectroscopy investigation of variously shaped Pt-decorated gold nanoplasmonics and associated three-dimensional models utilized to photoregenerate NADH cofactor molecules: (A) stars; (B) nanorods; and (C) nanocubes.

Source: Reprinted from Sánchez-Iglesias, A., et al., *Plasmonic substrates comprising gold nanostars efficiently regenerate cofactor molecules.* Journal of Materials Chemistry A, 2016. **4**(18): pp. 7045–7052. Copyright © 2016 with permission from The Royal Society of Chemistry.

So far, there have only been a limited number of organometallic complexes based on Rh that have been suitable mediators. Using plasmonic nanostructures in gold provides a simpler and more efficient solution. Researchers have found that gold nanorods surrounded by Pt domains (Figure 2.7 B) is an effective plasmonic cofactor photoregenerator, which chains a plasmonic structure (Au nanorods) and a catalytic active site (Pt) to reduce NADH cofactor molecules. The photoresponse of Au nanostars with epitaxial grown Pt domains was later discovered to be much better (Figure 2.7A), which was linked to the massive light-harvesting capabilities of star-shaped plasmonic structures. This discovery was made by the same team of researchers. It is exciting to consider the idea of creating appropriate mediators that might assist control of various physiological processes through light-induced inputs by combining these plasmonic heterojunctions with other semiconductors.

2.4.2 Artificial Enzymes

The exploration of new artificial enzymes is linked to a field of significant biotechnological interest, particularly for potential nanoplasmonic applications. Enzymes that are produced naturally are well-recognized biocatalysts that control all biochemical methods in living things. However, discovering novel nanomaterials that can imitate the action of natural enzymes as extremely

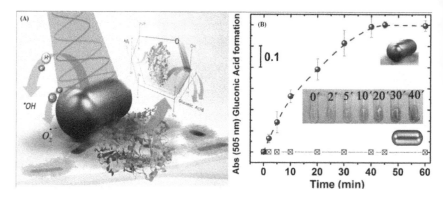

Figure 2.8 (A) Plot showing the titania-coated gold nanorods plasmonic photocatalyst's glucose-oxidase mimicking activity when exposed to NIR light. (B) Colorimetric detection of glucose when it is specifically transformed into gluconic acid by the production of a Fe-hydroxamate complex. Comparatively speaking to coated Au nanorods, uncoated Au nanorods display minor photo-oxidation characteristics. We performed photocatalytic studies using an 808 nm laser.

Source: Reprinted from Ortega-Liebana, M., et al., *Titania-coated gold nanorods with expanded photocatalytic response. Enzyme-like glucose oxidation under near-infrared illumination.* Nanoscale, 2017. **9**(5): pp. 1787–1792. Copyright © 2017 with permission from The Royal Society of Chemistry.

efficient and accurate catalysts, without encountering the limitations associated with natural enzymes, such as limited stability due to denaturation, susceptibility to changes in their optimal conditions, difficulties in productive recovery and recycle, and high prices resulting from numerous synthesis and purification stages, has recently become a critical area of research.

A viable approach to overcoming these significant difficulties is the development of inorganic/organic nanoparticles as a secure and affordable substitute. One of the first glucose oxidize substitute NIR-activated enzyme-like plasmonic photocatalysts has been created and evaluated by Santamaria's team (see Figure 2.8).[60] This biomimetic technology based on glucose oxidase enables the transition to biomass, targeted detection of blood glucose at trace levels, and internal metabolic control/monitoring in cells. Ortega-Liebana et al. used plasmonic gold nanorods with exceptional NIR absorption capabilities in combination with an outer titanium nanoshell, resulting in an extended response to the entire visible-NIR spectrum, further thermal stability, and photoactivity for effective and selective glucon oxidation of glucose (as shown in Figure 2.8). The thin TiO_2 shell created a strong Schottky barrier at the $Au\text{-}TiO_2$ contact, according to the researchers. Additionally, they noticed that the exposed Au nanorods showed limited photoresponse to the oxidation of glucose, indicating that the titanium shell component of the semiconductor is crucial for effective sugar molecule oxidation by facilitating active radical generation.[60]

2.5 SUMMARY

With the wavelength irradiation range being extended to cover the visible and NIR areas, plasmonic-based nanomaterials offer tremendous promise as potential replacements for present photocatalysts and processes. This will maximize the use of solar energy. It is possible to forecast a large range of prospective areas of action, including those specified in this chapter and some other options, comprising the creation of biofuels from water or biomass, photo-electrocatalysis, solar cells, or photovoltaics. Additionally, it is conceivable to utilize these catalysts to precisely target cell pathways during treatments including light-triggered therapy. In conclusion, it is anticipated that plasmonic utilization will manifest in the upcoming years with many exciting achievements.

REFERENCES

1. Zou, Z., et al., *Direct splitting of water under visible light irradiation with an oxide semiconductor photocatalyst.* Nature, 2001. **414**(6864): pp. 625–627.
2. Li, X., J. Zhu, and B. Wei, *Correction: Hybrid nanostructures of metal/two-dimensional nanomaterials for plasmon-enhanced applications.* Chemical Society Reviews, 2016. **45**(14): pp. 4032–4032.
3. Xiao, Q., et al., *Visible light–driven cross-coupling reactions at lower temperatures using a photocatalyst of palladium and gold alloy nanoparticles.* ACS Catalysis, 2014. **4**(6): pp. 1725–1734.
4. Fasciani, C., et al., *High-temperature organic reactions at room temperature using plasmon excitation: Decomposition of dicumyl peroxide.* Organic Letters, 2011. **13**(2): pp. 204–207.
5. Hallett-Tapley, G.L., et al., *Plasmon-mediated catalytic oxidation of sec-phenethyl and benzyl alcohols.* Journal of Physical Chemistry C, 2011. **115**(21): pp. 10784–10790.
6. da Silva, A.G., et al., *Plasmonic nanorattles as next-generation catalysts for surface plasmon resonance-mediated oxidations promoted by activated oxygen.* Angewandte Chemie International Edition, 2016. **55**(25): pp. 7111–7115.
7. Trinh, T.T., et al., *Visible to near-infrared plasmon-enhanced catalytic activity of Pd hexagonal nanoplates for the Suzuki coupling reaction.* Nanoscale, 2015. **7**(29): pp. 12435–12444.
8. Liu, Z., et al., *Selective reduction of nitroaromatics to AZOXY compounds on supported Ag-Cu alloy nanoparticles through visible light irradiation.* Green Chemistry, 2016. **18**(3): pp. 817–825.
9. Fujisima, A., and K. Honda, *Photolysis-decomposition of water at surface of an irradiated semiconduction.* Nature, 1972. **238**: pp. 37–38.
10. Manthiram, K., and A.P. Alivisatos, *Tunable localized surface plasmon resonances in tungsten oxide nanocrystals.* Journal of the American Chemical Society, 2012. **134**(9): pp. 3995–3998.
11. Huang, Q., et al., *MoO$_3$–x-based hybrids with tunable localized surface plasmon resonances: Chemical oxidation driving transformation from ultrathin nanosheets to nanotubes.* Chemistry–A European Journal, 2012. **18**(48): pp. 15283–15287.

12. Yan, J., et al., *Tungsten oxide single crystal nanosheets for enhanced multichannel solar light harvesting.* Advanced Materials, 2015. **27**(9): pp. 1580–1586.

13. Zhang, Y., et al., *Biomolecule-assisted, environmentally friendly, one-pot synthesis of CuS/reduced graphene oxide nanocomposites with enhanced photocatalytic performance.* Langmuir, 2012. **28**(35): pp. 12893–12900.

14. Joseph, C.G., et al., *Photocatalytic treatment of detergent-contaminated wastewater: A short review on current progress.* Korean Journal of Chemical Engineering, 2022. **39**(3): pp. 484–498.

15. Bian, W., et al., *The intermediate products in the degradation of 4-chlorophenol by pulsed high voltage discharge in water.* Journal of Hazardous Materials, 2011. **192**(3): pp. 1330–1339.

16. Jiang, R., et al., *Metal/semiconductor hybrid nanostructures for plasmon-enhanced applications.* Advanced Materials, 2014. **26**(31): pp. 5274–5309.

17. Fan, W., and M.K. Leung, *Recent development of plasmonic resonance-based photocatalysis and photovoltaics for solar utilization.* Molecules, 2016. **21**(2): pp. 180.

18. Xiao, M., et al., *Plasmon-enhanced chemical reactions.* Journal of Materials Chemistry A, 2013. **1**(19): pp. 5790–5805.

19. Meng, X., and Z. Zhang, *Synthesis and characterization of plasmonic and magnetically separable Ag/AgCl-Bi$_2$WO$_6$@ Fe$_3$O$_4$@ SiO$_2$ core-shell composites for visible light-induced water detoxification.* Journal of Colloid and Interface Science, 2017. **485**: pp. 296–307.

20. Xu, H., et al., *Microwave-assisted synthesis of flower-like BN/BiOCl composites for photocatalytic Cr (VI) reduction upon visible-light irradiation.* Materials & Design, 2017. **114**: pp. 129–138.

21. Chen, F., et al., *Novel ternary heterojunction photococatalyst of Ag nanoparticles and g-C$_3$N$_4$ nanosheets co-modified BiVO$_4$ for wider spectrum visible-light photocatalytic degradation of refractory pollutant.* Applied Catalysis B: Environmental, 2017. **205**: pp. 133–147.

22. Wan, Y., et al., *Fabrication of graphene oxide enwrapped Z-scheme Ag$_2$SO$_3$/AgBr nanoparticles with enhanced visible-light photocatalysis.* Applied Surface Science, 2017. **396**: pp. 48–57.

23. Ao, Y., et al., *A novel heterostructured plasmonic photocatalyst with high photocatalytic activity: Ag@ AgCl nanoparticles modified titanium phosphate nanoplates.* Journal of Alloys and Compounds, 2017. **698**: pp. 410–419.

24. Huang, Y., et al., *Facile synthesis of Bi/Bi$_2$WO$_6$ nanocomposite with enhanced photocatalytic activity under visible light.* Applied Catalysis B: Environmental, 2016. **196**: pp. 89–99.

25. Song, S., et al., *Structure effect of graphene on the photocatalytic performance of plasmonic Ag/Ag$_2$CO$_3$-rGO for photocatalytic elimination of pollutants.* Applied Catalysis B: Environmental, 2016. **181**: pp. 71–78.

26. Han, J., et al., *Efficient visible-light photocatalytic heterojunctions formed by coupling plasmonic Cu 2-x Se and graphitic carbon nitride.* New Journal of Chemistry, 2015. **39**(8): pp. 6186–6192.

27. Guo, W., et al., *Morphology-controlled preparation and plasmon-enhanced photocatalytic activity of Pt–BiOBr heterostructures.* Journal of Hazardous Materials, 2016. **308**: pp. 374–385.

28. Friedmann, D., et al., *Heterogeneous photocatalytic organic synthesis: State-of-the-art and future perspectives.* Green Chemistry, 2016. **18**(20): pp. 5391–5411.

29. Tsubogo, T., T. Ishiwata, and S. Kobayashi, *Asymmetrische Kohlenstoff-Kohlenstoff-Kupplungen unter kontinuierlichen Durchflussbedingungen mit chiralen Heterogenkatalysatoren.* Angewandte Chemie, 2013. **125**(26): pp. 6722–6737.

30. Chen, J., et al., *The application of heterogeneous visible light photocatalysts in organic synthesis.* Catalysis Science & Technology, 2016. **6**(2): pp. 349–362.

31. Chen, J.-R., et al., *Visible light photoredox-controlled reactions of N-radicals and radical ions.* Chemical Society Reviews, 2016. **45**(8): pp. 2044–2056.

32. Cheng, H., et al., *Harnessing single-active plasmonic nanostructures for enhanced photocatalysis under visible light.* Journal of Materials Chemistry A, 2015. **3**(10): pp. 5244–5258.

33. Zhao, X., et al., *Pd–Ag alloy nanocages: Integration of Ag plasmonic properties with Pd active sites for light-driven catalytic hydrogenation.* Journal of Materials Chemistry A, 2015. **3**(18): pp. 9390–9394.

34. Xiao, Q., et al., *Catalytic transformation of aliphatic alcohols to corresponding esters in O_2 under neutral conditions using visible-light irradiation.* Journal of the American Chemical Society, 2015. **137**(5): pp. 1956–1966.

35. Wang, Y., et al., *Photocatalytic esterification of aldehydes and degradation of organic dye over gold anchored zinc oxide nanoparticles under irradiation of visible light.* Nanoscience and Nanotechnology Letters, 2019. **11**(10): pp. 1395–1403.

36. Du, P., J.A. Moulijn, and G. Mul, *Selective photo (catalytic)-oxidation of cyclohexane: Effect of wavelength and TiO_2 structure on product yields.* Journal of Catalysis, 2006. **238**(2): pp. 342–352.

37. Guo, X., et al., *Copper nanoparticles on graphene support: An efficient photocatalyst for coupling of nitroaromatics in visible light.* Angewandte Chemie International Edition, 2014. **53**(7): pp. 1973–1977.

38. Tokarek, K., et al., *Green synthesis of chitosan-stabilized copper nanoparticles.* European Journal of Inorganic Chemistry, 2013. **2013**(28): pp. 4940–4947.

39. Ren, H., et al., *Photocatalytic materials and technologies for air purification.* Journal of Hazardous Materials, 2017. **325**: pp. 340–366.

40. Chen, X., et al., *Visible-light-driven oxidation of organic contaminants in air with gold nanoparticle catalysts on oxide supports.* Angewandte Chemie, 2008. **120**(29): pp. 5433–5436.

41. Dinh, C.T., et al., *Three-dimensional ordered assembly of thin-shell Au/TiO_2 hollow nanospheres for enhanced visible-light-driven photocatalysis.* Angewandte Chemie, 2014. **126**(26): pp. 6736–6741.

42. Sellappan, R., et al., *Effects of plasmon excitation on photocatalytic activity of Ag/TiO_2 and Au/TiO_2 nanocomposites.* Journal of catalysis, 2013. **307**: pp. 214–221.

43. Inoue, T., et al., *Photoelectrocatalytic reduction of carbon dioxide in aqueous suspensions of semiconductor powders.* Nature, 1979. **277**(5698): pp. 637–638.

44. Liu, H., et al., *Conversion of carbon dioxide by methane reforming under visible-light irradiation: Surface-plasmon-mediated nonpolar molecule activation.* Angewandte Chemie, 2015. **127**(39): pp. 11707–11711.

45. Upadhye, A.A., et al., *Plasmon-enhanced reverse water gas shift reaction over oxide supported Au catalysts.* Catalysis Science & Technology, 2015. **5**(5): pp. 2590–2601.

46. Shown, I., et al., *Highly efficient visible light photocatalytic reduction of CO_2 to hydrocarbon fuels by Cu-nanoparticle decorated graphene oxide.* Nano Letters, 2014. **14**(11): pp. 6097–6103.
47. Yuliati, L., H. Itoh, and H. Yoshida, *Photocatalytic conversion of methane and carbon dioxide over gallium oxide.* Chemical Physics Letters, 2008. **452**(1–3): pp. 178–182.
48. Tahir, B., M. Tahir, and N.S. Amin, *Gold–indium modified TiO_2 nanocatalysts for photocatalytic CO_2 reduction with H_2 as reductant in a monolith photoreactor.* Applied Surface Science, 2015. **338**: pp. 1–14.
49. Vaiano, V., et al., *Steam reduction of CO_2 on Pd/TiO_2 catalysts: A comparison between thermal and photocatalytic reactions.* Photochemical & Photobiological Sciences, 2015. **14**: pp. 550–555.
50. Tan, L.-L., et al., *Noble metal modified reduced graphene oxide/TiO_2 ternary nanostructures for efficient visible-light-driven photoreduction of carbon dioxide into methane.* Applied Catalysis B: Environmental, 2015. **166**: pp. 251–259.
51. Putri, L.K., et al., *Enhancement in the photocatalytic activity of carbon nitride through hybridization with light-sensitive AgCl for carbon dioxide reduction to methane.* Catalysis Science & Technology, 2016. **6**(3): pp. 744–754.
52. Zhang, X., et al., *Product selectivity in plasmonic photocatalysis for carbon dioxide hydrogenation.* Nature Communications, 2017. **8**(1): p. 14542.
53. Yui, T., et al., *Photochemical reduction of CO_2 using TiO_2: Effects of organic adsorbates on TiO_2 and deposition of Pd onto TiO_2.* ACS Applied Materials & Interfaces, 2011. **3**(7): pp. 2594–2600.
54. Zhai, Q., et al., *Photocatalytic conversion of carbon dioxide with water into methane: Platinum and copper (I) oxide co-catalysts with a core-shell structure.* Angewandte Chemie International Edition, 2013. **52**(22): pp. 5776–5779.
55. Baldoví, H.G., et al., *Understanding the origin of the photocatalytic CO_2 reduction by Au- and Cu-loaded TiO_2: A microsecond transient absorption spectroscopy study.* Journal of Physical Chemistry A, 2015. **119**(12): pp. 6819–6827.
56. Maciá-Agulló, J.A., A. Corma, and H. Garcia, *Photobiocatalysis: The power of combining photocatalysis and enzymes.* Chemistry, 2015. **21**(31): pp. 10940–10959.
57. Sánchez-Iglesias, A., et al., *Plasmonic substrates comprising gold nanostars efficiently regenerate cofactor molecules.* Journal of Materials Chemistry, 2016. **4**(18): pp. 7045–7052.
58. Sánchez-Iglesias, A., A. Chuvilin, and M. Grzelczak, *Plasmon-driven photoregeneration of cofactor molecules.* Chemical Communications, 2015. **51**(25): pp. 5330–5333.
59. Lin, Y., J. Ren, and X. Qu, *Catalytically active nanomaterials: A promising candidate for artificial enzymes.* Accounts of Chemical Research, 2014. **47**(4): pp. 1097–1105.
60. Ortega-Liebana, M., et al., *Titania-coated gold nanorods with expanded photocatalytic response. Enzyme-like glucose oxidation under near-infrared illumination.* Nanoscale, 2017. **9**(5): pp. 1787–1792.

Chapter 3

Synthesis Methods for Photocatalytic Materials

ABSTRACT

In this chapter, basically different methods for synthesis of photocatalytic materials have been described. The three most well-known processes used to generate nanomaterials are mechanical, chemical, and physical. Different interpretations exist with regards to the production and synthesis of nanomaterials. A number of research teams have recently suggested producing nanoparticles using biological systems. Because biological methods of synthesizing nanoparticles allow for physiological pH, temperature, and pressure while also requiring very little money, they can help eliminate harsh processing conditions. The chemical and physical techniques are very expensive. Chemical vapor deposition, colloidal dispersal, sol-gel, hydrothermal way, micro-emulsions, polymer course, and further precipitation procedures are examples of chemical procedures. Responsive milling, automated alloying, ball milling with high energy, and machine-driven crushing are examples of mechanical techniques. These methods have the advantage of being simple, requiring very little equipment, and allowing for the creation of small particles in any precipitate, similar to a bristle feed. The manufacturing of nanostructured constituents including basic or doped semiconductor photocatalysts is of significance. For example, a large number of materials have been published about the synthesis, characterization, as well as evaluation of photocatalytic activity of undoped and doped (metal or non-metal) titanium dioxide. It has been shown that the preparation method utilized in the manufacturing of semiconductors mostly determines their physicochemical features and photocatalytic activity. Many synthetic techniques, such as the sol-gel process, hydrothermal and solvothermal techniques, direct oxidation reactions, microwave-assisted method, sonochemical method, chemical vapor deposition method, physical vapor deposition, and electrodeposition method, can be used to prepare photocatalytic semiconductors in the arrangement of precipitates, threads, and films.

DOI: 10.1201/9781003403357-4

3.1 SYNTHESIS METHODS FOR PHOTOCATALYTIC MATERIALS

There are numerous synthesis techniques that can be used for photocatalytic materials. Among many of them have been reported about the synthesis and characterization of titanium dioxide that also includes evaluation for doped and undoped semiconductor materials. The techniques used for the synthesis of semiconductor materials are mainly responsible for photocatalytic activity and physiochemical properties of the semiconductor materials. Diverse synthetic approaches include the sol-gel process; chemical and electrodeposition methods; reactions methods, for example, direct oxidation; microwave method; and synthesis techniques like hydro- and solvothermal and many other methods. In this chapter, we mainly discuss the basic principles that we can use for various synthesis methodologies.

3.2 SOL-GEL PROCESS

The sol-gel process is a more chemical (wet chemical) way to produce various nanostructures, especially metal oxide nanoparticles. This process involves dissolving the molecular precursor (often metal alkoxide) in alcohol or water, heating it, and stirring it until it forms a gel. Depending on the intended usage and desirable properties of the gel, it must be dried suitably because the gel produced during the hydrolysis/alcohololysis process is wet or moist. For example, if the solution contains alcohol, the drying process is finished by burning the alcohol. After the drying phase, the produced gels are crushed and calcined. Due to the low reaction temperature configuration of the components, the sol-gel technique is inexpensive and allows for good chemical control.

In many different applications, ceramics can be produced using the sol-gel process as a moulding material and as a layer that acts as a transition between thin metal oxide coatings. The materials derived from the sol-gel technique find applications in optical, electrical, energy, surface engineering, biosensor, and chromatography, among other separation technologies. The sol-gel method is a widely used and useful technique for producing nanoparticles with different chemical compositions. The precursors are combined to create a homogenous sol, which is then the foundation of the sol-gel process, turning it into a gel. After the gel's solvent is extracted from the gel structure, the remaining gel is dried. The properties of the dried gel are significantly influenced by the drying process. Stated differently, the selection of the "removing solvent method" is contingent upon the intended purpose of the gel. Dried gels are used in a variety of industries, including as surface coating, building insulation, and the production of specialty clothing. It is significant to remember that gel can be crushed in specialized mills to synthesize nanoparticles.

3.2.1 Construction

The sol-gel process is used at low temperatures—typically below 100°C—while the material is liquid. Of course, the final product is a solid; this solid is produced during the polymerization process, which involves bonding the metal atoms in the raw materials together to form M-OH-M or M-O-M (where M is the metal atom). The sol-gel method for creating aerogels consists of the following two phases:

i. During the first stage, discrete, colloidal solid particles with nanoscale dimensions are produced.
ii. Colloidal particles mix with the solvent in the second phase to form a gel.

3.2.1.1 Basics Behind Sol-Gel Method

The hydrolysis and condensation reactions (sol-gel process) are influenced by a number of variables, including the temperature, type of solvent, alkali metal activity, water/alkoxide ratio, pH of the solution, and additive used. The regular injection of catalysts to control the volume and rate of hydrolysis and condensation processes is another important consideration. It is possible to alter these processing settings to create materials with different microstructures and surface chemistries. Ceramic materials in a range of shapes can be produced by processing the "sol" further. One can spin coat or dip coat a piece of substrate to form thin films. Casting the "sol" into a mould will result in wet "gel" forming. Dense glass or ceramic particles can be produced by heating and drying the "gel" further. When a wet gel liquid is extracted under supercritical conditions, a highly porous, extremely low-density material known as an "aerogel" is created. Polysilicate gel and particles are produced by silicate hydrolysis and condensation, as demonstrated by a variety of natural systems, including opals and agates. $Si-(OC_2H_5)_4$ can be considered the earliest "precursor" for glassy materials. Ebelmen synthesized the first metal alkoxide from $SiCl_4$ and alcohol and discovered that the product gelled upon exposure to the atmosphere.[1]

3.2.2 Working Principle of Sol-Gel Method

The sol-gel production methods are based on the formation of the homogeneous solution. For this purpose, metal alkoxide in organic solvents would be dissolved, and as a result we find the homogeneous solution with the accumulation of the chelating causes. We can easily synthesize these inorganic thin films by the process of coatings on the surface of the substrate with the help of the synthesis processes for instance, drying, annealing, and thermal decomposition. In Figure 3.1, the synthesis mechanism of the sol-gel procedure has been shown.

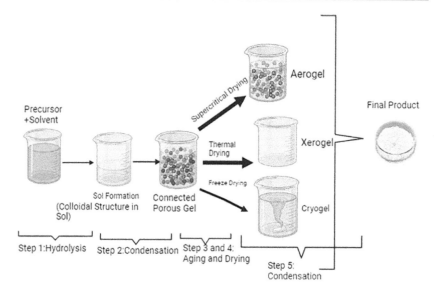

Figure 3.1 Schematic of different phases of sol-gel process from antecedent to sol-gel.

3.2.3 Pros and Cons of Sol-Gel Method

The sol-gel synthesis method consumes some benefits which are as follows:

  i. Sol-gel method can yield a thin coating so that the outstanding linkage among the substrate and the upper layer can be assured
 ii. Little temperature synthesis
iii. Creation of highly pure products
 iv. Extremely high fabrication effectiveness
  v. Creation of photosensitive mechanisms with intricate forms
 vi. Creation of composite oxides from homogenous chemicals
vii. Probability of generating uniform chemical compositions through composition design
viii. Possibility of applying the material in thin layers and employing it as fibers, aerogels, and external coverings
 ix. Chance of generating materials using this approach in an unstructured format
  x. Making materials with altered physical characteristics, such as those with high optical limpidity, a small thermal expansion coefficient, and a short UV absorption
 xi. Manufacture absorbent materials that enable carbon-based and polymeric chemical enrichment
xii. Due to the procedure in solution, precursors have a high chemical reactivity

xiii. A material assembly process with unique characteristics that could transform the constraints maintaining the development of tuberculosis (TB) and provide a net formation with a low initial cost.[1]

xiv. To provide safety against corrosion, it can also produce a thick protection

xv. Has a unique property of being sintered at low temperature at ranges from 200°C to 600°C

xvi. Sol-gel method is also simple, inexpensive as compared to other methods, and also efficient as it produces coverage in high quality

Sol-gel technique has some draw backs which are explained below:

i. Need to use organic solutions that can be harmful
ii. Requires long time for processing
iii. During processing, contraction can occur
iv. Waste of hydroxyl or carbon groups that can be obtained

3.3 HYDROTHERMAL METHOD

Hydrothermal technique for the combination of the nanomaterials is a very general technique based on a solution method. We can synthesize a wide range of nanomaterials by using the hydrothermal process at very high temperature as well as room temperature. For the aim to control the morphology of the nanomaterials, certain conditions can be employed on the primary arrangement in the proposed response either at the small and great temperature.

According to Byrappa and Yoshimura,[2] the word "hydrothermal" usually mentions a slightly heterogeneous reaction occurring below high pressure and temperature circumstances in the existence of aqueous diluters or mineralizers. This synthetic process is often carried out in autoclaves. The reaction is conducted in autoclaves, steel pressure containers with or deprived of Teflon liners, at a precise temperature and/or stress, with the response taking place in aqueous explanations. The pressure can be reached at temperatures higher than the water's boiling point or to the saturation of vapor. The degree of solution added to the mixture and its internal pressure generated by the autoclave are substantially determined by it. The investigational temperature and pressure settings and the erosion opposition in that pressure-temperature variety in a certain flush or hydrothermal unsolidified are the most crucial factors for choosing an appropriate autoclave. The corrosion resistance is a key consideration when choosing an autoclave material if the reaction is occurring inside the vessel. High-strength amalgams, such as 300 sequence (austenitic) stainless steel, ferrous steel, nickel, cobalt-based superalloys, titanium, and the situation of alloys, are the most effective materials aimed at inhibition of corrosion.[2]

The hydrothermal process has been extensively used by researchers to formulate nanoparticles with photocatalytic characteristics. For instance, quick peptization of a titanium predecessor can be hydrothermally treated with water to create TiO_2 nanoparticles. Tetra alkyl ammonium hydroxide (peptizer), titanium butoxide, and other precursors were used to create the precipitates. By increasing the dimension of the alkyl restraint, the particle size under the same peptizer concentration dipped. According to Chen and Mao,[3] the peptizers' concentrations had an impact on the particles' shape. In addition, TiO_2 nanotubes have drawn more and more interest[4]; a thorough study was published of hydrothermally produced TiO_2-based nanotubes. This work discusses the creation mechanism, structural modifications, and photocatalytic uses of TiO_2 nanotubes. Additionally, TiO_2 nanotubes have drawn more and more interest[4]; a thorough study was published of hydrothermally produced TiO_2-based nanotubes. The work discussed the formation, procedure, changes to the structure, and photocatalytic uses of TiO_2 nanotubes, evaluating the properties of the TiO_2 predecessor, hydrothermal temperature, and reaction time, as well as auxiliary techniques (such as ultrasonication and warm-up assistance) and support action (such as acid coating and calcination) on the development of titanate nanotubes.

3.3.1 Construction

As the title of the process indicates, it involves the role of water in its synthesis process. We dissolve the required precursors in the defined amount of distilled water. In recent years,[5] hydrothermal methods have been used to create phosphorus-doped TiO_2 (P-TiO_2) with a mesoporous morphology. The titanium source utilized was tetrabutyl titanate, while the dopant preparation used was the acid phosphoric.

The hydrothermal process was performed over a period of 12 hours at 200°C in a stainless steel autoclave coated with Teflon. P-doped materials were additionally produced for comparison using a similar great temperature for calcination of the sol-gel method. Beneath reproduction illumination from a xenon table lamp, the photocatalytic bustle of the produced illustrations was assessed utilizing as assessment responses the degradation of methylene blue (MB) and the disintegration of formaldehyde aqueous solution. Both hydrothermal and sol-gel techniques were used to create well-crystallized mesoporous P-doped titania nanoparticles. When associated to individuals organized by sol-gel procedure or Degussa-P25 substance, the treatment has better photocatalytic activity, a insignificant average hole size, a uneven crystallite extent, and a large surface space.[5] In addition, hydrothermal conduct generates additional lattice defects in the P-TiO_2 semiconductor. These lattice imperfections are vigorous in capturing photo-generated electrons and enhancing catalytic activity.[6]

In another investigation, it was said that WO_3 was produced by the hydrothermal method for the photoelectrochemical process in conjunction with solar water oxidation, as well as for the breakdown of an aqueous solution of the element rhodamine B (RhB) under the influence of visible light. By utilizing composites like ammonium meta tungsten hydrate as a precursor of tungsten in the acidic medium (HCl), this reaction is carried out as the hydrothermal response in an apparatus of stainless steel Teflon lined autoclave at temperature of 180°C designed for the time of 4 hours without the addition of the process of stirring. In addition, hierarchical hydrated WO_3 assemblies were manufactured by means of HNO_3 or H_2SO_4 rather than HCl. Manufactured precipitates were strengthened for 30 minutes airborne at 500°C. The aqueous phase of amitriptyline (AMT) was hydrothermally kept at 180°C for the purpose of comparison. Without the need of a tiny template or organic intermediary, a distinct three-dimensional (3D) hierarchical structure of WO_3 hydrates with a flower-like shape was created in the presence of the different acids. The morphology of the final products was not affected by the existence of the anions present in the acids. Nevertheless, it has been seen that when the result was accomplished in the non-existence of the acerbic, the final manufactured goods that we acquired were of irregular shape, insignificant in quantity, and also needed to be agglomerated. When it came to the photocatalytic performance on the breakdown of carbon-based pollutants, hierarchical WO_3 hydrates exceeded commercial WO_3 with regard to photocatalytic activity. (It is utilized as a bulk alternative.) Compared to 82% RhB degradation, 84% was obtained. With commercial oxide, 40% dye degradation was achieved. Additionally, when WO_3 film-coated fluorine-doped tin oxide conductor was exposed to recreated solar light, WO_3 microflowers displayed the synthetic enhanced activity for solar light-driven water absorption.[7]

As an alternative, ternary Bi_2WO_6 photocatalyst consumes into spheres of micro size using a sol-gel and hydrothermal method combination. The forerunner solutions of $(NH_4)_6W_7O_{24}.6H_2O$ and $Bi(NO_3)_3.5H_2O$ were present in EDTA. The produced gel, which had a pH of 3.0, was aged for the whole night until being placed in an autoclave that was continued at temperature of 220°C for 24 hours. Conversely, Bi_2WO_6 was produced straight by a hydrothermal response deprived of using the sol-gel technique. While hydrothermal (H)Bi_2WO_6 has an inconsistent platelike framework, sol-gel-hydrothermal (SH)Bi_2WO_6 was made up of monodispersed ranked nanoparticles that were porous in the centre.

According to the results of the degradation of methylene blue (MB) in the occurrence of the various Bi_2WO_6 catalysts, SH-Bi_2WO_6 has better photocatalytic activity than H-Bi_2WO_6. One explanation for SH-Bi_2WO_6 increased photocatalytic activity is the unique hierarchical structure that was created by combining sol-gel and hydrothermal processes.

3.3.2 Working Principle of Hydrothermal Method

The hydrothermal method lies within the broad category of liquid-phase chemical processes. This one is defined as a chemical response occurring in a closed container at great temperatures and pressures using water that assists as deionized solvent (solvothermal is the term used for alternative solvents). The hydrothermal process produces a crystal with fewer internal flaws, less thermal stress, and crystal planes. The element size of the creation can be accurately regulated during the hydrothermal amalgamation, which is easy and suitable. The hydrothermal method also has drawbacks, including expensive equipment needs (extraordinary temperature and pressure-unaffected steel, impervious to corrosion coating), mechanical challenges (severe temperature and pressure regulator), typically lengthy reaction times (numerous hours), and inadequate security performance.[8] Under some circumstances, hydrothermal procedures can reduce a metal alloy into an ultrafine precipitate. These procedures can also work for a solution covering different metal ions to create crystalline precipitate at great temperatures and pressures. Technology has advanced, and it is now widely used in many different industries. The route of synthesis mechanism of the hydrothermal method is as shown in Figure 3.2.

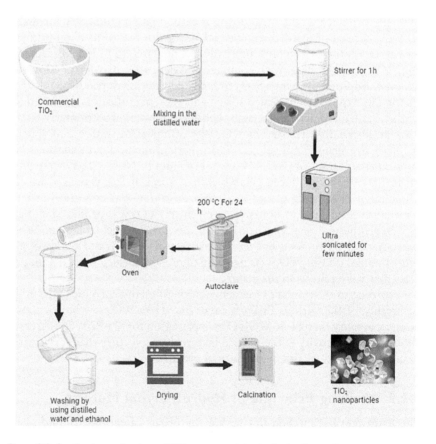

Figure 3.2 Synthesis mechanism of TiO$_2$ nanoparticles by hydrothermal route.

3.3.3 Pros and Cons of Hydrothermal Method

Compared to other approaches, the hydrothermal synthesis process has several advantages.

    i. In essence, it employs a reaction technique based on solutions.
    ii. Hydrothermal synthesis allows for the synthesis of nanomaterials at a wide range of temperatures, from very low to very high.
    iii. The morphology of the materials to be synthesized can be controlled under low-pressure or high-pressure circumstances, depending on the vapor pressure of the primary component in the reaction.
    iv. Numerous types of nanomaterials have been successfully synthesized using this technology.
    v. At high temperatures, hydrothermally synthesized nanomaterials might not be stable.
    vi. High vapor pressure nanomaterials can be produced with the hydrothermal method with the least amount of material loss.

The compositions of the nanomaterials may be precisely controlled using hydrothermal synthesis.[9] The hydrothermal method has the ability to produce crystals on a large scale with high quality.

The hydrothermal method of synthesis also has several disadvantages which are as follows:

    i. Expensive apparatus required for the process of synthesis i.e. autoclave.
    ii. In some reactions, high temperature and pressure are required to synthesize the materials which is sometimes difficult to achieve.
    iii. If we used steel tube to grow the crystal, it is impossible to observe that specific crystal.

3.4 SOLVOTHERMAL METHOD

Using a non-aqueous solvent, the solvothermal method is a hydrothermal technique. The reaction temperature may be raised to considerably higher levels than in the hydrothermal approach due to the large spectrum of organic solvents with high boiling points that can be utilized. The solvothermal technique often enables better control over the size, shape, dispersion, and crystallinity of semiconductor particles. TiO_2 nanoparticles and nanorods have been made using the solvothermal process, either with or without the application of surfactants. It has been demonstrated to be a workable method for producing a variety of nanoparticles with a restricted size distribution and dispersity.[10]

In all cases, the solvent greatly affects the crystal growth process. Solvents with a range of physical and chemical characteristics can influence the solubility, reactivity, and diffusion behaviour of the reactants; in particular, the polarity and coordinating power of the solvent can influence the morphology.[11]

Medina-Ramírez et al. described the final products' crystallization behaviour.[12] In addition to changing the solvent's polarity, the ethanol occurred at extraordinary concentrations. Non-aqueous solvothermal techniques are often more straightforward to control the size, crystallinity, and agglomeration behaviour of the nanoparticles. For instance, large specific surface area is often a feature of mesoporous TiO_2 microspheres with rough outer surfaces. This is important for semiconductor activity because it creates more active sites for photocatalytic processes. Different sizes of mesoporous TiO_2 samples have been produced using solvothermal methods.[13–15] Hierarchical mesoporous materials were created using tetrabutyl titanate as a precursor in a polyethylenimine solution diluted with 100% ethanol.[13–15] TiO_2 microspheres with great crystallinity and extraordinary BET specific surface capability have been manufactured in this context.

Hierarchical mesoporous TiO_2 microspheres with high crystallinity and high BET specific surface area were created by using tetrabutyl titanate as a precursor in a polyethylenimine solution mixed with 100% ethanol.

The activity of the produced TiO_2 microspheres was measured in the aqueous solution degradation of methyl orange (MO) and phenol under simulated solar light irradiation. The results showed a narrow range of pore diameters that are centred at 2.4 and 10.1 nm, respectively, and a high specific surface area of 118.3 m_2 g^{-1}. The TiO_2 microspheres exhibit superior photodegradation activities on both MO and phenol in comparison to P25.[15]

The solvent's physicochemical behaviour parameters, such as the dielectric constant, are modified under solvothermal conditions, which involves high pressure. Over the past few decades, solvothermal synthesis has developed dramatically and has become a flexible method for achieving the morphological control of well-defined metal nanostructures. The most current developments in solvothermal metal nano-crystallization synthesis and use are covered in this review.

3.4.1 Materials and Preparation

The anhydrous calcium chloride ($CaCl_2$, AR), sodium hydroxide (NaOH, AR), sodium dihydrogen phosphate dihydrate ($NaH_2PO_4.2H_2O$, AR), and ethanol (AR) were supplied by Sinopharm Chemical Reagent Co. Ltd. (Shanghai, China). Oleic acid was purchased from Shanghai Aladdin Bio-Chem Technology Co. Ltd. in China. All of the chemical reagents were utilized precisely as supplied, without any further purification.

The hydroxyapatite products were synthesized using a solvothermal method, as previously mentioned. First, a mixture was made in a closed system at room temperature by mixing 12 g of oleic acid, 11 g of ethanol, and 10 g of deionized water. Subsequently, 20 mL of NaOH aqueous solution were added, dropwise, to the previously described mixture to generate reaction solution B. The range of concentrations utilized was 1.0 to 2.0 M. In this process, sodium oleate is produced as a result of the reaction between oleic acid and NaOH. We also supplied the molar concentrations with the molar ratio of NaOH to oleic acid. Following preparation, 20 mL of an aqueous $CaCl_2$ solution (0.2 M) and 20 mL of an aqueous $NaH_2PO_4.2H_2O$ solution (0.19 M) were added separately into the previously described reaction solution B to create the final combination C. All the previous processes were carried out with magnets agitating very quickly. The generated mixture C was placed in a 100 mL stainless steel autoclave coated with Teflon and heated to 180°C for 23 hours as the last stage. The white product was collected for further use by centrifugation at 2000 rpm after the autoclave had cooled to ambient temperature and was disinfected three times with ethanol and deionized water.[16]

3.4.2 Working Principle of Solvothermal Technique

"A chemical reaction in a closed system in the presence of a solvent (aqueous and non-aqueous solution) at a temperature higher than that of such a solvent's boiling point" is a description of a solvothermal method. A solvothermal manner subsequently involves crucial compressions. A chemical response that happens in a solvent at a temperature better than the solvent's boiling point in a wrapped vessel is identified as a "solvothermal synthesis." By varying the response circumstances, this method suggests a fine regulator over the dissemination of constituent part size and procedure, as shown in Figure 3.3.

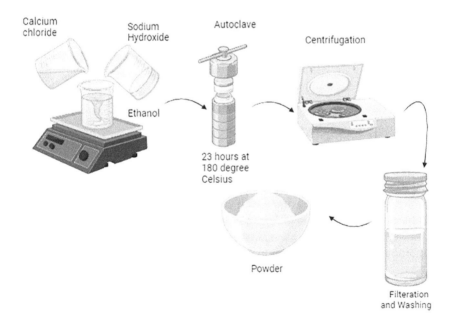

Figure 3.3 Synthesis mechanism of solvothermal method.

3.4.3 Pros and Cons of Solvothermal Technique

The solvothermal method has many advantages over other techniques of synthesis:

i. This method's main advantage is its ability to be closely controlled using liquid-phase or multiphase chemical processes, which result in high vapor pressures and little nanomaterial loss.
ii. For the production of ferrite materials with enhanced physical and chemical properties suited to both industrial and biological fields, the solvothermal synthesis approach is helpful.
iii. Aqueous or non-aqueous solvents can be employed in the solvothermal synthesis method to create ferrite materials with exact control over the size distribution, shape, and crystalline phases.
iv. In comparison to the normal solvothermal synthesis that uses joule-heating as the heat source, the microwave-assisted solvothermal technique may produce samples in a matter of minutes because of the extremely effective heating given by the microwave irradiation.

The solvothermal method has some disadvantages which are explained as follows:

i. We are able to identify the extensive reaction stages, the discontinuity of the procedure, and the dispersion of the boiler as some of the key problems of the solvothermal strategy. Therefore, a wide range of additional technologies are employed together with this procedure.
ii. High temperatures and the need for expensive equipment are some of its drawbacks.
iii. There are safety issues in the reaction process, and it is impossible to observe the products.

3.5 DIRECT OXIDATION METHOD

3.5.1 Construction

The direct oxidation method is the interaction of dry gases with metals that results in the surface production of oxides or other compounds; it is most visible at high temperatures. When synthesizing TiO_2 nanoparticles, titanium metal can also be directly oxidized with the addition of oxidants or during anodization. Utilizing hydrogen bleach to rust a titanium metal platter, crystalline TiO_2 nanorods have been produced. When a washed Ti plate is submerged in 50 mL of a 30 wt% H_2O_2 resolution at temperature of $353°K$ for the time of 72 hours, TiO_2 nanorods on a Ti plate are produced. A suspension drizzle instrument causes the production of crystalline TiO_2.

It is possible to modify the transparent stage of TiO_2 nanorods by including inorganic NaX salts ($X_{14}F^-$, Cl^-, and SO_4^{2-}). While Cl is added for use in the manufacture of pure anatase, F^- and SO_4^{2-} have been introduced to assist its synthesis. On the other hand, the presence of Cl promotes the growth of rutile, which in turn produces pure anatase. A promising approach to generate precipitated TiO_2-based photocatalysts is anodic oxidation, which may contribute to the regulated formation of restrained self-organized TiO_2 on differently structured Ti surfaces.

The electrochemical process of anodic oxidation is used to create an oxide coating on a metallic substrate. Although the substrate remains absorbed in an acid steam bath, an electrical bias is placed on it at relatively low currents. The films can have a range of microstructural characteristics and be extremely durable and thick.

3.5.2 Working Principle of Direct Oxidation Technique

Through the gas transparent barrier, the oxidant (air) penetrates the container from the opposite side. The reaction causes aluminum and oxygen to combine to form the increasing layer of the oxide matrix. When the reaction front touches the protective coating, the process is completed. Figure 3.4 (A, B) shows the direct and indirect methods of oxidation.

3.5.3 Pros and Cons of Direct Oxidation Technique

The direct oxidation method has many advantages. Some of them are as follows:

   i. Low loss of volume
  ii. Economical and straightforward apparatus

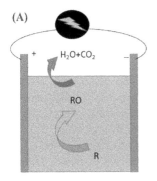 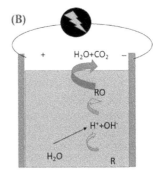

Figure 3.4 Mechanism of direct oxidation method: (A) direct oxidation and (B) indirect oxidation.

iii. Inexpensive raw materials
iv. Good mechanical qualities at high temperatures (such as creep strength) because of the lack of impurities or combustion aids; minimal remaining porosity

This method also has some disadvantages which are as follows:

i. A growth rate of approximately 0.04″/hour (1 mm/hour) demonstrates low productivity.
ii. Two to three days is far inadequate for manufacture.
iii. Aluminum residual (non-reacted) could be found in the oxide matrix.

3.6 SONOCHEMICAL METHOD

Sonication is the technique of employing sound waves to agitate the particles in liquids. These disruptions are used to combine the solutions, accelerate the pace at which a solid dissolve in a liquid, and extract the liquids' dissolved gases.

3.6.1 Construction

Pulsed ultrasound has recently attracted more attention for its potential application across an abundant larger variety of chemistry and dispensation, which has been commonly referred to as sonochemistry. Nearly all of these usages are contingent on auditory cavitation being produced in fluid mediums. Ultrasound has proven to be very helpful in the mixture of several different types of nanostructured resources, for example transition from high surface of metals, amalgams, oxides, colloids, carbides, and so forth. By applying ultrasonic technology under certain circumstances, it is possible to quickly create nanocomposites. Time, mild conditions, air, and no calcination are required.

There is no obvious correlation between the chemical effects of ultrasonography and the interaction between molecular species. Instead, audile cavitation—the development, expansion, and implosive breakdown of foams in a fluid—is the source of sonochemistry. Specific hot adverts with temporary temperatures of around 10,000 K, pressures of around 1000 atm or higher, and freezing speeds exceeding 109 K s¹ are produced when bubbles from cavitation burst. Many nanostructured materials, as well as metals, alloys, oxides, and biomaterials, can be successfully manufactured with the appropriate element size dissemination under such high circumstances, resulting in a variety of chemical reactions and physical changes.[15] The distinctive feature of the sonochemical technique is the catalyst for the chemical response; it is a viable option for reducing energy consumption throughout the production of nanoparticles. Simple necessary ways to obtain this kind of material include hydrothermal and sonochemical approaches,[17] transformed commercial Ti powder into mesoporous TiO_2 nanorods utilizing an entirely novel sonochemical method which is template free. The process of research was approved at a low temperature of 70°C and normal atmospheric pressure. Under UV light, benzene was subjected to photocatalytic deprivation in training to measure the movement of the produced samples. The photocatalytic movement was greater than P25 TiO_2 and a proposed method aimed at the development of ultrasonic mesoporous TiO_2 nanorods. The sample's higher exterior BET of 91.4 m^2g^{-1}, greater hole bulk size of about 0.55 cm^3g^{-1}, and mesoporous arrangement can be used to explain its superior activity.[17]

There are numerous reports in using sonochemical methods, non-metal–doped TiO_2 was produced.[17–19] As an illustration,[18] nanocrystalline N-doped TiO_2 was produced by sonicating the solution. Tetraisopropyl titanium and urea have been heated for 150 minutes at 80°C in water and isopropyl alcohol because the sonication period affects the synthesis of N-doped TiO_2 nanoparticles. The sample's crystallinity became more prominent at (60, 120, 150, and 180 minutes) higher sonication times; nevertheless, 150 minutes was the ideal sonication period during which TiO_2 crystallization was finished. TiO_2 can be synthesized with nanostructures employing a template-free method.

A comprehensive investigation into the impact of ultrasound on the stage arrangement, assembly, and functioning of unadulterated and nobbled TiO_2 nanocatalysts was carried out in another work.[20] Both a single-step sonochemical approach and a conservative approach were used for generating cerium and Fe-doped TiO_2 nanocatalysts with dissimilar quantities of doping fundamentals at room temperature. Both conventional and sonochemical methods for synthesis used 2-propanol as the solvent and cerium (III), ferric, and titanium (IV) nitrate by way of the predecessors. Using a direct absorption titanium buzzer in the sonication compartment, the ultrasonic procedure's sonication was carried out.

The titanium isopropoxide in 2-propanol explanation was introduced to the ultrasound reactor along with explanations of various meditations of cerium/iron predecessors, and the combination was then sonicated for a limited time of 30 minutes. While the precipitate formed during the sonochemical fabrication within 30 minutes, the conventional synthesis needed a 4-hour reaction time. The cavitation effect, which causes a faster hydrolysis response, is what caused the reaction time to be lower than it would have been during conventional synthesis. The preparation technique also affects the samples' optical characteristics.

In this particular case, it was discovered that the preparation process as well as the type of dopant had an effect on the rise in absorption in the detectible expanse. As a consequence, the sonochemically produced particles displayed greater absorption. Studies on the breakdown of crystal purplish-blue dye below UV radiation were done on demand to associate the photocatalytic action of produced catalysts. In arrears to the subordinate element size of the catalyst obtained using the sonochemical technique, the doped TiO_2 catalysts exhibited better activity when compared with catalyst created using the conventional method. This was clarified by the higher surface range for the reaction. The disintegration of the TiO_2 atoms was due to the enhanced high-velocity antiparticle crashes between the divisions.

The production of titania crystals and hydrolysis may be accelerated by ultrasonic irradiation. Furthermore, according to Huang, the sonication process may equally scatter the metallic ions throughout the TiO_2 rock crystal lattice. Silica-coated ZnO nanoparticles are additionally produced using the sonochemical method. A mixture consisting of ZnO, tetraethoxysilane (TEOS), and based on ammonia was distributed in an ethanol-water solution mixture before being exposed to ultrasound. The first TEOS/ZnO dosage that is of 0.8 for the time of 60 minutes by irradiating with ultrasonic rays caused the development of the silica coating layer. HRTEM images showed that the covering was identical and prolonged up to a range of 3 nm from the ZnO superficial. According to Siddiquey et al.,[21] silica-coated ZnO nanoparticles substantially decreased the photocatalytic activity against the photodegradation of MB peroxide in an aqueous explanation.

In recent years, ultrasound was used to generate TiO_2/WO_3 nanoparticles at room temperature. The precursors include titanium isopropoxide $(Ti[OC_3H_7]_4)$ and sodium tungstate dihydrate $(Na_2WO_4 .2H_2O)$.[22] The process of synthesis has been carried out at an intense level for 2.5 hours at room temperature in an argon environment utilizing an ultrasonic instrument. The TiO_2/WO_3 sample was made up of particles with a combination of square and hexagonal shapes, measuring 8 to 12 nm in diameter. MB degradation was explored for the photocatalytic movement of TiO_2/WO_3 nanoparticles during irradiation with visible light. The consequences demonstrated that under similar conditions, TiO_2/WO_3 nanoparticles degrade faster than bare TiO_2 nanoparticles.[23]

3.6.2 Working Principle of Sonochemical Method

The sonication method makes use of ultrasonic sound waves. Hundreds of microscopic vacuum bubbles are created in the solution throughout the procedure as a result of the pressure being applied. The produced bubbles explode into the solution during the cavitation process. When bubbles burst in the cavitation field, waves are produced that release a massive quantity of energy. As a result, the molecular bonds holding the water molecules together are severed. When the molecular bonds weaken, the particles begin to separate, and the mixing process may proceed. From heating, friction is created in the solution as a result of the sound waves' energy release. Ice

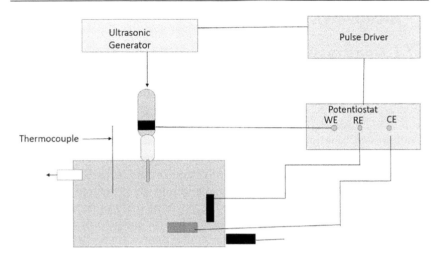

Figure 3.5 General schematic diagram of sonochemical method.

cubes are used both before and following sonication in order to stop the sample throughout the procedure[24] as shown in Figure 3.5.

3.6.3 Pros and Cons of Sonochemical Method

Some benefits of the sonochemical method are as follows:

i.   Sonochemistry is now a developed part of investigation in both carbon-based chemistry and pharmaceutical production.
ii. Ultrasound irradiation establishes an important instrument in organic combination due to its numerous benefits such as reduced reaction times, higher crops, lower budgets, and minor circumstances.
iii. This technique improves the reaction rate.
iv. In a short period of time it involves a high amount of energies and pressure.
v.   There is no need for any additives.
vi. This method also lessoned the number of steps of reactions.

The sonochemical technique of synthesis also has some disadvantages associated with it which are as follows:

i.   This process does not have the capacity to provide high efficiency of production.
ii. The amount of provided energy is not sufficient.
iii. There is a low yield.
iv. There is a delay in problems.

3.7 MICROWAVE METHOD

3.7.1 Construction

For well-organized, "green" amalgamation using little-boiling diluters at high temperature in sealed chambers, uninterrupted and lot microwave apparatuses were generated. New chemical reactions and procedures have been made easier by characteristics for quick space heating and cooling, simultaneous heating system and cooling, and discrepancy heating. On the basis of these discoveries, commercial microwave systems are available. Conventional reactions frequently require two to three instructions of fewer seconds. Green strategies have additionally generated efficiencies that are frequently greater than average, fewer or without catalyst, effortlessly recyclable diluters, or broad casting. For the purpose of determining the ideal conditions, complementary interactive software was developed. In the years before 1990, concurrent methodologies like gas chromatography-mass spectrometry and high-performance liquid chromatography-mass spectrometry, along with the use of computers and the Fourier transform, significantly enhanced the methods used for the separation and recognition of organic molecules. On the other hand, chemical synthesis and manufacture remained to be performed using outdated equipment as well as frequently inefficient procedures that had not been modified considerably over time. A great deal of the improvements had been gradual, such as rotational evaporated water, that corresponds ground-glass couplings, particularly adapters, and barriers for glassware in place of the corks or latex bungs. A high level of importance was not being paid to environmental concerns. Usually, productivity and income were given greater importance than the environmental impact or safety of chemical processes and products. Synthetic procedures frequently employ low temperatures and methods that were indirect.[25]

We searched for technologies that would allow for direct, effective, and environmentally inconspicuous synthesis due to increasing concern over the possible environmental impacts of chemicals and chemical processes. Our attempts have mostly concentrated on techniques and technology for controlled heating. Hardware (unremitting and consignment microwave reactors), software (analytical machinery for identifying the best reaction circumstances), and "green" thermal artificial techniques, counting novel reactions and required sources, are all incorporated in the services. We were intrigued in the primary microwave-assisted carbon-based syntheses, which were published in 1986. Common chemical reactions have much shorter reaction times than usual, including esterification, hydrolysis, etherification, addition, and modification. Nevertheless, the crude machinery used had insufficient controls, which led to dangers like explosion.

This suggested that if microwave heating could be performed properly, it would be beneficial to synthesis. After establishing that the method was inconsistent with organic solvents, scientists looked towards situations absent of solvent, specifically uses of "dry media" in containers that were open. The approach they took was easily put into practice and has received a lot of scrutiny. In general, reasonable internal microwave ovens were in employment deprived of temperature dimension or reaction inspiring though small variations in the scope, figure, similarity, dielectric characteristics, and smoothing of the sample in the oven had an influence on its results. Due to the varied densities of flux across the microwave chamber and the usage of cycles of duty for providing the power, uniform heating can be difficult to achieve in household ovens. These impressions seemed to have helped in the identification of unusual "microwave effects," instances of which proved problematic to reproduce.[26]

We approach in another way that is centred on our predictions that temperature measurement was essential for reproducible reactions across laboratories and that solvents made from organic matter were going to keep playing significant functions in artificial chemistry despite the emergence of solvent-free approaches. Consequently, in our opinion, for microwave machinery to be broadly advantageous, it must be allowed to function in a safe, accessible, and measurable manner. Accurately and consistently, carbon-based solvents are preferred. Studies have been carried out into solvent-free methods concurrently with others, and two separate subfields of microwave chemistry have been developed. The distinctions depended on whether closed or open containers were used, usually but not necessarily with or without solvent. The secure application of microwaves in organic synthesis was the goal of all the techniques. The whole thing illustrated how different perspectives on the same problem can lead to remarkably different yet useful solutions. As we should say about it, open-vessel electromagnetic chemistry and closed-vessel electromagnetic chemistry usually have unique benefits and drawbacks rather than common ones, and as previously mentioned, they are considered to be more supportive than conflicting.[27]

3.7.2 Working Principle of Microwave Method

In the stove, a magnetron electron pipe produces microwaves. Foodstuff absorbs the microwaves after they are caught by the stove's metallic inner. Food is cooked with microwaves because they generate heat by disturbing the water molecules. The dielectric material used in the process known as microwaves can be treated utilizing dynamism in the manner of electromagnetic waves which is high frequency. The majority of microwave heating system frequencies range from 900 to 2450 MHz.[28] Figure 3.6 shows the general synthesis mechanism of the microwave-assisted method.

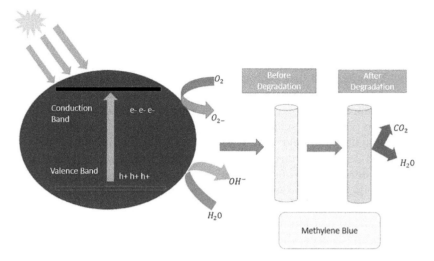

Figure 3.6 Synthesis mechanism for nanoparticles using microwave-assisted method.

3.7.3 Pros and Cons of Microwave Method

The microwave-assisted method has many advantages associated with it which are as follows:

   i. Acceleration in the rate of reaction
   ii. Requires a short period of time for reaction
   iii. Versatility in the applied conditions of reactions
   iv. High value of production
   v. Heating can be selective
   vi. Excellent value of reproducibility
   vii. Easy to handle equipment

There are some disadvantages of the microwave-assisted method which are as follows:

   i. Expensive apparatus
   ii. Not suitable for scale-up
   iii. Unfeasible to monitor the reaction

3.8 CHEMICAL VAPOR DEPOSITION

3.8.1 Construction

Chemical vapor deposition (CVD) has historically been used to create thin films based on heterogeneous materials. However, the development of specialized techniques such as plasma-enhanced CVD (PECVD), which enhances chemical vapor deposition, has made this statement approach more applicable to a wider range of polymerization. Because PECVD may disrupt sensitive resource polymers, biocompatible polymers, and biomaterials, it can potentially combine biotic and abiotic systems. The main topics include the existing applications of conventional PECVD thin films of organic and inorganic materials in biological contexts, as well as the mechanics of low-pressure PECVD thin-film development. The last half of this chapter explores the recently found applicability of low-pressure PECVD to biological materials.

Another technique that has been used for producing reinforced titanium dioxide compounds is CVD. TiO_2 nanorods that are mesoporous substances, composite substances (SiO_2/TiO_2, activated carbon/TiO_2, amines polymer/TiO_2), thin films of (which is backed in glass, metallic materials, or semiconducting substrate), besides others, have all been produced through CVD. A technique to manufacture materials at the crossroads of physical and chemical processes is chemical vapor coating. The most frequently utilized techniques to produce TiO_2 materials are chemical based because they are relatively simple to regulate, have a limited size distribution for nanoparticles, are inexpensive, and have dependable maintenance of nanoparticles in the classification. Physical approaches have attracted fewer attention, principally for significant material developed facilities since they demand extremely specialized tools and individuals for the development of material things.[29]

Highly pure materials can be generated via the CVD process, which also removes the necessity for post-heating for improving crystallinity. Nevertheless, to accomplish material depositing, ultra-high vacuity and predecessors with high suspension pressure are essential.

Depending on the kind of precursor used, the kind of substrate, the required level of thin-film homogeneity, and the reaction conditions, gas deposition procedures have been developed with minor alterations.[30] Examples of these techniques include atmospheric pressure chemical vapor deposition (APCVD), PECVD, metal-organic chemical vapor deposition (MOCVD), which is based on metal-organic precursors, and hybrid physical chemical vapor deposition. CVD stands for one of the most precise methods available when a material needs to be coated symmetrically. Consequently, CVD offers several benefits in the creation of lightweight, controllable nanostructured films with a range of constituents.[30] Heterogeneous molecular substrates have to be produced by CVD as the source for film development. Precursor particles travel in a flow of carrier gas, which may be either reactive or inert to chemicals. They become transformed into a thin solid layer of the determined substances by chemical processes that take place in the gas stage closest to the surface. The deposition process is stimulated by hot carrier gases in CVD techniques. The main chemical processes associated with the development of thin films utilize the CVD technological advances. Others are from numerous CVD processes used for fabricating thin-film nanostructured substances. There are three primary approaches for generating photocatalytic substances. The quantity of predecessors, temperature, and deposition time were modified while the films were being formed on the 85 substrates utilized. Due to modifications in the conditions of experimentation, various shapes of the films were produced. SEM was performed on films that were grown at 400°C. A compressed structure is obtained for skinny films (fewer than 400 nm) placed on uniform materials, although columnar morphology is shown when the film thickness is raised.[31] Conversely, non-flat substrates (such as glass microfibre) generate column expansion regardless of film thickness. The decomposition of pollutants in an aqueous solution for example malic acid, imazapyr, and orange G and the gas phase (toluene) has been evaluated for the films. Due to their high porosity, column shape, and high specific surface area, films placed on fibrous substrates exhibited higher photocatalytic effectiveness.

There have been observations regarding CVD-based non-metallic (B, C, N)-doped TiO_2 materials during the synthesis process. Boron-doped TiO_2 nanotubes were manufactured with visible light photocatalytic properties. It was carried out via electrochemical anodization to generate TiO_2 nanotubes. Following being anodized, the TiO_2 nanotube electrode received CVD annealing in boron-containing vapor in order to perform boron doping. The option selected for the boron availability was boric acid. The nanotubular layer's strength was compromised by boron doping. B-doped specimens, nevertheless, exhibited more intense UV and visible absorbance. Now the breakdown of methyl red, the B-doped TiO_2 nanotube electrode shown remaining photoelectrocatalytic effectiveness and good stability (more than 10 times reproducibility). The development of assisted visible light reactive TiO_2 materials has been attracting a lot of interest. In conjunction with this, Sn^{+4}-TiO_2, CNT/TiO_2, TiO_2/Ag, carbon-doped TiO_2, nitrogen-doped TiO_2, and Sn^{+4}-TiO_2 have been generated, and their environmental implications have been studied.[32]

It may be concluded from the findings mentioned later and some further examples discovered in the literature that the increased catalytic activity of CVD-demonstrated TiO_2 materials is attributable to the excellent TiO_2. Foster and coworkers presented details on the atmospheric pressure microwave CVD fabrication of TiO_2/Ag and TiO_2/CuO small films. Particularly towards methicillin-resistant *Staphylococcus aureus* and several additional infections, the films exhibited good microbicidal activity. Employing an altered CVD method, nanocomposite of V_2O_5/TiO_2 nanoparticles and films have been produced. The raw materials have been thoroughly explained and evaluated for MB decomposition.

Through the use of CVD, ZnO along with additional chalcogenide films have been produced. The components are constructed around the application of materials from only one source. ZnO, zinc silicate, and zinc selenium thin films are extensively deposited using MOCVD technology and aerosol-assisted chemical vapor deposition, based on the physico-chemical properties of these elements. Because of their restricted chemical resistance, zinc silicate and zinc selenium have been less extensively studied than ZnO, a substance that has implications for degradation in the environment.

In the final analysis, CVD is a helpful technique to generate a number of photocatalytic substances as thin films on a multitude of substrates. The substances can be grown at the atmosphere's pressure according to the work of the organometallic precursor. Although the potential that utilizing CVD techniques for manufacturing photocatalytic substances on an extensive basis will be costlier than utilizing different techniques of manufacturing, the physical characteristics are associated with the manufacture of the savings.

3.8.2 Working Principle of Chemical Vapor Deposition

The "bottom-up" technological advances group comprises CVD. The fundamental concept underlying it is to surround the process in a chamber with the vapor of a gas or liquid reactant which includes the film's components and other elements needed to perform the reaction. From the end of the nineteenth century, CVD has been investigated. It is mainly used in the preparation and synthesis of essential materials.

There are four primary phases to CVD:

i. The reaction occurs when the gas dissipates to the material's surfaces.
ii. The outermost layer of the material has an adsorbed reactions gas.
iii. On the outermost layer of the material, a reaction of chemicals begins in the third step.
iv. The outermost layer of the material has been separated from the by-products that are gaseous.[33] Figure 3.7 shows a general sketch of CVD.

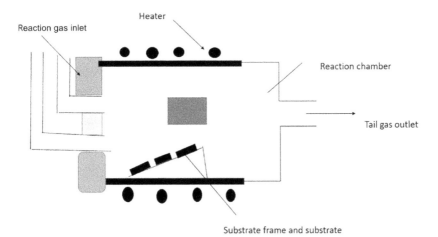

Figure 3.7 Chemical vapor deposition method.

3.8.3 Pros and Cons of Chemical Vapor Deposition

For many reasons, CVD is encouraged in numerous manufacturing processes:

i. It is not limited to a direct line of accumulation, which is a frequent characteristic of sputtering, the process of evaporation.
ii. As a result, CVD has an impressive breaking power.
iii. The rate of deposition is substantial, thick coverings can be easily acquired, the method is usually competitive, and it is comparatively easier to initiate chemical reactions on the surface of the substrate than physical modifications.
iv. Three-dimensional surfaces in nature arrangements like spaces and cracks may be coated with relatively little effort.
v. The procedure is adaptable for developing, and novel technologies may be invented on the foundation of functioning producing units.
vi. Because of its non-line-of-sight nature, it has the capacity to similarly cover components with complicated shapes and the capacity to produce extremely pure and solid films or tiny particles at relatively rapid deposition rates.
vii. Through the use of CVD, multiple metallic, ceramic, and conducting thin films are being produced.

The CVD method is not, however, an ideal choice for coating. It has a number of disadvantages:

i. The intense heat supply is the first drawback. The precursor substance, which is the procedure's particle substance, frequently possesses a high evaporating temperature. The precursors usually vaporize at 600°C, a temperature at which surfaces are fragile. Only altering the energy source can make this disappear on extra material.
ii. The temperature is the initial difficulty; nearly all dynamically motivated MOCVD activities operate at temperatures around 300°C and 800°C which remain significantly more than the outside temperature.
iii. It is challenging to synthesize multicomponent nanomaterials by CVD due to variations in pressure of vapor, the nucleation process, as well as development rates during the process of gas-to-particle transformation, all of which contribute to the heterogeneous structure of the particles.
iv. In thermal-activated CVD, there are also no highly volatile, environmentally friendly, or pyrophoric materials.

3.9 PHYSICAL VAPOR DEPOSITION

3.9.1 Construction

Physical vapor deposition (PVD) is a different approach to deposition technique that can be employed for generating materials both with and without TiO_2 dopants. In contrast to the procedure known as CVD, the

PVD technique does not require a change in chemical composition from precursors to final product preceding the development of thin films from the phase of gas. Sputtering and evaporation are the two main groups into which the basics of PVD mechanisms occur. Through the development of electron-beam heating foundations, extraordinarily high rates of deposition have been attained. For the formation of thin films, several PVD procedures are possible. Among these include magnetron and arcs methods, magnetron sputtering (also known as sputtering ion coating), and cathodic arc vapor formation. The method in which the raw materials undergo evaporation and the plasma parameters used during the depositing process differ among these PVD procedures. Spark evaporates include rapidly propagating an extremely high energy arc over a target area on the metal's surfaces with the goal to evaporate a tiny, constrained cathodic area. Metal vapor that has become significantly charged forms the ionized plasma that is produced. In the situation of the sputter, energetic neutral atoms or ions strike an object and forcefully discharge atoms. Due to advancements in sputter methods which allow the formation of various lightweight films at low temperatures, polymer substrates are currently being studied for use in industrial processes.[25]

By utilizing nitrogen ion irradiation at various nitrogen pressures along with the minerals rutile and nitrogen ion destruction, coworkers produced N-doped TiO_2 coatings. The materials' photo activity was evaluated for their capacity to break down MB. In contrast to the films' structure and nanometric dimensions, the process of photocatalytic degradation required a long time, for example time for reaction can be 60 hours. On glassware and plastic substrates, dc responsive magnetic sputtering was used to produce iron-doped TiO_2 layers. RhB's photodegradation was evaluated, and it was found that films established on polymeric surfaces demonstrated a greater rate of degradation by sunlight; subsequently, W-doped TiO_2 thinner sheets' fabrication, characterization, and photoelectrochemical properties were given. Considering the material can be utilized as the photocatalyst as well as a support for the catalyst in proton-exchange membrane fuel chambers, the use of tungsten-doped TiO_2 is of commercial significance. [34]

Employing conductive Ti and W objectives, the thin layers were applied by using magnetron sputtering on overheated (001) silicon substrates and fluorinated tin oxide-coated glassware. To produce oxidized crystalline films, the sheets were subjected to annealing in air for a period of 2 hours at 550°C. The highest concentration of tungsten which could be soluble was determined to be 33%; additionally, contamination prompted the establishment of WO_3 separated crystallized stage. The thin layers with a structure of $Ti0.92$ and $W0.08O_2$ showed the greatest photoelectrochemical efficiency for H_2 production. By oxidation annealing deposited TiN coverings, N-TiO_2 coatings are generated and then placed on metallic surfaces. Enhanced resistance to wear in the combined materials N-TiO_2@SS renders it an appropriate choice for application in medical and environmental uses.[22]

Dynamic magnetic sputtering was used to generate layers of iron oxynitride. The development of hydrophobic coatings that show potential for safeguarded, applied, and resistance to corrosion applications was made feasible by optimizing the formation variables (which include nitrogen velocity, formation time, and sputtered intensity). Sputtered silver-based TiO_2 coatings synthesis was just recently described. It was fascinating to learn that silver slowed the process of crystallization of the TiO_2 substrate. The development of acetate composites that had better tribological attributes was made easier by heating the films at high temperatures (500°C). For the production of TiO_2 and ZnO films, the PVD method and the cryogenic cooling of substrate were evaluated. When ZnO was coated at room temperature, crystallized films with a substance called structure were produced. The films of TiO_2 had an arranged shape of the surface and had the properties of a heterogeneous material, irrespective of the formation temperature. Purified acetate materials have been produced by annealing the thin films at 600°C. Microblasting is a simple, fast, and inexpensive powder coating method that was employed by Mc Donnell and colleagues to deposit TiO_2. It is significant to note that this process enables the removal of cracked films on a diversity of materials, include fluorine tin oxide–coated metallic, glass, and polymer compounds, at atmospheric temperatures and pressures while using compressed air as a carrier gas.[35]

Flexible films and angled heterostructures electron-beam deposition was employed to generate tiny films, inclined nanorods, and perpendicular nanorods of WO_3/TiO_2. In an attempt to synthesize the materials, they were brought up to between 300°C and 400°C. Although annealing at 300°C leads to a WO_3 substance with an arbitrary structure (TiO_2 in acetate structure), annealing at 400°C generates materials with orthorhombic (WO_3) and acetate (TiO_2) crystal structures. The two-layer vertical TiO_2/WO_3 tiny rods annealing at 300°C, which displayed the photodegradation ratio almost 10 times higher than the single-layer TiO_2 vertical nanorod arrays, proved to have enhanced photocatalytic efficacy for the breakdown of the colour MB. The rise in photo activity is caused by the crystalline WO_3 having a higher conductivity band level. A stage that is easier for transferring charge carriers than TiO_2 and is less distant from it.[36]

Dynamic explosive deposition, a PVD-derived method, was just recently issued. By implementing this method, it is feasible to drastically change the film's structure and constitution by changing a single deposition variable. Surfaces dispersion is limited during the development of reactive ballistic deposition through either cryogenically freezing the substrate's surface or through the existence of sturdy contacts among substrate and implanted molecule or atoms, enabling the growth of exteriors that are kinetically restricted. The substance requires evaporation using pulsing laser or charged particle sources. A high vacuum is needed to make the evaporant average free path larger than that observed in the SEM pictures which demonstrate how the shape and structure of films of TiO_2 established at various temperatures are affected by temperature. This method has been utilized to produce an increasing number of photocatalytic substances, including TiO_2, $-Fe_2O_3$, Sn- and Ti-doped $-Fe_2O_3$, bismuth vandate, S- and C-doped TiO_2, and Mo- and W-doped bismuth vandate.[26]

In the final analysis, various kinds of raw and doped TiO_2 fragile films have been extensively produced utilizing PVD methods for a wide range of uses, comprising durability against wear, the fabrication of cleaning themselves materials, and remediation of the environment. Sputtering is also one of the multiple PVD procedures that are frequently employed for spreading films throughout wide areas. In addition, as reported by numerous authors, the use of the sputtering handle allows exact control over the film's thickness, size of grains, structure, and morphology. Sputtering technique usage had previously been restricted by costly expense and complex reaction-condition management. By generating films at ambient pressure and temperature, multiple teams of researchers have been able to eradicate the problems mentioned earlier with this method of production. On the contrary, the extensive use that can be covered by thin films employing the pulsed-light deposition is restricted by the films' pollution as a consequence of particle growth throughout the film formation process.[35]

3.9.2 Working Principle of Physical Vapor Deposition

Vacuum-deposited methods that are able to produce thin films and coatings on substrate like ceramics, metals, glass, and plastics are referred to as the process of PVD, occasionally referred to as physical vapor transport (PVT). PVD is characterized by a procedure in which the substance changes repeatedly among a thin-film condensation phase and a vapor phase. Sputtering and evaporation are the two most common PVD methods. PVD is utilized in the fabrication of substances which involve squeaky films for mechanical, chemical, electrical, sound, or optical functions. By a physical process (high-temperature pressure or gas plasma), the substance that is to be deposited undergoes transformation into a liquid. The vapor proceeds to move from the source to a region with a low pressure.[29] Figure 3.8 shows a general sketch of PVD.

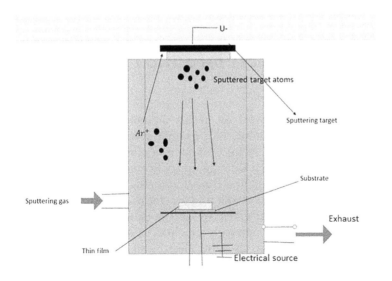

Figure 3.8 Schematic diagram of physical vapor deposition.

3.9.3 Pros and Cons of Physical Vapor Deposition

The PVD method has several advantages in contrast to other methods. Some of them are as follows:

   i. At times, PVD coatings are more durable and impervious to degradation than coatings produced via electroplating methods.

  ii. The vast majority of coatings have outstanding resistance to abrasion at high temperature and exceptional impact resistance and are so resistant that protective topcoats are infrequently required.

 iii. With an extensive selection of techniques, PVD coatings can use nearly any type of artificial and some endogenous coating materials on a broad range of materials and surfaces.

 iv. According to more traditional coating processes like painting and the electroplating process, PVD techniques are often more environmentally friendly.

  v. There are multiple methods to deposit a certain film.

The PVD method is also associated with some disadvantages which are as follows:

  i. Some methods can place limitations; for example, most PVD coating methods often restrict access for a direct line of transfer, but certain methods provide for comprehensive covering of complicated geometries.

 ii. Some PVD techniques require specific attention from operating individuals and, in certain instances, a cooling water supply to eliminate substantial thermal loads. These kinds of equipment work at high temperatures and vacuums.[7]

3.10 ELECTROCHEMICAL DEPOSITION

3.10.1 Construction

Nanoparticles are deposited on metals or conductive surfaces employing electrochemical synthesis. This technique is additionally possible for producing progressive films for example epitaxial, super lattice, nanospot, and nanoporous films. In wide-ranging applications, such as a measureable that is powdery circulates in a fluid containing binary parallel-arranged electrodes: the opposite electrode and the one that conducts substrate. The resulting solution is then given an electric potential, and nanoparticles proceed to travel in the opposite direction of the conducting surface to form the film. In addition, changing electrolysis limitations like thickness of current, temperature, and pH render it simple to control the films' distinct states. While reports of the electrode placement of films of TiO_2 by a number of titanium compounds, include TiF_4 and $(NH_4)_2TiO(C_2O_4)_2$, the application of the titanium in the form of inorganic salts in solution form has not been explored. Both a solution of acid and an environment deprived of oxygen are required for electroplating. Non-aqueous fluids provide a potential response to this problem. In accordance with multiple sources, the production of pure, doped, and composite supportive materials of TiO_2 that have different morphologies (nanoparticles, such as nanotube and tiny wires) are produced electrochemically. Amadelli et al. reported the synthesis of composite PbO_2/TiO_2 films.[36] The substances efficiently decomposed both oxalic acid and benzyl alcohol by light electrocatalysis. For the development of one-dimensional TiO_2 nanotubes and nanoparticle arrays, electrochemical depositing methods have been shown to be especially useful. Since comparisons demonstrate that directed TiO_2 nanotube arrays offer superior properties than particles of TiO_2 for a wide range of different uses (such as detecting, electrochemical splitting of water, remediation of the environment, solar cell dye sensitization, delivery of drugs, and other medical applications), these substances have caught the attention of multiple groups of researchers in the past 10 years. While nanotubes of TiO_2 have been produced via a wide range of synthetic approaches, anodizing on titanium in an electrolyte based on fluoride enables perfect oversight of the shape and dimension of the materials that are produced. TiO_2 nanotube panels with a broad range of shapes and features have been generated through modifying and enhancing the electrochemical deposition components. There are examples of multiple nanotube array morphologies generated through titanium anodization. It is apparent that alterations in deposition situations have an enormous effect on the material's building design, transforming it from well-separated, independent nanotubes to tightly arranged arrays. The thickness of the wall of a typical nanotube ranges from 5 to 30 nm, the size of the pores is between 20 and 350 nm, and their length is between 0.2 and 1000 mm.[12]

The manufacturing process of the employed conductor for solar compartments sensitive to dye has used TiO_2-supported nanomaterials; nevertheless, annealing of the physical reduces its electrochemical material goods. It has been proven that the electrode position delivers a quick and modest-cost approach to manufacturing films of TiO_2 on a wide range of substrates at lower temperatures. Electrochemically, the nanocrystalline films of TiO_2 developed. Cetyltrimethyl ammonium bromide, which stands to serve as the deposition promoting an agent, employing $TiCl_3$ as the initial material. Films that are thin of TiO_2/SiO_2 composite electrodes have been generated and documented. It is well known that the efficiency of TiO_2 materials is diminished by the combination of electrons and holes generated by bandgap electrical stimulation. Metallic nanoparticles have been introduced into the TiO_2 mixture since they can serve as acceptors of electrons, limiting the process of a recombination reaction. In the context of this, composites which include sequences of nickel nanoparticles attached to arrays of TiO_2 nanotubes have been developed using an electrochemical method. Another study describes the development of very flexible TiO_2 nanotube hybrids with a wide range of conductive materials, including polymer compounds, inorganic electronic components, and metals (Ni and Au). A broad range of materials have been electrodeposited to produce several forms of nanocomposites.[28]

3.10.2 Working Principle of Electrochemical Deposition

Electrochemical deposition is a technique for synthesizing metals that involves depositing a solid metal coating from ion solutions on an electrode-like layer that carries electricity. When sufficient electricity is passed through the mixture employing the three electrodes which are needed (working, workspace, and reference), metal ions decrease at the electrode that is working. It is an approach in which a metallic ion can solidify into a metal nanoparticle and settle on the cathode surface if sufficient electric current passes through the solution of electrolyte (Figure 3.9).

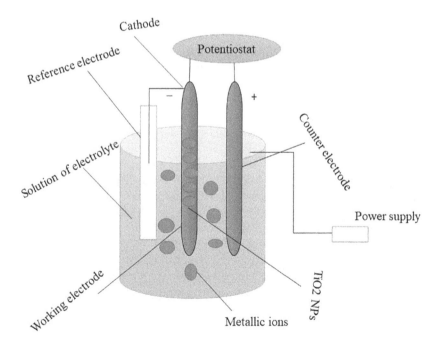

Figure 3.9 Schematic diagram of the electrochemical deposition.

3.10.3 Pros and Cons of Electrochemical Deposition

Electrochemical deposition has a number of benefits for equipment. These specific advantages of electrochemical deposition involve the following:

i. The surface of the substrate acquires a barrier throughout electroplating to protect it from damage from its surroundings. On certain occasions, this layer of protection may be sufficient to prevent corrosion carried by the atmosphere. Materials with this quality benefit significantly since they can withstand harsher circumstances for longer periods of time and need to be changed more rarely.

ii. Miniature precious metal coatings are frequently put on exterior components to improve their shine and appeal to the eye. Attractive parts may be offered for a lower price since the coating enhances their visual appeal without requiring major monetary investments. The silverware displayed is frequently electroplated in order to avoid tarnishing, which over time enhances both endurance and aesthetic appeal. Silver and copper electroplating encourages electrical conductivity in sections and provides a useful, affordable solution to improve conductivity in electronics and electrical parts.

iii. Heat susceptibility:
The potential of the substrate to endure damage from heat is enhanced by the heat susceptibility of certain metals, particularly Au and Zn-Ni. In consequence, this could improve the plated items' life expectancy.

iv. Improvement in hardness:
Electroplating is frequently employed to reinforce substrate materials, minimizing how susceptible they are to harm from strain or harsh usage. This characteristic may extend the durability of plated items while decreasing the need for maintenance.
Metals only provide some of the benefits. For instance, nickel plating can help to reduce friction, which reduces wear and tear and lengthens the item's lifespan. On the opposite hand, utilizing Zn-Ni compositions throughout production assists to avoid cutting edges that might cause damaged components. Copper is also frequently employed specifically as an undercoating because it makes it easier for different metal coatings to cling to, boosting the quality of the final product.

The various advantages of electroplating have been considered, but it is only right to also take into account its disadvantages. Let's examine the negative aspects of electroplating:

i. The main and biggest disadvantage of electroplating is pollution. The electroplating process produces a variety of hazardous wastes. However, with good waste management, this might be avoided.

ii. Another disadvantage of electroplating that is commonly noted is the cost. A full electroplating system is quite pricey, as you might anticipate. You will need to buy a lot of expensive materials, reagents, and other equipment before you can start. Thankfully, the cost of plating equipment is decreasing over time.

iii. The final disadvantage of the electroplating procedure relates to time. Since the metal layer is applied extremely slowly during electroplating, the process can be time-consuming and demanding of a great deal of tolerance. Perseverance will eventually pay off, even when components require multiple layers.

iv. The trash generated during the electrodeposition is difficult to remove from the environment since it poses a health risk.

v. The process of coating the metal in numerous layers is time consuming.

vi. Electroplating calls for expensive equipment.

REFERENCES

1. Bokov, D., et al., *Nanomaterial by sol-gel method: Synthesis and application.* Advances in Materials Science and Engineering, 2021. **2021**: pp. 1–21.
2. Byrappa, K., and M. Yoshimura, *Handbook of Hydrothermal Technology.* William Andrew, 2012.
3. Chen, X., and S.S. Mao, *Titanium dioxide nanomaterials: Synthesis, properties, modifications, and applications.* Chemical Reviews, 2007. **107**(7): pp. 2891–2959.
4. Liu, N., et al., *A review on TiO$_2$-based nanotubes synthesized via hydrothermal method: Formation mechanism, structure modification, and photocatalytic applications.* Catalysis Today, 2014. **225**: pp. 34–51.
5. Guo, S., et al., *Synthesis of phosphorus-doped titania with mesoporous structure and excellent photocatalytic activity.* Materials Research Bulletin, 2013. **48**(9): pp. 3032–3036.
6. Wang, Z., et al., *Facile construction of carbon dots via acid catalytic hydrothermal method and their application for target imaging of cancer cells.* Nano Research, 2016. **9**: pp. 214–223.
7. Biswas, S.K., and J.-O. Baeg, *A facile one-step synthesis of single crystalline hierarchical WO$_3$ with enhanced activity for photoelectrochemical solar water oxidation.* International Journal of Hydrogen Energy, 2013. **38**(8): pp. 3177–3188.
8. Li, Y., et al., *Microwave-assisted hydrothermal synthesis of copper oxide-based gas-sensitive nanostructures.* Rare Metals, 2021. **40**: pp. 1477–1493.
9. Gan, Y.X., et al., *Hydrothermal synthesis of nanomaterials.* Journal of Nanomaterials, 2020. **2020**: pp. 1–3.
10. Wu, Z., et al., *Progress in the synthesis and applications of hierarchical flower-like TiO$_2$ nanostructures.* Particuology, 2014. **15**: pp. 61–70.
11. Xiazhang, L., et al., *Ce$_{1-x}$Sm$_x$O$_2$–δ-attapulgite nanocomposites: Synthesis via simple microwave approach and investigation of its catalytic activity.* Journal of Rare Earths, 2013. **31**(12): pp. 1157–1162.

12. Medina-Ramírez, I., A. Hernández-Ramírez, and M.L. Maya-Trevino, *Synthesis methods for photocatalytic materials*. Photocatalytic Semiconductors: Synthesis, Characterization, and Environmental Applications, 2015: pp. 69–102.

13. He, F., et al., *Solvothermal synthesis of mesoporous TiO_2: The effect of morphology, size and calcination progress on photocatalytic activity in the degradation of gaseous benzene*. Chemical Engineering Journal, 2014. **237**: pp. 312–321.

14. Tanasković, N., et al., *Synthesis of mesoporous nanocrystalline titania powders by nonhydrolitic sol-gel method*. Superlattices and Microstructures, 2009. **46**(1–2): pp. 217–222.

15. Zhu, L., et al., *Solvothermal synthesis of mesoporous TiO_2 microspheres and their excellent photocatalytic performance under simulated sunlight irradiation*. Solid State Sciences, 2013. **20**: pp. 8–14.

16. Wang, Y.-C., et al., *Investigation on [OH−]-responsive systems for construction of one-dimensional hydroxyapatite via a solvothermal method*. New Journal of Chemistry, 2021. **45**(1): pp. 358–364.

17. Guo, S., et al., *Synthesis of mesoporous TiO_2 nanorods via a mild template-free sonochemical route and their photocatalytic performances*. Catalysis Communications, 2009. **10**(13): pp. 1766–1770.

18. Wang, X.-K., et al., *A novel single-step synthesis of N-doped TiO_2 via a sonochemical method*. Materials Research Bulletin, 2011. **46**(11): pp. 2041–2044.

19. Yu, C., et al., *Sonochemical fabrication of novel square-shaped F doped TiO_2 nanocrystals with enhanced performance in photocatalytic degradation of phenol*. Journal of Hazardous Materials, 2012. **237**: pp. 38–45.

20. Shirsath, S., et al., *Ultrasound assisted synthesis of doped TiO_2 nano-particles: Characterization and comparison of effectiveness for photocatalytic oxidation of dyestuff effluent*. Ultrasonics Sonochemistry, 2013. **20**(1): pp. 277–286.

21. Siddiquey, I.A., et al., *Sonochemical synthesis, photocatalytic activity and optical properties of silica coated ZnO nanoparticles*. Ultrasonics Sonochemistry, 2012. **19**(4): pp. 750–755.

22. Anandan, S., T. Sivasankar, and T. Lana-Villarreal, *Synthesis of TiO_2/WO_3 nanoparticles via sonochemical approach for the photocatalytic degradation of methylene blue under visible light illumination*. Ultrasonics Sonochemistry, 2014. **21**(6): pp. 1964–1968.

23. Carneiro, J., et al., *Study of the deposition parameters and Fe-dopant effect in the photocatalytic activity of TiO_2 films prepared by DC reactive magnetron sputtering*. Vacuum, 2005. **78**(1): pp. 37–46.

24. Carneiro, J., et al., *Iron-doped photocatalytic TiO_2 sputtered coatings on plastics for self-cleaning applications*. Materials Science and Engineering: B, 2007. **138**(2): pp. 144–150.

25. Arin, J., S. Thongtem, and T. Thongtem, *Single-step synthesis of ZnO/TiO_2 nanocomposites by microwave radiation and their photocatalytic activities*. Materials Letters, 2013. **96**: pp. 78–81.

26. Bilecka, I., I. Djerdj, and M. Niederberger, *One-minute synthesis of crystalline binary and ternary metal oxide nanoparticles*. Chemical Communications, 2008. (7): pp. 886–888.

27. Roberts, B.A., and C.R. Strauss, *Toward rapid, "green", predictable microwave-assisted synthesis*. Accounts of Chemical Research, 2005. **38**(8): pp. 653–661.

28. Abadias, G., A. Gago, and N. Alonso-Vante, *Structural and photoelectrochemical properties of $Ti_{1-x}WxO_2$ thin films deposited by magnetron sputtering*. Surface and Coatings Technology, 2011. **205**: pp. S265–S270.

29. Bilecka, I., P. Elser, and M. Niederberger, *Kinetic and thermodynamic aspects in the microwave-assisted synthesis of ZnO nanoparticles in benzyl alcohol*. ACS Nano, 2009. **3**(2): pp. 467–477.

30. Shan, A.Y., T.I.M. Ghazi, and S.A. Rashid, *Immobilisation of titanium dioxide onto supporting materials in heterogeneous photocatalysis: A review*. Applied Catalysis A: General, 2010. **389**(1–2): pp. 1–8.

31. Afzaal, M., M.A. Malik, and P. O'Brien, *Preparation of zinc containing materials*. New Journal of Chemistry, 2007. **31**(12): pp. 2029–2040.

32. Ângelo, J., et al., *An overview of photocatalysis phenomena applied to NO_x abatement*. Journal of Environmental Management, 2013. **129**: pp. 522–539.

33. Yoshimura, M., and K. Byrappa, *Hydrothermal processing of materials: Past, present and future*. Journal of Materials Science, 2008. **43**: pp. 2085–2103.

34. An, H.-J., et al., *Cationic surfactant promoted reductive electrodeposition of nanocrystalline anatase TiO_2 for application to dye-sensitized solar cells*. Electrochimica Acta, 2005. **50**(13): pp. 2713–2718.

35. Anwar, N.S., et al., *Synthesis of titanium dioxide nanoparticles via sucrose ester micelle-mediated hydrothermal processing route*. Sains Malaysiana, 2010. **39**(2): pp. 261–265.

36. Amadelli, R., et al., *Composite PbO_2–TiO_2 materials deposited from colloidal electrolyte: Electrosynthesis, and physicochemical properties*. Electrochimica Acta, 2009. **54**(22): pp. 5239–5245.

Chapter 4

Material Characterization Techniques

ABSTRACT

In this chapter, characterization procedures have been described for examining characteristics such as morphology and size of nanomaterials and to categorize the quantity of constituent chemicals and materials employed in the synthesis processes. It uses a range of techniques, such as microscopy, spectroscopy, and chromatography. Many approaches that have been utilized for analysis were introduced in the past, and new ways are now being developed. To identify the different properties of nanomaterials, different characterization techniques have been utilized. For purity, evaluation of the crystallinity and chemical structure of synthesized material has been done by utilizing X-ray diffraction. Further, scanning electron microscopy, scanning tunneling microscopy, and transmission electron microscopy have been applied to check the morphology of desired samples. UV-visible spectroscopy and atomic force microscopy have been used for optical properties and topographical analysis of nanomaterials. Additionally, to identify the surface composition, X-ray photoelectron spectroscopy analysis have been used. In this chapter, various characterization techniques are described.

4.1 SCANNING ELECTRON MICROSCOPE

On the surface of solid objects, the scanning electron microscope (SEM) works with a concentrated stream of high-energy electrons to generate a variety of signals. The morphology, chemical properties, crystallinity degree, and alignment of the sample's components are all revealed by the signals generated by electron-sample interactions. It works by regularly scanning the specimen with a beam of focused electrons. An electron source is used to produce the beam of electrons, whereas electromagnetic lenses basically used to focus the beam, just like in the transmission electron microscope (TEM).

DOI: 10.1201/9781003403357-5

The specimen's surface emits high-energy backscattered electrons and low-energy secondary electrons because of the electron beam's effect. SEM does not require complicated procedures for specimen preparation, so even large specimens can be supported.[1]

To get a clear picture, it is best to make the specimen electrically conductive. In a scanning mode with 20× to nearly 30,000× magnification and a spatial resolution of approximately 50 to about 100 nm, conventional SEM techniques can be utilized to photograph areas with widths between 1 cm and 5 microns. Additionally, SEM can focus on specific points on the specimen. Using an energy-dispersive x-ray spectroscopy (EDX), this method is mainly useful for qualitative or quantitative evaluations of chemical compositions, crystal structure, and orientations of crystals. In terms of structure and operation, the SEM and electron probe micro-analyzer (EPMA) share many similarities, and their capabilities are extensive.[2]

4.1.1 Fundamental Principles of Scanning Electron Microscopy

In SEM, the accelerated electrons have a lot of kinetic energy, and a range of signals are emitted as the incident electrons slow down in the solid specimen. Backscattered electrons can be used to highlight compositional differences in multiphase samples, while secondary electrons help to indicate the morphology and topography of samples. Both types of electrons can be used to image samples. The sample emits X-rays because of inelastic collisions among incoming electrons and electrons in particular atomic orbitals.

When the accelerated electrons return to states with lower energy, they produce X-rays with a specific wavelength. Consequently, distinct X-rays are generated by all specimen elements that are "excited" by a beam of electrons. SEM examination is regarded as "non-destructive", allowing for repeated examination of the same materials.[3]

4.1.2 Instrumentation and Working of Scanning Electron Microscopy

Instrumentation of SEM includes the following:

- Electron source
- Anode
- Condenser lens and condenser aperture
- Scanning/detection coils
- Objective lens and objective aperture
- Sample stage
- Detectors for all types of emitted electrons
- Display output devices

The sample's surface is exposed to a concentrated electron beam (Figure 4.1). Unless the microscope is designed to operate at low vacuums, the specimen is placed on a sample stage, and a series of pumps are used to produce vacuum for optimized results. The microscope's construction will affect the vacuum pressure. The position of the electron beam adjusted on the specimen can be changed with the help of scanning coils that are placed directly above the objective lens. Due to this phenomenon of beam scanning, data can be collected on a particular point on the specimen's surface. Hence, the interaction between the electron and the sample results in a number of signals. After that, the proper detectors are used to locate these signals.[4,5]

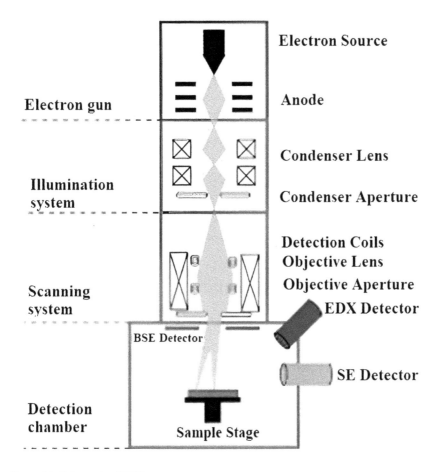

Figure 4.1 Schematic of SEM.

4.1.3 Strengths and Limitations

Following are the strengths of SEM:

i. Without certainty, the SEM has more applications for studying solid materials than any other instrument. In order to characterize solid materials, SEM plays a significant role for different kinds of domains.
ii. The majority of SEMs have extremely user-friendly "intuitive" edges. Sample preparation is minimal for most of the applications. Many applications require quick data capture (less than 5 minutes per image for solid electrolyte interphase, back-scattered electron, and spot EDX studies).
iii. Latest development of SEMs give output in digital form that is extremely portable.[6,7]

Following are limitations of SEM:

i. The samples must be solid or powdered. While the maximum limit for horizontal dimensions should be typically less than 10 cm and rarely exceeds 40 mm, the limit in the case of vertical dimensions is comparatively much larger.
ii. For the majority of devices, a vacuum of 10–5 to 10–6 torr is required.
iii. Standard SEMs should not be used to study samples that are likely to degrade at low pressures, such as rocks that are saturated with hydrocarbons, "wet" samples like coal and organic compounds, or those that are likely to outgas in the range of low pressure.
iv. SEM EDX-based detectors are limited for some elements and cannot work for elements with low atomic numbers like H, He, and Li.[7]

4.2 SCANNING TUNNELING MICROSCOPE

The scanning tunneling microscope (STM) is a device that made it possible to observe surfaces at the atomic level for the first time. STM was first developed by Gerd Binning and Heinrich in 1981 (Figure 4.2) at the IBM laboratory in Zurich, and it was given the Noble Prize in physics in 1986.[8,9] The preferred tool for observing and presenting nanostructures based on quantum tunneling is STM. Basically, STM creation led to the groundbreaking advancement known as nanotechnology. The electron microscope has long been a popular imaging tool for studying nanoscience. STM is a type of electron microscope that produces three-dimensional images of the sample. In the STM, a probe tip that raster-scans the surface from a set distance images the structure of a surface.

Due to its high resolution and clear imaging, Ms became a valuable tool for nanotechnology in the 1980s. The development of many nanostructures, including corrals, was monitored using STM in the 1990s. Additionally, Friedel oscillations were restricted to the nano-objects and enhanced by adding contaminants to an electron gas.[10-12] With the use of STM, lateral resolution

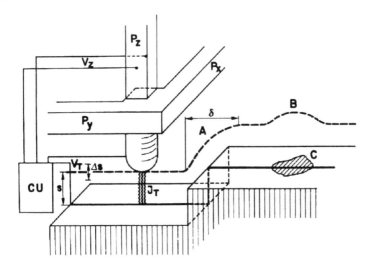

Figure 4.2 Schematic of first STM.

Source: Reprinted with permission from Binnig, G., et al., *Surface studies by scanning tunneling microscopy*. Physical Review Letters, 1982. **49**(1): p. 57. Copyright © 1982 American Physical Society.

of 0.1 nm and depth resolution of 0.01 nm can be achieved. Along with other outstanding qualities, it has the rare ability to function anywhere. It provides versatility for working in a variety of gas or liquid environments, including ultra-high vacuum, water, air, and other gases or liquids, at temperatures ranging from almost 0 K to over 1000°C.[13-16] Spin-polarized STM (SP-STM) was developed at the beginning of the twenty-first century and used to research the formation of magnetic layers.

4.2.1 Basic Principle of Scanning Tunneling Microscopy

STM works on the quantum tunneling theory. Particle tunneling through a wall that would be impassable in classical physics is referred to as quantum tunneling. Early in the twentieth century, the effect was predicted, and by the middle of the century, it had been recognized as a general physical phenomenon. In the phenomenon of tunneling, electrons behave like waves and pass through potential barriers.[17] Quantum tunneling occurs in STM between the probe tip and the sample. The very tiny STM tip creates an electron tunnel between the sample and probe surfaces (Figure 4.3). Calculating tip height and sample work function for imaging while maintaining a constant tunnel current.[18]

In 1982, the first paper on the application and operation of STM was published.[19] A metal probe attached to a support made consisting of the piezoelectric P_X, P_Y, and P_Z elements. When a voltage is applied across piezoelectric materials, they alter in size. While P_X and P_Y are used to perform a raster-scan by the probe across the sample surface, P_Z is used to

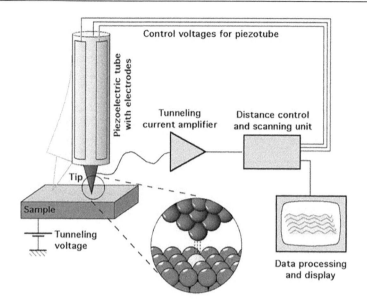

Figure 4.3 The components and working of STM.

Source:Adapted with permission from under the terms of the Attribution 2.0 International (CC BY ShareAlike 2.0) AT (https://creativecommons.org/licenses/by-sa/2.0/at/chem.libretexts. org).Michael Schmid,TuWien,https://www2.iap.tuwien.ac.at/www/surface/stm_gallery/stm_ schematic; https://chem.libretexts.org/Bookshelves/Physical_and_Theoretical_Chemistry_ Textbook_Maps/Supplemental_Modules_%28Physical_and_Theoretical_Chemistry%29/ Quantum_Mechanics/02._Fundamental_Concepts_of_Quantum_Mechanics/Tunneling.

maintain probe height. Following a set level of tunnel current, the tip is moved across the sample's surface to scan it in a raster pattern, and it is then halted. Through a feedback electrical system, tunnel current is kept constant to control the height of the probe during the scan. The height of the tip may alter during the scan due to work or surface texture.[20]

This remarkable atomic scale precision was utilized in ultra-high vacuum and at high temperatures to produce photographs of the complex surface that arises from heating and slowly cooling the silicon (111)* structure. Because of the symmetry of the pattern created when low-energy electrons are diffracted from this rebuilt silicon surface, it is referred to as the silicon "7 7" surface.

4.2.2 Construction and Working of Scanning Tunneling Microscopy

The operational amplifiers (OP AMP) in integrated circuits 1, 2, and 3 have high gain values. The input impedance of these amplifiers is very high. This occurs because these OP AMPs are configured to function in negative feedback. The result is produced with a zero-voltage difference between the positive and negative terminals. A sample biased at sample volts with respect to ground is depicted in Figure 4.4.

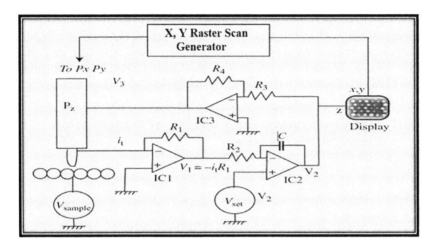

Figure 4.4 Schematic circuit of the height control circuit of an STM. The triangles represent "operational amplifiers," very high gain amplifiers that are operated by feeding the output back to the input so as to keep the "+" and "–" input terminals at the same potential.

The inverting input of IC1 is linked to the STM probe tip, and the non-inverting input is connected to ground. The tip of the probe needs to be on virtual ground. The output for this configuration comes from IC1, and V_1 increases to a level equivalent to the negative product of the feedback resistance (R_1) and tunnel current (It). Consequently, IC1's output voltage depends on tunnel current.

The height of the tip is under the control of IC2. The non-inverting terminal is applied with a predetermined voltage (V_{set}), and the feedback capacitor in the feedback loop measures the difference between V_{set} and V_1. The difference between the observed output tunnel current and the predetermined tunnel current directly affects V_2. The error signal is sent by IC3 to a high-voltage amplifier, which also controls the P_Z element.[20,21]

The feedback process is kept entirely negative throughout. The probe is brought back from the surface if the tunnel current increases. The height of the sample is also represented by the control signal V_2, which is used to modify the display's brightness. The display's *x-y* position and the probe's *x-y* position are synchronized.

4.3 TRANSMISSION ELECTRON MICROSCOPY

Similar to the light microscope, TEM is based on optical principles. In addition, TEM offers higher resolution. We can study the ultrastructure of organelles, viruses, and macromolecules with this increased resolution. TEM can also be used to examine samples of specially prepared materials. When conducting research, the light microscope and the TEM are frequently used in conjunction with one another. Microscopes called TEM utilize a beam of electrons in order to see samples and produce an image with high magnification. Up to 2 million times can be magnified with TEMs. It is not surprising that TEMs have grown to be so useful in the biological and medical fields.[5,22]

4.3.1 Instrumentation and Working of Transmission Electron Microscope

TEMs consist of the following components:

- An electron source i.e. thermionic gun
- Condensers
- Sample stage
- Objective diaphragm
- Intermediate lens
- Projector lens
- Phosphor or fluorescent screen
- Electron energy-loss spectrometer
- Computer to display output

The TEM is a type of electron microscope that is divided into three categories: The first category is an electron gun that creates a particular beam of electrons along with a condenser system that directs it toward the target. Second, there is a system for making images that entails an objective lens, a sample stage for the specimen, and intermediate projector lenses (Figure 4.5). These lenses are used to focus the beam of electrons on the specimen to make a real, high-magnification image. Last, there is an apparatus for recording images that typically includes a digital camera and a fluorescent screen for displaying the image. It also gives information about loss of electrons through electron energy-loss spectrometer on the display. Additionally, a vacuum system consisting of pumps, gauges, and valves, as well as power supplies, is needed.[23,24]

4.3.1.1 The Electron Gun and Condenser System

The source of the incident electrons is a V-shaped tungsten filament that looks like a sharply pointed pole made of lanthanum hexaboride. The central aperture of the control grid, also referred to as a Wehnelt cylinder, that surrounds the filament is positioned so that the cathode's apex lies at, just above, or just below it. The isolated instruments such as cathode and grid have a negative potential that should be equivalent to the accelerating voltage. The electron cannon's final electrode is the anode, which is a disc with an axial hole. If the high voltage has been sufficiently stabilized, electrons will pass through the centre aperture with constant energy as soon as they

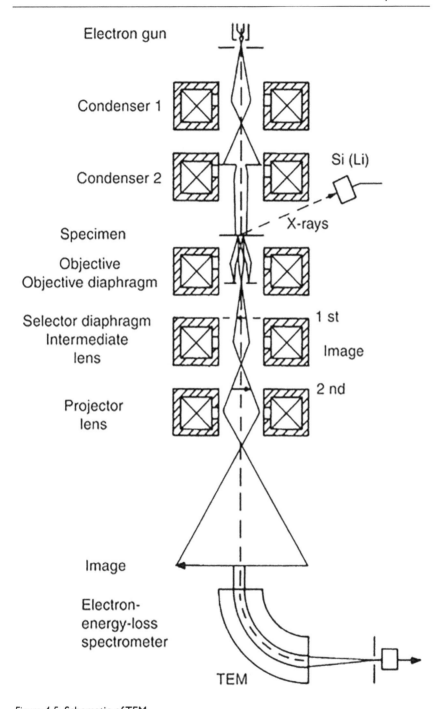

Figure 4.5 Schematic of TEM.

Source: Reprinted from Arnold, W. *Encyclopedia of nanotechnology*. In *Encyclopedia of Nanotechnology*. Springer, 2012. Copyright © 2012 with permission from Springer.

leave the shield and cathode. Control and alignment are crucial for the electron gun to function effectively.

On the other hand, a condenser lens system basically regulates the beam's brightness and angular aperture. The beam can be focused onto the target using a single lens, although a double condenser is more frequently utilized. Here, the second lens projects the source's reduced image onto the object after the first lens, which is powerful, generates a reduced picture of the source. Small spot sizes are used to reduce specimen disruptions brought on by heating and radiation.[25]

4.3.1.2 The Image-Producing System

In a mobile specimen stage, a little holder holds the specimen grid. The projector lens or lenses further enlarge the real intermediate picture produced by the objective lens, which typically has a modest focal length of approximately 1 to 5 mm. A single projector lens may have a 5:1 magnification range, and a greater range of magnifications may be attained by using the projector's interchangeable pole components. The latest instruments include two projector lenses for high magnification and optimized results.

For purposes of picture stability, a high magnification of 1000 to 250,000 is frequently offered by the microscopes at the output. Enlarging a photo or digital image can result in a higher final magnification. The electron microscope's final image quality is greatly influenced by the precision of aligning the lenses accordingly. The incident sources used by the lenses need to be extremely stable in order to achieve the highest possible resolution. Electronic stabilization is required to a level greater than one part in a million. Computers operate the latest electron microscopes, and specialized software is also easily accessible.[25]

4.3.1.3 Image Recording

The monochromatic electron picture must be transformed into something the human eye can see, either by leaving the incident electrons on a screen attached to the microscope or by digitally recording the view for projection on a computer screen display. Prior to publishing, computerized photographs can be edited or processed using a format like TIFF or JPEG. It is possible to add artificial colours to a monochrome image by identifying sections of the image or pixels with specific properties. This may turn a raw image into a visually appealing image, which can help with visual interpretation and education.[25,26]

4.3.2 Strengths and Limitations

TEM methods are extremely effective in material analysis and offer a number of benefits, including the following:

  i. TEMs allow for the creation of high-quality pictures.

  ii. High spatial resolution TEM applications exist in a variety of sectors, including science, education, and industry.

  iii. Information on elements and atomic bonds may be obtained using TEMs in conjunction with EDX and electron energy loss spectroscopy.

  iv. TEMs offer the maximum magnification compared to other microscopes.

  v. TEMs can deal with specimen's surface characteristics, shape, size, and structure.[27]

TEMs do, however, have a few drawbacks:

  i. The instruments are pricey and quite huge.

  ii. Because TEMs are sensitive to mechanical vibration, fluctuations in electromagnetic fields, and changes in cooling water, they need housing and maintenance.

  iii. Making samples out of bulk materials often takes a long time.

  iv. Sample preparation can produce potential artefacts.

  v. Specialized training is required for data analysis and tool operation.

  vi. The types of TEM samples that work under vacuum are limited.[27]

4.4 X-RAY DIFFRACTION

In 1912, Max von Laue observed that crystalline materials behave similarly to crystal lattice plane spacing as 3D diffraction gratings at different wavelengths of X-rays. X-ray diffraction (XRD) is now commonly utilized to study atomic distances and crystal structures. If monochromatic X-rays are incident on a crystalline specimen, constructive interference occurs which has led to a foundation of XRD. The monochromatic X-rays come from a cathode and must be filtered, collimated, and then incident on the specimen.

This technique basically determines the crystallographic structure of materials by measuring the intensities and scattering inclination of the incident X-rays. It also helps to examine the phase, composition, and impurities in a sample for comparatively small-sized specimens. The identification of materials by their diffraction pattern is one of the primary applications for XRD analysis. In addition to phase identification, XRD offers the ability to calculate the internal stresses that differentiate between the actual structure and others.[28]

4.4.1 Working of X-Ray Diffraction

Crystals represent the regular arrangements of atoms; however, X-rays are basically electromagnetic (EM) waves. In order to scatter incident X-rays,

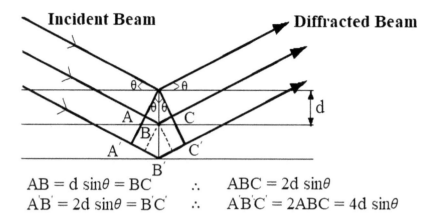

$AB = d \sin\theta = BC$ $\therefore$ $ABC = 2d \sin\theta$

$A'B' = 2d \sin\theta = B'C'$ $\therefore$ $A'B'C' = 2ABC = 4d \sin\theta$

Figure 4.6 Bragg's law of XRD.

crystal atoms primarily interact with their electrons. In this case of elastic scattering, the electron acts like a scatterer. These waves cancel the effect of each other and interfere destructively in different directions, but in some, they add positively.

Bragg's Law:

$$n\lambda = 2d\sin\theta \tag{4.1}$$

where n, d, and λ represent an integer, distance, and beam wavelength, respectively. Wavelengths of X-ray beam is an optimal choice due to its comparable wavelength with the spacing in the specimen's atom (Figure 4.6). In consequence, this spacing will have a significant effect on diffraction angle. The sample is then subjected to X-rays, which "bounce" off the structure's atoms.[29,30]

4.4.2 Data Interpretation in X-Ray Diffraction

The obtained results of XRD can be interpreted by observing the peak of intensity for a variety of signals at specific diffraction angles. The 2θ sites for pure ZnO, which stand for a specific separation between the crystals in the specimen, are examined by the diffraction angle from incoming X-ray beam transmitted on the surface of the specimen (Figure 4.7). The peak's intensity depends on the quantity of molecules in that phase or at that distance apart. Peak intensity leads to an increase in the number of crystals or molecules with that specific spacing.

Crystal size and peak width are inversely related. A narrower apex is linked to a larger crystal. A peak that is wider indicates the existence of a small size crystal, a defect in the structure, amorphous materials show lack

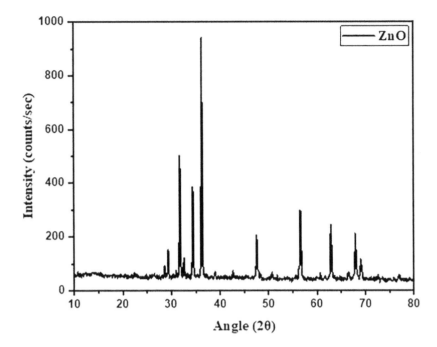

Figure 4.7 Interpretation of XRD spectra of pure ZnO.

in crystallinity. For smaller samples, the composition of a sample can be determined using the patterns that are found during XRD analysis. A huge database of these substances stores the diffraction patterns of elements, compounds, and minerals. The elemental analysis can be interpreted by comparing its peak values and positions with literature.

4.4.3 Applications of X-Ray Diffraction

Following are the useful applications of X-ray powder diffraction:

  i. Crystalline material characterization
 ii. Detection of optically challenging fine-grained minerals and quantitative analysis
iii. Calculating the size of unit cells and multilayered epitaxial structures
 iv. Sample's purity determination and use of Rietveld refinement to identify structural analysis
  v. Measurement of lattice mismatching, density dislocations, and optimized film quality on substrate
 vi. Helpful for analyzing the polycrystalline specimen i.e. grain orientation[30,31]

4.4.4 Strengths and Limitations

Following are the strengths of XRD analysis;

    i. It is a powerful and quick (<20 min) approach for identifying unknown minerals.
    ii. Most of the time, it offers an unambiguous mineral diagnosis.
    iii. It requires little sample preparation.
    iv. XRD equipment is widely available.
    v. The process of interpreting data is straightforward.

Following are the limitations of XRD analysis:

    i. It is limited to only homogeneous and single-phase materials.
    ii. It is helpful only for powdered samples having weight of tenths of a gram.
    iii. The percentage of threshold detection is only 2%.
    iv. Unit cell measurements require challenging indexing patterns for non-isometric crystal systems.
    v. Peak overlap is a possibility and gets worse for high angle "reflections".[32]

4.5 ATOMIC FORCE MICROSCOPE

Binnig, Quate, and Gerber led to the development of atomic force microscopy (AFM) and the Nobel Prize–winning STM, the phenomenon of scanning probe microscopy has grown significantly at nanometer scale further than the use of interatomic forces to examine the topography. The possibility of seeing atoms and measuring interactions between molecules intrigues scientists. Because the material characteristics and surface topology frequently do not correlate, topography imaging is not enough to give researchers the necessary information.[33] Because of these factors, sophisticated imaging techniques that give quantitative information on different surfaces have been developed. AFM techniques can now be used to measure a vast range of material properties, including electrical forces, surface potential, capacitance, and magnetic properties.

4.5.1 Why Atomic Force Microscopy?

Since STM is only limited to conducting and semiconducting surfaces, in order to overcome these limitations, the development of AFM was addressed. The advantage of the AFM is that it can examine virtually any surface as well as biological materials, polymers, composites, glass, and ceramics. The AFM was developed in 1985 by Binnig, Quate, and Gerber.

A gold foil strip and a diamond shard served as their initial AFM. When the diamond tip came into direct contact with the surface, the interaction mechanism was interatomic van der Waals forces. The STM above the cantilever served as the second tip for detecting the cantilever's vertical movement.[33]

4.5.2 Working of Atomic Force Microscope

Similar to STM, a feedback loop is used to raster-scan a sharp tip over a surface to change the settings required to take a surface picture. Unlike STM, the AFM does not require a conducting specimen. In this case, the tip of the cantilever basically exerts a force on sample atoms (Figure 4.8). Virtually each measurable force like van der Waals, electrical, magnetic, and thermal, can be examined using this technique. The term "scanning probe microscopy" is frequently used to describe these methods. Programming adjustments and tweaked tips are expected for a portion of the further developed techniques. Additionally, along with feedback loop device, AFM frequently incorporates deflection and force measurement.[34,35]

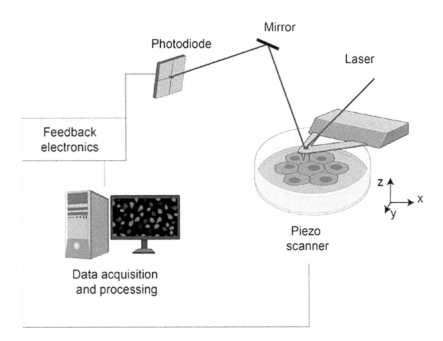

Figure 4.8 The components and working of the AFM.

Source: Reprinted with permission from Müller, D.J., et al., *Atomic force microscopy-based force spectroscopy and multiparametric imaging of biomolecular and cellular systems.* Chemical Reviews, 2020. 121(19): pp. 11701–11725. Copyright © 2021 American Chemical Society.

4.5.2.1 Probe Deflection in Atomic Force Microscopy

Most AFMs have traditionally employed a laser beam deflection technique, in which the laser beam is reflected onto a position-sensitive detector from the back of the reflecting AFM lever. Micro-fabrication using Si or Si_3N_4 has been done for cantilever and tip of AFM whose radius is about few to tens of nanometers.[33]

4.5.3 Modes of Operation

The following sections discuss the modes of operation (Figure 4.9) for AFM.

4.5.3.1 Contact Mode

By keeping a constant tip deflection, the contact mode approach uses a steady force for tip-sample interfaces. In order to maintain the cantilever's initial deflection, the scanner moves the entire probe while the tip transmits feedback loops containing information about the probe's interactions with the surface. The force is computed rather than directly measured with the cantilever's stiffness and the lever's deflection measured.[36,37]

According to Hook's Law:

$$F = -kz \tag{4.2}$$

where F is the applied force, k is the lever's stiffness, and z is the length.

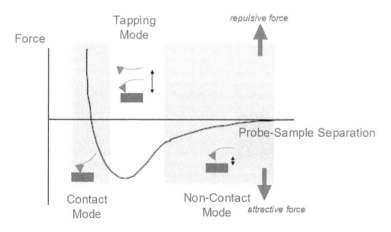

Figure 4.9 Different modes of operation of AFM.

4.5.3.2 Tapping Mode

The cantilever oscillates externally at the frequency of its intrinsic resonance when it is in tapping mode. When the probe starts scanning the specimen's surface, the frequency of the oscillation is controlled by mounting a piezo-electric on apex of the cantilever. Interactions between the probe and the surface are what lead to changes in frequency of oscillations. These modifications can be used to deduce specifics about the sample's surface or materials. Because the tip is not dragged across the surface, this technique is more delicate than contact AFM, but scanning takes more time comparatively. Additionally, it frequently offers lateral resolution that is superior to that of contact AFM.[38]

4.5.3.3 Non-Contact Mode

The cantilever oscillates in the non-contact mode slightly above its resonance frequency, and when the tip approaches the surface, the forces produced by the material are observed. If the oscillation frequency or amplitude is consistent, the average tip-to-sample distance may be utilized to scan the surface. The method's little stress on the sample helps to prolong the life of the tip. However, unless put in a powerful vacuum, it often does not offer very excellent resolution.[37]

4.5.4 Mechanism of Feedback Loop

In AFM, laser deflection is used in a feedback loop to counter the force and specific location of the tip (Figure 4.8). As seen, a cantilever with an AFM tip reflects a laser from its underside. As the tip connects with the surface utilizing the laser position of the photodetector, the feedback loop scans and measures the surface.

4.5.5 Limitations

AFM images frequently contain artefacts, which are distortions of the actual topography. These issues are generally brought about by issues with the probe or image processing. The AFM gradually scans the surface, making it susceptible to changes in the outside environment that cause thermal drift. If the tip is damaged or blunt, it can also result in deceptive pictures and images with low contrast.

Noise, which manifests as streaks or bands in the picture, can be produced by the movement of particles on the surface because of the cantilever's movement. The sample size varies depending on the equipment, but it is typically 8 mm wide by 8 mm high. AFM has difficulty with solid samples because the tip might move the material as it scans the surface. To apply a layer of material as uniformly as possible and obtain the most precise measurement of particle heights, solutions or dispersions work best. The typical method for accomplishing this is to spin-coat the solution over newly sliced mica, which enables the particles to adhere to the surface after drying.[39]

4.6 SURFACE COMPOSITION

X-ray photoelectron spectroscopy (XPS) is such a type of technique which is used for surface examination, sometimes stated as electron spectroscopy for chemical analysis (ESCA). It is a great place to start when examining surface structure since it gives both qualitative as well as quantitative data regarding the surface composition of a material.

4.6.1 X-Ray Photoelectron Spectroscopy

A surface-dependent, non-destructive procedure called XPS is used to calculate the topmost layer of about 10 nm (about 30 layers of atoms) of organic and artificial materials. By estimating the limiting energies of components, which are associated with the sort and strength of their chemical bonds, XPS examines the organization of material surfaces. It also examines the overall corresponding abundances of constituents on surfaces (semi-quantitative examination), and the chemically active polyvalent particles. The surfaces of numerous materials, including inorganic mixtures (minerals), semiconductors, natural mixtures, natural and artificial materials are described utilizing XPS. Research on processes including surfaces, like sorption, catalysis, redox, disintegration/precipitation, erosion, and dissipation/statement type responses, is upheld by XPS.[40]

4.6.2 Fundamental Principle and Working of X-Ray Photoelectron Spectroscopy

The photoelectric effect, which Einstein originally depicted in 1905 and was awarded the Nobel Prize for in 1921, is utilized in this technique, which includes particles emitting electrons in response to electromagnetic radiation. When the limiting energy of the material's electrons is surpassed by the photon energy, photoelectrons will be generated from the material for which the energy does not correspond to the intensity of EM radiation but the frequency (hv).

A sample is exposed to X-rays (photons), and after the sample's electrons have absorbed enough energy, they are expelled with a certain amount of kinetic energy. A detector measures the energy of those expelled electrons, and a plot of their energies and relative densities of electrons is created. Differentiating the electrons and creating the spectra are made possible by the varied pathways that electrons of different energy take through the detector (Figure 4.10).

Atoms in the chemical being evaluated by XPS are identified using the following equation:

$$E_{binding} = E_{photon} - E_{kinetic} - \phi \qquad (4.3)$$

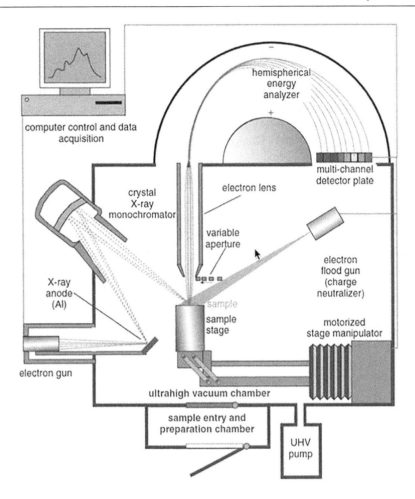

Figure 4.10 Schematic components and working of XPS.

Source: Adapted with permission under the terms of the CC BY-NC-SA license 3.0 (http://creativecommons.org/licenses/by-nc-sa/3.0/) X-Ray Photoelectron Spectroscopy (XPS; aka Electron Spectroscopy for Chemical Analysis, ESCA) David Mogk, Imaging and Chemical Analysis Laboratory, Montana State University (https://serc.carleton.edu/msu_nanotech/methods/xps.html).

Here, binding energy $E_{binding}$ is the force that pulls an electron toward a nucleus, photon energy E_{photon} is the force that drives the X-ray photons that the spectrometer uses, and kinetic energy $E_{kinetic}$ is the force that drives the electrons that are expelled from the sample. The correlation factor of the instrument deals with the work function ϕ that is basically the least amount of energy needed to evict an electron from an atom. The detector measures the kinetic energy, and the work function includes known quantities. The main remaining question is the binding energy, which may then be answered. The binding energy is lower for higher orbitals because ejecting electrons from orbitals farther from the nucleus requires less energy. The energies of the electrons in various subshells (s, p, d, etc.) also vary. XPS enables the determination of a material's composition by displaying the energy of electrons released from it. Additionally, XPS may be used to identify chemical changes. This is because binding energy depends on more than just the electron's shell.[41,42]

4.6.3 Data Interpretation

The fraction of electrons with a certain binding energy is shown as peaks in the XPS spectrum. Less electrons are indicated by shorter peaks (Figure 4.11). It can be understood that there were half as many electrons detected with the binding energy at peak B as there were at peak A if, for instance, peak B is half as tall as peak A. As a result, the peak intensities reveal a material's percentage makeup. As can be observed in the preceding image, the O (1s) has the biggest peak, indicating that oxygen has the highest atomic composition.

An electron is attracted to the nucleus more strongly depending on how much binding energy there is. For example, electrons in the first state will

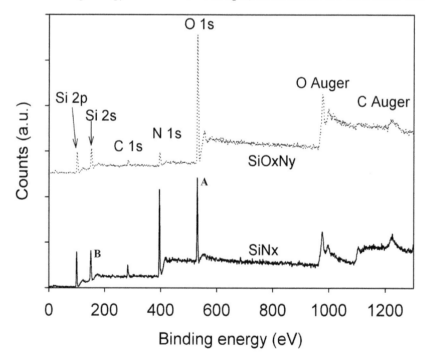

Figure 4.11 Interpretation of XPS spectra.

have peaks with a greater energy than electrons in the second state. The energy of electrons in 2s will be higher than that of 2p (Figure 4.11). While some instruments offer peak identification characteristics, it is still possible to identify peaks and lines on spectra by comparing them to standards made of various materials. For assistance in interpreting the spectra, Moulder (1992) provides examples of these standards.[43]

4.6.4 Quantitative Analysis of X-Ray Photoelectron Spectroscopy

Semi-quantitative analysis of elements can be integrated by evaluating the area under the peak and using the proper elemental sensitivity aspects. Concentration of the desired element in quantitative analysis is obtained as follows:

$$C_x = \left(I_x / S_x\right)\left(\Sigma I_i / S_i\right) \tag{4.4}$$

where C_x, I_x, and S_x represent concentration intensity and sensitivity factor of elements, respectively. Additionally, I_i / S_i is the total of the studied element's proportions of intensities divided by their respective sensitivity factors. Atomic concentrations may often be calculated to within 10%.[44]

4.6.5 Limitations of X-Ray Photoelectron Spectroscopy

Following are the limitations of XPS:

i. The ultra-high vacuum chamber used by XPS (10–9 Torr) causes some materials to become unstable.

ii. The researched surface areas are much greater since X-ray beams cannot be concentrated like electron beams can. The studied region will typically be 1 mm × 1 mm or, at best, 10 to 100 microns wide. The latest XPS devices could be able to produce a "tiny spot," but doing so might require physical scaling which minimizes the count rate. A synchrotron source's X-rays can be used for small spot XPS.

iii. When lines or peaks overlap, it might be challenging to tell which atoms' electrons are being represented. It is critical to recognize auger lines with intricate patterns that are unrelated to the kinetic energy from electron ionization.

iv. X-ray ghost lines are tiny lines or peaks that are slightly off-centre due to X-ray sources other than the instrument itself. The behaviour at the surface and the bulk may differ significantly because just the surface has been examined.[45]

4.7 ULTRAVIOLET-VISIBLE SPECTROSCOPY

Ultraviolet (UV)-visible spectroscopy is an analytical method that investigates the discrete wavelengths of either UV or visible light that are absorbed or transmitted. The composition of the sample has a considerable impact on this aspect and reveals the types and quantities of by-products present in the specimen. As this spectroscopic method relies on the use of light, the properties of light are crucial in determining the results. Therefore, understanding the characteristics of light is essential in UV-visible spectroscopy.[46]

4.7.1 Instrumentation and Working of UV-Visible Spectroscopy

The most efficient variant of the UV-visible spectrophotometer, which includes significant key features as depicted in Figure 4.12, comprises various components. These components are as follows:

  i. Light source
 ii. Wavelength selector (monochromator)
iii. Splitter and reference cell
 iv. Optical lens
  v. Sample cell
 vi. Detector
vii. Amplifier
viii. Computer for signal processing and output

Since this spectrophotometer utilizes light-based technology, it necessitates an elevated source of light that can operate across a range of wavelengths. Among other light sources, a xenon lamp is considered the most efficient high-energy light source for UV and visible wavelengths.[46,47]

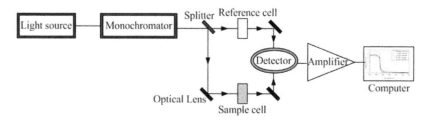

Figure 4.12 Instrumentation and working of UV-visible spectrometer.

Xenon lamps, in comparison to tungsten and halogen lamps, are less stable and more expensive. The next step involves selecting suitable wavelengths from the range emitted by the light source for sample preparation and analyte monitoring during inspection. Several techniques are available for this, such as monochromators, absorption filters, interference filters, bypass filters, and cutoff filters. The most popular method is the monochromator because of its adaptability. To further limit the light wavelengths utilized for accurate measurements and to improve the signal-to-noise ratio, filters and monochromators are frequently used.

In a spectrophotometer, the light passes through the sample using the chosen wavelength selection technique. It is essential to measure a reference sample, also called a "blank sample," for all experiments. This reference sample is typically a cuvette filled with the same solvent used to extract the sample. The baseline is the aqueous solution devoid of the target chemical when an aqueous buffered solution is used for analysis. When analyzing bacterial cultures, a sterile culture medium is employed as the guide.[48] The spectrophotometer then employs the reference sample signal automatically to help acquire the accurate absorption spectra of the substances. After passing through the sample, a detector transforms the light into a visible electrical signal. Typically, detectors are based on photoelectric layers or semiconductors. The detector generates an electric current, and the resulting signal is identified and sent to a computer or screen.

4.7.2 Data Interpretation

UV-visible spectroscopy generates output in the form of a graph that can depict wavelength-dependent absorbance, optical density, or transmittance. However, the most used form of data presentation is a graph with wavelength along the x-axis and absorbance along the y-axis. This graph is often referred to as an "absorption spectrum".

4.7.2.1 Beer-Lambert's Law

If a linear relationship can be shown using a calibrated set of typical solutions covering the same substance, Beer-Lambert's Law is particularly helpful for calculating the amount of a substance. This law describes the absorbance (A) which is directly related to the concentration of sample (c), molar absorptivity (ε) and path length (L):

$$A = \varepsilon L c \qquad (4.4)$$

This is referred to as Beer-Lambert's Law. Beer-Lambert's Law describes the relationship between the concentration of a sample and the absorbance of light at a given wavelength. The sensitivity of the system depends on the calibration curve, which is obtained from a plot of absorbance relative to concentration. This curve is useful for data analysis in this technique. The slope of the curve represents the sensitivity of the system. The system's sensitivity rises as the slope gets steeper. The capacity to recognize small differences in sample concentration is known as sensitivity. The molar absorptivity (ε) can be used to estimate the sensitivity of the system from Beer-Lambert's Law.[49,50]

4.7.3 Strengths and Limitations

UV-visible spectroscopy, like any other technique, has its limitations. However, it offers several advantages that contribute to its popularity. These advantages are as follows:

i. The non-destructive nature of the method allows the sample to be reused or subjected to further processing and analysis.
ii. Rapid measurements enable easy incorporation into experimental procedures.
iii. The instrument is user-friendly and requires minimal training.
iv. Data analysis is generally straightforward and does not require extensive processing, making user training less complicated.
v. The affordability of the instrument makes it accessible to many laboratories.

Despite these significant advantages, there are a few limitations associated with the technique. These include stray light, light scattering, interference from multiple absorbing species, and geometrical limitations.[50,51]

REFERENCES

1. Vernon-Parry, K.D., *Scanning electron microscopy: An introduction*. III-Vs Review, 2000. **13**(4): pp. 40–44.
2. Mohammed, A., and A. Abdullah, *Scanning electron microscopy (SEM): A review*. In *Proceedings of the 2018 International Conference on Hydraulics and Pneumatics—HERVEX, Băile Govora*, 2018.
3. Hafner, B., *Scanning electron microscopy primer*. In *Characterization Facility* (p. 1–29). University of Minnesota, 2007.
4. Ul-Hamid, A., *A Beginners' Guide to Scanning Electron Microscopy* (Vol. 1). Springer, 2018.
5. Inkson, B.J., *Scanning electron microscopy (SEM) and transmission electron microscopy (TEM) for materials characterization*. In *Materials Characterization Using Nondestructive Evaluation (NDE) Methods* (pp. 17–43). Elsevier, 2016.
6. Sargent, J., *Low temperature scanning electron microscopy: Advantages and applications*. Scanning Microscopy, 1988. **2**(2): p. 19.

7. Afanasyev, S., T. Kychkina, and L. Savvinova, *Scanning electron microscope (advantages and disadvantages)*. Colloquium-Journal, 2019. **2-2**(26): pp. 25–27.

8. Binnig, G., et al., *Tunneling through a controllable vacuum gap*. Applied Physics Letters, 1982. **40**(2): pp. 178–180.

9. Binnig, G., et al., *7× 7 reconstruction on Si (111) resolved in real space*. Physical Review Letters, 1983. **50**(2): p. 120.

10. Eigler, D.M., and E.K. Schweizer, *Positioning single atoms with a scanning tunnelling microscope*. Nature, 1990. **344**: pp. 524–526.

11. Crommie, M.F., C.P. Lutz, and D.M. Eigler, *Confinement of electrons to quantum corrals on a metal surface*. Science, 1993. **262**(5131): pp. 218–220.

12. Heller, E., et al., *Scattering and absorption of surface electron waves in quantum corrals*. Nature, 1994. **369**(6480): pp. 464–466.

13. Decker, R., et al., *Local electronic properties of graphene on a BN substrate via scanning tunneling microscopy*. Nano Letters, 2011. **11**(6): pp. 2291–2295.

14. Binnig, G., et al., *Scanning tunneling microscope combined with a scanning electron microscope*. Review of Scientific Instruments, 1986. **57**(2): p. 221.

15. Manoharan, H., C. Lutz, and D. Eigler, *Quantum mirages formed by coherent projection of electronic structure*. Nature, 2000. **403**(6769): pp. 512–515.

16. Wiesendanger, R., et al., *Observation of vacuum tunneling of spin-polarized electrons with the scanning tunneling microscope*. Physical Review Letters, 1990. **65**(2): p. 247.

17. Choi, B.-Y., et al., *Conformational molecular switch of the azobenzene molecule: A scanning tunneling microscopy study*. Physical Review Letters, 2006. **96**(15): p. 156106.

18. Lauritsen, J., et al., *Chemistry of one-dimensional metallic edge states in MoS_2 nanoclusters*. Nanotechnology, 2003. **14**(3): p. 385.

19. Campbell, T., et al., *Dynamics of oxidation of aluminum nanoclusters using variable charge molecular-dynamics simulations on parallel computers*. Physical Review Letters, 1999. **82**(24): p. 4866.

20. Stroscio, J.A., and D. Eigler, *Atomic and molecular manipulation with the scanning tunneling microscope*. Science, 1991. **254**(5036): pp. 1319–1326.

21. Pan, J., T. Jing, and S. Lindsay, *Tunneling barriers in electrochemical scanning tunneling microscopy*. Journal of Physical Chemistry, 1994. **98**(16): pp. 4205–4208.

22. Winey, M., et al., *Conventional transmission electron microscopy*. Molecular Biology of the Cell, 2014. **25**(3): pp. 319–323.

23. Kohl, H., and L. Reimer, *Transmission electron microscopy*. Springer Series in Optical Sciences, 2008. **36**.

24. Schorb, M., et al., *Software tools for automated transmission electron microscopy*. Nature Methods, 2019. **16**(6): pp. 471–477.

25. Zuo, J.M., and J.C. Spence, *Advanced Transmission Electron Microscopy*. Springer, 2017.

26. Langford, R., and A. Petford-Long, *Preparation of transmission electron microscopy cross-section specimens using focused ion beam milling*. Journal of Vacuum Science & Technology A: Vacuum, Surfaces, and Films, 2001. **19**(5): pp. 2186–2193.

27. Williams, D.B., et al., *The Transmission Electron Microscope*. Springer, 1996.
28. Warren, B.E., *X-Ray Diffraction*. Courier Corporation, 1990.
29. Bunaciu, A.A., E.G. Udriștioiu, and H.Y. Aboul-Enein, *X-ray diffraction: instrumentation and applications*. Critical Reviews in Analytical Chemistry, 2015. 45(4): pp. 289–299.
30. Artioli, G., *X-Ray diffraction*. In *Encyclopedia of Geoarchaeology* (pp. 1–7). Springer, 2022.
31. Ameh, E., *A review of basic crystallography and x-ray diffraction applications*. International Journal of Advanced Manufacturing Technology, 2019. 105: pp. 3289–3302.
32. Li, J., and J. Sun, *Application of X-ray diffraction and electron crystallography for solving complex structure problems*. Accounts of Chemical Research, 2017. 50(11): pp. 2737–2745.
33. Eaton, P., and P. West, *Atomic Force Microscopy*. Oxford University Press, 2010.
34. Giessibl, F.J., *Advances in atomic force microscopy*. Reviews of Modern Physics, 2003. 75(3): p. 949.
35. Rugar, D., and P. Hansma, *Atomic force microscopy*. Physics Today, 1990. 43(10): pp. 23–30.
36. Yang, C.-W., et al., *Imaging of soft matter with tapping-mode atomic force microscopy and non-contact-mode atomic force microscopy*. Nanotechnology, 2007. 18(8): p. 084009.
37. Tello, M., and R. García, *Nano-oxidation of silicon surfaces: Comparison of noncontact and contact atomic-force microscopy methods*. Applied Physics Letters, 2001. 79(3): pp. 424–426.
38. Putman, C.A., et al., *Tapping mode atomic force microscopy in liquid*. Applied Physics Letters, 1994. 64(18): pp. 2454–2456.
39. Meyer, E., *Atomic force microscopy*. Progress in Surface Science, 1992. 41(1): pp. 3–49.
40. Fadley, C.S., *X-ray photoelectron spectroscopy: Progress and perspectives*. Journal of Electron Spectroscopy and Related Phenomena, 2010. 178: pp. 2–32.
41. Van der Heide, P., *X-ray Photoelectron Spectroscopy: An Introduction to Principles and Practices*. John Wiley & Sons, 2011.
42. Stevie, F.A., and C.L. Donley, *Introduction to x-ray photoelectron spectroscopy*. Journal of Vacuum Science & Technology A: Vacuum, Surfaces, and Films, 2020. 38(6): p. 063204.
43. Sobol, P., et al., *Single crystal CuInSe₂ analysis by high resolution XPS*. Surface Science Spectra, 1992. 1(4): pp. 393–397.
44. Powell, C.J., and P. Larson, *Quantitative surface analysis by X-ray photoelectron spectroscopy*. Applications of Surface Science, 1978. 1(2): pp. 186–201.
45. Andrade, J.D., *X-ray photoelectron spectroscopy (XPS)*. Surface and Interfacial Aspects of Biomedical Polymers: Surface Chemistry and Physics, 1985. 1: pp. 105–195.
46. Penner, M.H., *Basic principles of spectroscopy*. Food Analysis, 2017: pp. 79–88.
47. Vitha, M.F., *Spectroscopy: Principles and Instrumentation*. John Wiley & Sons, 2018.

48. Perkampus, H.-H., *UV-VIS Spectroscopy and Its Applications*. Springer Science & Business Media, 2013.

49. Rocha, F.S., et al., *Experimental methods in chemical engineering: Ultraviolet visible spectroscopy—UV-Vis*. The Canadian Journal of Chemical Engineering, 2018. **96**(12): pp. 2512–2517.

50. Akash, M.S.H., and K. Rehman, *Ultraviolet-visible (UV-VIS) spectroscopy*. Essentials of Pharmaceutical Analysis, 2020: pp. 29–56.

51. Tom, J., *UV-Vis spectroscopy: Principle, strengths and limitations and application*. Technology Networks Analysis, 2021: pp. 1–20.

Chapter 5

Fundamentals of Green Photocatalysis

ABSTRACT

The availability of pure and potable water is a crucial prerequisite for human existence and the growth of society. Using green, sustainable chemistry, photocatalysis is an artificial means of photosynthesis technology that addresses energy and environmental problems. A wastewater treatment system with cost-effectiveness and good performance is presently in high demand. By changing the form, dimension, and characteristics of nanomaterials, the function of nanotechnology in water purification has substantially advanced. This chapter discusses current advancements in environmentally friendly photoactive nanostructures, such as plasmonic photocatalytic materials for water splitting and metals, metal oxides, and metal-doped metal oxides. The discrepancies between the energy ideas for photosynthesis and photocatalysis, as well as a newly developed heterogeneous photocatalyst-based Z-scheme method, are all thoroughly examined.

5.1 INTRODUCTION

The world is facing a severe environmental issue known as water pollution, as stated.[1,2] According to Ma et al. and Bhateria and Jain, the consequences of modernity, the disposal of industrial waste, contamination in seas, and the buildup of plastic trash in water resources have all contributed to human contamination of the majority of water resources (2016). Obtaining clean or even slightly polluted water for domestic purposes can require individuals to walk long distances regularly, leading to a global water crisis. In certain cases, consuming untreated or polluted water can cause malnutrition or disease, particularly during droughts, floods, or insufficient sanitation. Nevertheless, as observed by the World Health Organization (2001) and United Nations Environmental Programme (2001), resolving issues connected to water distribution and utilization is complicated by a lack of expertise, financing, and awareness (2015).[3–6]

DOI: 10.1201/9781003403357-6

The solution to problems with water can be found in ecological sustainability, a crucial idea with three main components: economic, social, and environmental. The green economy and legal mechanisms for ecological sustainability, both of which are focused on Rio+20, are crucial in this respect. The notions and goals of a "green economy" and "sustainable development" only hold weight when the sea is fully incorporated. Sustainable development has been defined as "growth that meets the needs of the present without compromising the ability of future generations to meet their own needs".[7] The global community has embraced and supported sustainable development as a method of addressing the economic, social, and ecological problems that the globe has encountered in recent decades. Nonetheless, the issues still exist and have even gotten worse in maritime areas. Because government agencies are failing to react to institutional changes, coastal communities are having a difficult time coping with the current problems.[8,9]

In addition to environmental and global shifts, human activities including industrialization, agriculture, and deforestation are the main causes of water pollution. Due to the ongoing buildup of dangerous toxins and chemicals in water bodies, the quality of water resources is continuously declining. Industrial effluents are the most dangerous inorganic water pollutants because they contain both inorganic and organic pollutants, including heavy metal ions such as mercury (Hg^{2+}), lead (Pb^{2+}), arsenic, cadmium (Cd^{2+}), chromium (Cr^{6+}), and nickel (Ni^{2+}).[10–12] Additionally, a wide range of detrimental organic contaminants have been detected in diverse aquatic environments. These include but are not limited to organic dyes, colouring agents, pesticides, oils, phenols, fertilizers, pharmaceutical waste, hydrocarbons, and detergents.[13–16]

In 2013, cities generated an estimated 1.3 billion tonnes of solid trash, according to the World Bank. Due to the present rate of urbanization, this amount is anticipated to rise by 70% to 2.2 billion tonnes annually by 2025.[17–19] As a result, trash management will cost more money. The easiest method to address this problem is to alter how garbage is used and recycled. There are several programmes in place to reuse solid waste. This approach involves converting solid waste into a valuable product, like sorbent. Sorbents made from solid waste have been successfully utilized in gas separation processes. Furthermore, researchers are actively pursuing the improvement of low-cost and high-capacity sorbents to eliminate toxic substances, and this topic has been the subject of extensive research. Due to their maximum adsorption capacity, simplicity of operation, and convenience of design, adsorbents have been well recognized for their effectiveness in removing pollutants. Furthermore, there is a growing demand for the development of adsorbents derived from alternative materials that are readily accessible and cost-effective compared to expensive commercially manufactured coal-based adsorbents. In order to solve this issue, there is a lot of research interest in creating adsorbents from a more accessible and sustainable precursor.[20–23]

5.2 GREEN ENERGY

Humans derive their primary energy from the sun, which emits radiation waves that continuously flow to the earth's surface (Figure 5.1). Despite the tremendous amount of energy that is emitted across the solar system, the planet only absorbs a very small amount of it. Photosynthesis has proven to be the most effective method of converting solar energy into a form that can be used, and it is essential to human survival through forestry and agriculture. Solar radiation has a total energy of 384.6 yotta-watts (3.846×1026 W). The sun radiates this significant quantity of energy equally in all directions. At a range of 150 million kilometers, the planet receives around 1368 Wm^2 in exposed places.

On the earth's surface, 1000 Wm^2 of renewable radiation is acquired everywhere thanks to air reflection, which results in around 30% of the solar energy being reflected. The crucial difficulty is to efficiently gather and utilize solar energy, despite the sun's abundance of energy.[24]

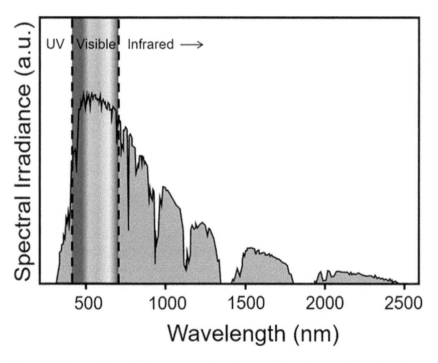

Figure 5.1 Observation of solar spectrum through spectral irradiance and wavelength.

Source: Reprinted with permission from Schreck, M., and M. Niederberger, *Photocatalytic gas phase reactions*. Chemistry of Materials, 2019. **31**(3): pp. 597–618. Copyright © 2019 American Chemical Society.

5.3 HARNESSING SOLAR ENERGY: PHOTOSYNTHESIS AND PHOTOCATALYSTS

An excitonic chemical inverter refers to a system that utilizes light to induce the generation of electrons and holes, facilitating a chemical reaction. Contrarily, photocatalysis describes the speeding up of a chemical process as an outcome of the combination of photons and a catalyst. Moreover, a more precise description that distinguishes between photosynthetic and photocatalytic control system design on the thermodynamics of the two-step reaction:

> Photoelectrochemical cells can be categorized as photosynthetic or photocatalytic. In the former, radiant energy provides Gibbs energy to drive a reaction, and electrical or thermal energy can later be recovered by permitting the reverse, spontaneous reaction to occur. In a photocatalytic cell, photon absorption triggers a reaction with $\Delta G < 0$, so there is no net storage of chemical energy, but the radiant energy accelerates a sluggish reaction.[25,26]

Fujishima and Honda are credited with the introduction of photocatalysis in the 1970s, which has since been commonly referred to as the Honda-Fujishima effect. The process of water splitting (Figure 5.2), in which nanoscale catalysts, such as nanoparticles, react with sunlight to separate water molecules into hydrogen and oxygen atoms, produces hydrogen gas as a clean energy source, is one notable use of photocatalysis.[27]

The efficiency of this mechanism relies predominantly on nanoscale catalysts that harness sunlight and convert photon energy into chemical energy to facilitate the transformation of water molecules, considering water's transparency to sunlight. Water splits can be summed up in the following manner:

$$H_2O + Light\ Energy \rightarrow 1/2O_2 + H_2 \qquad (5.1)$$

Water splitting requires a minimum of 1.23 eV of light energy, which corresponds to a wavelength of approximately 1 mm. Although most photons in sunlight can potentially trigger water splitting, water itself is transparent to the entire spectrum of sunlight, and a catalyst is needed to present photon energy into the water molecules. This catalyst converts solar energy into water by first absorbing it.[28] Solar energy is transformed into chemical energy that is contained in certain kinds of molecules through the process of photosynthesis. By dissolving carbon dioxide (CO_2) and water (H_2O), plants employ photosynthesis to create carbohydrates ($C_6H_{12}O_6$) and oxygen (O_2).[29]

$$6CO_2 + 12H_2O + Light\ Energy \rightarrow C_6H_{12}O_6 + 6O_2 + 6H_2O \qquad (5.2)$$

It is reported that photosynthesis consists of two basic phases: the reaction of light (photosystem [PS] II; Hill reaction) and the dark reaction.

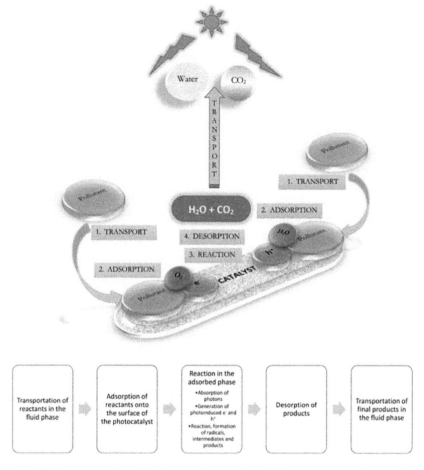

Figure 5.2 Represents the role of nano-photocatalyst in heterogeneous photocatalytic reactions.

Source: Reprinted from Bora, L.V., R.K. Mewada, and S.E. Reviews, *Visible/solar light active photocatalysts for organic effluent treatment: Fundamentals, mechanisms and parametric review.* Renewable and Sustainable Energy Reviews, 2017. **76**: pp. 1393–1421. Copyright © 2017 with permission from Springer Nature.

Photosynthesis is a process that transforms light energy into chemical energy (PS I; Calvin cycle). The Z-scheme is a visual representation that illustrates the process by which photosynthetic organisms, such as chloroplasts, elevate the energy state of electrons and effectively capture energy as the electrons traverse through an electron transport chain as depicted in Figure 5.3. Even if many stages of the process happen simultaneously, by examining the figure, a linear sequence may be made. In the Z-scheme diagram, electron energy is represented by the *y*-axis.[30–32]

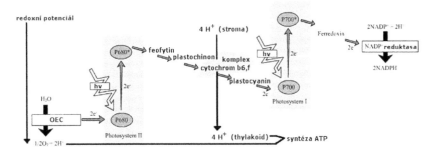

Figure 5.3 Z-scheme analysis in photosynthesis process.

Source: Adapted with permission under the terms of the Attribution-ShareAlike 4.0 International (CC BY-SA 4.0).

The following steps occur during the initial stage of photosynthesis:

i. When a photon enters Photosystem II (PS II), its energy is absorbed and distributed until it reaches the reaction centre located at P680, facilitated by the photosystem.

ii. An electron from a chlorophyll molecule undergoes energization and is transferred to the primary electron acceptor of Photosystem II (PS II). The positioning of the electron in the primary acceptor of PS II at a higher level in the diagram signifies that it possesses greater potential energy compared to the initial electrons present in the non-energized chlorophyll of PS II.

iii. Within Photosystem II (PS II), the oxygen-evolving complex facilitates the extraction of electrons from a water molecule, allowing them to enter the primary electron acceptor (P680) while the chlorophyll electron is received. This series of reactions generates protons and oxygen as by-products.

iv. The energy released in step 4 drives cellular work, specifically the conversion of adenosine diphosphate and Pi to adenosine triphosphate.

v. The electron from PS II reaches PS I with low energy, but as it travels, it receives energy from photons striking PS I, causing its electrons to become energized. The electrons' energy is transmitted to P700, which raises it before the principal electron acceptor in PS I captures them. The electrons arriving from PS II, which are ultimately sourced from water, replenish these lost electrons. It should be noted that the electron enhanced from PS I raises the energy of these electrons to the highest possible level.

vi. PS I's electron transport mechanism, which does not inject protons, receives its powered electrons from the main electron acceptor. It sends PS I's energetic electrons in the direction of NADP+ reductase.

vii. NADP+ reductase changes NADP+ from a low-energy oxidized form to a high-energy reduced state known as NADPH by using the energy of the incoming electron.

The conversion of energy from electron-hole pairs, along with the utilization of photocatalysis and photosynthetic technologies, collectively contribute to the conversion of light energy into chemical energy. In systems that involve solid-gas or solid-liquid interfaces, these processes generate photo-generated charge carriers, which subsequently give rise to reactive species as illustrated in Figure 5.4a. Both photocatalytic and photosynthetic mechanisms must fulfil two essential conditions in order to do this: (i) light absorption to produce electron and hole pairs and (ii) stimulation of redox reactions at the interface of the compounds by photo-generated charge carriers.[26]

The thermodynamics of surface chemical processes largely distinguishes photocatalysis from photosynthetic devices (Figure 5.4b). Photocatalysts use light energy to propel a chemical process that has a favourable thermodynamic outcome. The thermodynamically unfavourable processes ($G > 0$) that occur in photosynthesis systems, on the other hand, demand more energy from the light source for the photochemical reaction and conversion to take place. Therefore, the reverse process of photosynthesis is thermodynamically favourable ($\Delta G < 0$) due to the higher free energy content of the products compared to the reactants. Hence, photosynthetic systems must possess the capability to accomplish three fundamental tasks: (i) facilitating the forward photosynthesis process, (ii) performing the required tasks as mentioned earlier, and (iii) inhibiting the reverse photosynthesis process.

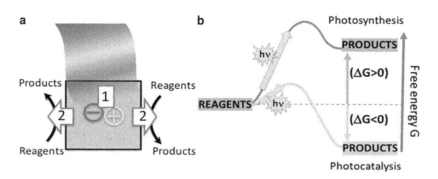

Figure 5.4 (a) Basic principles in photocatalysis and artificial photosynthesis with respect to energy and (b) Energy of photosynthesis and photocatalysis.

Source: Reprinted with permission from Osterloh, F.E., *Photocatalysis versus photosynthesis: A sensitivity analysis of devices for solar energy conversion and chemical transformations.* ACS Energy Letters, 2017. 2(2): pp. 445–453. Copyright © 2022 American Chemical Society.

5.4 ADVANCEMENTS IN GREEN PHOTOCATALYSIS

5.4.1 Solar-Powered Photocatalysis

In the field of material science, the quest for appropriate metal oxide semiconductors that can serve as photocatalysts for the solar-driven detoxification or removal of pollutants from water bodies is a significant research objective. The ideal photocatalytic material should possess specific bandgap properties, enabling efficient absorption of a broad range of solar radiation, successful water molecule splitting, and sustained stability in aqueous environments during the reaction processes. Additionally, the material should be economical, easily processed, readily available, and environmentally nontoxic. Over the past few decades, numerous nanoassemblies based on metal oxide semiconductors have been developed and utilized as catalytic materials for water remediation applications driven by solar energy. Extensively studied semiconductors have shown promising efficiency, particularly those incorporating transition or d-block metal ions, making them promising candidates for advanced photocatalytic materials.[33]

The photocatalytic process is influenced by several factors, including the following:

i. *Dye's concentration*: This plays a critical role. The catalyst's ability to degrade a specific amount of dye is essential. During stimulated light conditions, the dye quantity is adsorbed onto the photocatalyst's surface, which initiates the photocatalytic reaction. The amount of dye adsorbed on the photocatalyst's surface is directly proportional to the original dye concentration, making it a vital parameter that must be carefully monitored. Typically, the degradation percentage declines as the dye concentration rises; however, the necessary amount of photocatalyst must be maintained.[34]

ii. *Amount of dye and catalyst*: The amount of dye employed in the photocatalytic process matters greatly. An appropriate amount of catalyst is required to degrade an average quantity of the dye. Increasing the amount of photocatalyst in the heterogeneous photocatalytic process can improve the percentage of photodegradation of the dye by creating more active sites in the reaction, resulting in a greater number of reactive radicals being produced.[35]

iii. *The pH of the solution*: This is another important factor that influences the degradation process. The surface potential of the catalyst, such as metal oxide nanoparticles, may change when the pH of the solution is altered, which could impact the pollutant adsorption on the photocatalyst's surface, resulting in a change in the photodegradation rate.[36]

iv. *Morphology*: The surface morphology of the photocatalyst, including particle size and shape, is a critical factor that influences the photodegradation activity. Each morphology has a direct relationship with the surface of the catalyst and organic pollutant. The rate of photocatalytic activity is directly proportional to the number of photons striking the photocatalyst's surface, and photocatalysts with a variety of morphologies can promote faster reaction rates.[37,38]

v. *Area of surface*: To enhance the photocatalytic enactment, it is recommended to use materials with a larger surface area. These materials can create numerous active sites on the photocatalyst surface, resulting in the generation of more radical reactive species and efficient photodegradation activity.[39]

vi. *Effect of temperature*: To ensure optimal photocatalytic performance, it is recommended that the reaction temperature fall within the range of 0 to 80°C. Temperatures above 80°C can lead to the promotion of electron-hole pair recombination and consequently reduce photocatalytic activity. Therefore, controlling the reaction temperature is crucial for achieving efficient photocatalytic performance.[40,41]

vii. *Properties of pollutants*: The extent of photodegradation depends on the absorption and nature of pollutants present in the H_2O. For instance, TiO_2 may not be efficient in disinfecting pollutants at larger concentrations because it saturates the photocatalyst's surface and prevents the production of active radicals, which lowers the photocatalytic effectiveness.[42]

viii. *Irradiation time and light's intensity*: The duration and intensity of the irradiation have a big impact on how quickly pollutants photodegrade. Although the production of excitons is more prevalent at low light intensities and prevents the recombination of electron-hole pairs, high light intensities may have an adverse effect on the photodegradation percentage. However, it should be noted that intensifying the light irradiation can lead to the recombination of electron-hole pairs at the surfaces of photocatalysts, resulting in reduced catalytic activity within the reaction media.[43]

ix. *Degradation of dyes and dopants*: TiO_2 nanoparticles that can capture photons with low energy may be created using a number of different techniques. One strategy includes "bandgap engineering", which modifies the valence and conduction bands of photocatalytic materials by including metals and non-metals. Surface modification using semiconductors and organic compounds is another technique.[35]

5.4.2 Metal Oxides

Semiconductor metal oxide nanostructures have been extensively employed in photocatalytic processes, serving purposes such as wastewater purification and hydrogen fuel generation through oxygen and hydrogen splitting. The key criteria for an efficient photocatalytic material encompass an appropriate bandgap, favourable band-edge position, significant surface area, well-defined shape, chemical stability, and the ability to be reused. TiO_2, ZnO,

SnO_2, Cu_2O, and WO_3 are among the metal oxide semiconductors with similar photocatalytic properties, including light absorption, which generates photo-generated charge carriers that oxidize organic constituents.[44]

Semiconducting metallic oxide nanostructures are ignited during this reaction by ultraviolet (UV) light, visible light, direct sunshine, or a combination of visible and UV illumination. During this activation process, the photo-generated charge carriers experience excitation, causing their transition from the valence band to the conduction band, resulting in the generation of electron-hole pairs. The photocatalytic activity of semiconducting metal oxides stems from the synthesis of hydroxyl radicals through the oxidation of OH-anions and the production of superoxide radicals through the reduction of O_2. These radical reactive species can react with organic contaminants, disassembling the molecular chains and producing harmless by-products through disinfection or mineralization. Photocatalytic materials are therefore of great scientific significance for the energy, environmental, and hydrogen fuel sectors. These substances are frequently employed to clean wastewater by getting rid of bacteria and other dangerous impurities.[45]

Figure 5.5 displays the bandgap and band-edge positions of various semiconducting materials that are often employed in photocatalytic applications. Although certain of these materials, like Fe_2O_3, have appropriate bandgap

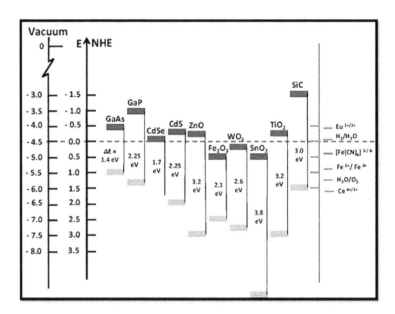

Figure 5.5 Displays band locations relative to vacuum and normal hydrogen electrode of several metal oxide semiconductors.

energies for visible light region utilization, they exhibit some limitations in terms of their efficiency as photocatalysts.[46] In order to address these constraints, scientists are investigating alternative materials. For instance, metal chalcogenide-coated semiconductors such as PbS and CdS have demonstrated photocatalytic capabilities. However, it is important to note that these materials are also recognized for their instability, toxicity, and susceptibility to photocorrosion. On the other hand, oxide semiconductors with good stability and resistance to photocorrosion in aqueous solutions include SnO_2, WO_3, and Fe_2O_3. These materials have conduction band edges located below the potential of the normal hydrogen electrode (NHE). In their 2011 experiment, it is shown how external voltage might be used to activate the charge transfer feature that causes hydrogen to evolve during water splitting. In contrast to TiO_2 and ZnO, it has been observed that Fe_2O_3 exhibits reduced photoactivity due to corrosion or the formation of transient metal-to-ligand or ligand-to-metal electrochemical reaction states. The disintegration and instability of ZnO in water over time after its formation is observed. On the other hand, TiO_2 nanoparticles have exceptional sustainability in an aqueous solution and are resistant to corrosion.[47–49]

5.4.3 Metal-Doped Oxides

According to Neamen, the process of doping involves adding impurity atoms to any semiconducting material's lattice structure. The characteristics of the substrate material are significantly impacted by these dopant atoms. A graphic illustration of the doping-related flaws in a lattice framework can be seen in Figure 5.6. Substitutional doping refers to the presence of an impurity or foreign atom as a substitute at the host atom's lattice positions. The following requirements must be followed in order to produce this form of doping: (i) the host and dopant metals must have the same crystalline phase, electron affinity, and solubility phase; and (ii) the difference in the atomic radii of the dopant and host metals must not be greater than 15%.[50]

Doping is a method for introducing impurity atoms into the lattice structure of a semiconductor, which can modify and impact the host material's

(a)   **(b)**   **(c)**

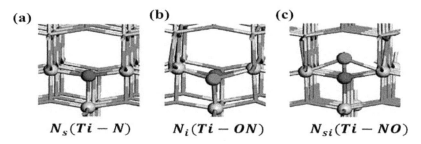

$N_s(Ti - N)$ $N_i(Ti - ON)$ $N_{si}(Ti - NO)$

Figure 5.6 Represents the doping mechanism in TiO_2: (a) substitutional, (b) interstitial, and (c) substitutional-interstitial doping.

properties. Substitutional doping takes place when atoms from impurities substitute host atoms at lattice sites. Substitutional doping can be achieved when the atomic radii of the dopant atoms do not exceed 15% of the difference in atomic radii between the host and dopant metals, and when they share the same crystalline phase, electronegativity, and solubility state. Interstitial doping is the incorporation of atoms of impurities among host lattice sites, which causes gaps between the host atoms. The radii of the host and interstitial atoms influence the probability of impurity atoms occupying interstitial sites. By observing the fluctuations in atomic radius, it is possible to pinpoint the specific position of dopant atoms in the interstitial site. According to the proportion of the ionic radius, the coordination numbers of the cations in the interstitial sites are as follows: 6 (octahedral), 4 (tetrahedral), etc. The ratio of the cation/anion (r+/r−) ionic radius measurements may be used to identify the cations present in the specific interstitial spots. The ratio of the ionic radius is between 0.225 and 0.414 in tetrahedral holes and between 0.414 and 0.732 in octahedral holes.[50,51]

Doping is a common technique for enhancing the electrical characteristics and absorption capacity of semiconducting photocatalysts. Using this technique, the characteristics of the host material are engineered by introducing appropriate molecular dopants, non-metals, or metals as shown in Figure 5.7 which represents the mechanism of tantalum-doped tungsten trioxide (Ta/WO_3). Several factors, including the oxidation state, surface chemistry, and ionic radii of the dopant, are among the variables that can influence the effectiveness of the doping process. Doping semiconductors with non-metal ions (such C, N, B, and S) to change their bandgap and band-edge location

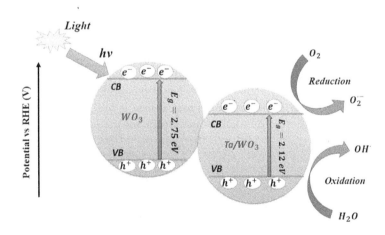

Figure 5.7 Photocatalytic mechanism for metal-doped semiconductor (Ta/WO_3).

has received a lot of attention lately to make them acceptable for visible-light photocatalysis.

Due to their large bandgap energy, semiconductor photocatalysts like TiO_2, SnO_2, and ZnO are often only active under UV light. Appropriate metal or non-metal ions can be doped into the semiconductor lattice to change the electrical structure and band-edge locations, extending their absorption range into the visible light area. This results in extraordinary photocatalytic activity when exposed to visible light.[52,53] Recent decades have seen a significant amount of research into the visible-light-active photodegradation of dangerous organic contaminants, pharmaceutical waste, and poisonous colours by metal-doped TiO_2 nanoparticles. From the early 1990s, doping with non-metals including nitrogen, carbon, fluorine, and sulphur has also received a lot of attention. As a consequence, photocatalysts with exceptional activity against a variety of contaminants when exposed to sunlight have been produced.[53]

5.4.4 Plasmonic-Enhanced Photocatalysis

Because of their superior charge transport capabilities, wide spectrum of sunlight absorption, and improved photodegradation efficiency when exposed to visible light, plasmonic photocatalysts have attracted the attention of many researchers. These materials may be produced by distributing noble metal nanoparticles onto semiconductors. A Schottky barrier and localized surface plasmon resonance (LSPR) are two different characteristics of this type of design. These characteristics contribute to the efficient separation and transfer of the charge carriers when exposed to visible light.

Due to their capacity to increase the photodegradation rate under visible light irradiation, wide range of sunlight absorption, and superior charge transport capabilities, plasmonic photocatalysts are the subject of intensive study. With this kind of material architecture, noble metal nanoparticles are scattered atop semiconductors, resulting in two distinctive properties: a Schottky barrier and LSPR. Under the influence of visible light, these features effectively separate and transmit charge carriers.[54]

The LSPR, which represents intense oscillation on the outermost layer of metal nanoparticles and semiconductor photocatalysts, is an essential component of plasmonic photocatalysis. The consequences of the plasmonic metal nanoparticles photocatalytic activity are shown in Figure 5.8. Because of its brief diffusion length and the interfacial charging effect at the heterojunction, the metal-semiconductor junction enables the separation of electron-hole pairs and transfers charge carriers quickly in plasmonic photocatalysts.

Based on the morphology, structure, and size of noble metal nanoparticles, plasmonic metals like Ag, Au, and Pt exhibit resonance oscillation at specific wavelengths. This phenomenon can shift the absorption limit of photocatalysts such as TiO_2 from UV light to the visible range. By strongly

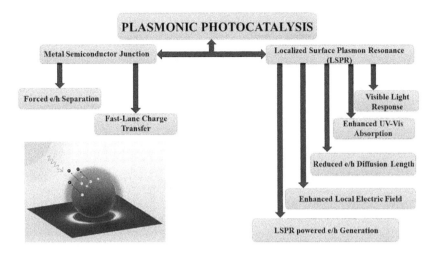

Figure 5.8 Consequences of photocatalytic activity performed by plasmonic metal nanoparticles.

absorbing incoming light in a very thin layer of metal nanoparticles, this surface plasmon resonance effect can significantly improve the visible-light absorption efficiency of low-bandgap semiconducting photocatalysts, such as Fe_2O_3, and enhance the electron transport characteristics of poor electron transport semiconductors. The transport characteristics of the charge carriers are improved during photon excitation as a result of the close proximity of the photo-generated electron/hole to the noble metal nanoparticle surface and the short diffusion length.

In this technique, the surface of the TiO_2 photocatalyst is partly encircled by Au nanoparticles. TiO_2 nanoparticles often display n-type properties as a result of oxygen defects and the extra electrons they produce in the material.[55] Au-TiO_2 nanostructures have been thoroughly compared to traditional TiO_2 photocatalysts to ascertain their photocatalytic activity. Figure 5.9 illustrates how the LSPR phenomenon enables a plasmonic Au nanoparticle on the surface of TiO_2 to absorb the full visible spectrum of electromagnetic energy. The reduction and oxidation reactions that enable the degradation of dangerous contaminants in an aqueous medium are made possible by this effect, which includes the absorption of collective oscillations of electrons and holes in TiO_2 and interfacial charge transfer mechanisms.

One of the most important factors impacting the photocatalytic properties of the semiconductor is the interfacial charge transfer mechanism in plasmonic photocatalysts, which prevents the recombination of electrons and holes. Au in plasmonic photocatalysts has two main advantages over traditional TiO_2 photocatalysts: it increases the visible light range's absorption capacity and reduces the rate of electron-hole pair recombination. As a result, plasmonic photocatalysts made of Au have the potential to perform photocatalysis more effectively than do TiO_2 photocatalysts.

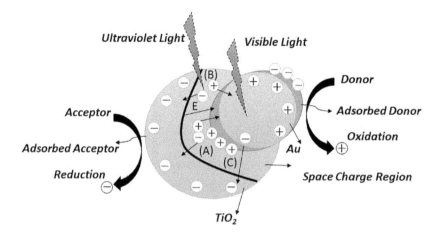

Figure 5.9 Comparison of charge transfer in plasmonic photocatalysts and semiconductor photocatalysts.

5.4.5 Carbon-Based Photocatalysts

In recent years, there has been significant research on carbon nanomaterials due to their unique physiochemical, structural, optical, and electronic properties. Researchers have explored the potential for developing nanocomposites that incorporate these favourable characteristics into conventional photocatalytic materials for effective light-based water remediation.[56]

The ability of different types of nanomaterials to purify water, absorb oil, and eliminate diseases from water bodies has been thoroughly studied. Among these substances, nanostructures made of carbon have extraordinary physiochemical qualities that enable them to remove organic, inorganic, and heavy metal substances from water. The market has a few commercially available carbon-based adsorbents. Moreover, due to their quick photodegradation and charge transport characteristics, fullerenes, graphene, and carbon nanotubes are now employed as co-catalysts in typical photocatalytic materials like TiO_2, ZnO, SnO_2 etc. Strong reduction capabilities may be provided by carbon-based nanoparticles, which aid in the breakdown of dangerous and complicated contaminants in water bodies.

It is reported that graphene decorated TiO_2 nanocomposites exhibited exceptional photocatalytic activity when illuminated with UV light. Various material combinations have been investigated to enhance the activity of these nanocomposites. Among the numerous carbon nanomaterials, graphene and its counterparts have been shown to exhibit exceptional physiochemical characteristics, robust electron-accepting capacities, work function modification, and electronic features that make them particularly effective

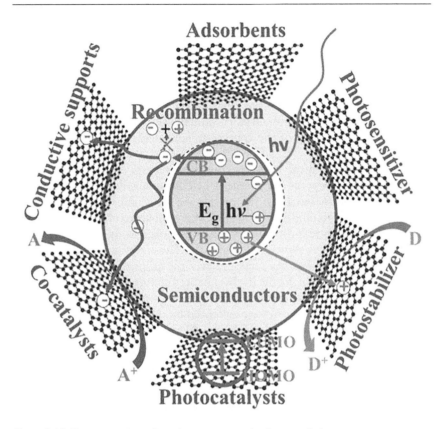

Figure 5.10 Demonstration of graphene structure in photocatalysis.

Source: Reprinted from Li, X., et al., *Graphene in photocatalysis: A review*. Small, 2016. 12(48): pp. 6640–6696. Copyright © 2016, with permission from Wiley (https://onlinelibrary.wiley. com/doi/10.1002/smll.201600382).

in removing various pollutants. Since their characteristics may be modified to affect the photocatalytic activity of semiconductors, graphene-based nanocomposites are thought to be promising prospects for photocatalytic applications as shown in Figure 5.10.

In order to integrate graphene with different morphologies of photocatalysts for photocatalytic technologies, either electrostatic interactions or chemical bonding has been used by researchers. For this aim, a variety of photocatalysts have been used, such as organic, inorganic, metal-organic compounds, semiconductors, plasmonic metals, non-metal plasmonic nanomaterials, and dyes.[57]

As they have better qualities than other common catalyst substrates like graphite, activated carbon, and soot, carbon nanotubes (CNTs) have been widely exploited as effective catalytic materials.[58] Recent studies have demonstrated that Cnts' large surface areas and the existence of additional active sites enable them to efficiently absorb some hazardous substances. Compared to virgin CNT and TiO_2, an outstanding photocatalytic material with increased activity has been created as a result of the integration of TiO_2 into the CNT matrix. These nanocomposites trap the electrons in the valence band and absorb a large range of visible light, which dramatically reduces the rate of electron-hole pair recombination. Moreover, it is possible to close the energy gap of the composite nanostructures to permit visible light.

Due to their chemical stability, cost-effectiveness, and ability to effectively handle a variety of environmental challenges, organic materials have shown to be advantageous in the energy and environmental sectors. Consequently, by combining graphene or CNT nanostructures, metal-organic compounds, graphitic carbon nitride, and other organic dyes have been used as photocatalysts,[59] as shown in Figure 5.11. The coupling method creates a heterojunction of the Schottky type between the graphene and organic semiconductor surfaces, which effectively enhances charge transport throughout the junction. A graphene-organic semiconductor's structural, electrical, and physiochemical characteristics are enhanced by the use of metal-free photocatalysts in a multilayer architecture, producing remarkable photocatalytic

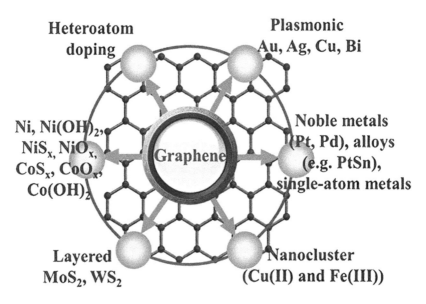

Figure 5.11 Represents the different configuration of catalysts, co-catalysts, and composites in graphene structure.

Source: Reprinted from Li, X., et al., Graphene in photocatalysis: A review. Small, 2016. 12(48): 6640–6696. Copyright © 2016, with permission from Wiley (https://onlinelibrary.wiley.com/doi/10.1002/smll.201600382).

performance. Graphene/gC₃N₄ nanocomposites were also used to show the combination of chemical reduction and impregnation techniques, which could create hydrogen by splitting into hydrogen and oxygen from water.

5.4.6 Z-Scheme Photocatalysis

By using the visible region of the electromagnetic spectrum for different photodegradation processes, the use of nanomaterials in the design and development of correctly constructed metal oxide-based semiconducting photocatalysts clearly shows promise in solving environmental challenges. The Z-development scheme in photocatalysis offers a number of beneficial processes, such as superior sunlight harvesting, quick charge separation, the production of active species for oxidation and reduction reactions, and high redox proficiency, all of which lead to increased photocatalytic activity as depicted in Figure 5.12. The Z-scheme is a photocatalytic system that employs two dissimilar semiconducting photocatalysts coupled with an appropriate redox mediator. It has advantages over conventional photocatalysis, such as more efficient utilization of sunlight and a lower energy requirement for activation. The Z-scheme can also be used for both water oxidation and reduction potential, depending on the photocatalyst utilized in the method.

Tada et al. evaluated the photocatalytic performance of CdS-Au-TiO₂ nanocomposites in terms of charge transfer processes. They also presented a technique for their fabrication. With the Au nanoparticles specifically positioned between the CdS and TiO₂ nanoparticles, significant photo-induced electron and hole transfers from TiO₂ to Au and from CdS to Au were produced. As a result, photoexcited holes on TiO₂ have powerful oxidation

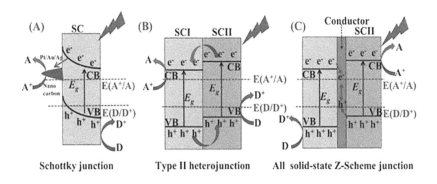

Figure 5.12 Three models of photocatalysis mechanism are shown in a band-edge diagram.

Source: Reprinted from Li, X., et al., *Graphene in photocatalysis: A review*. Small, 2016. 12(48): 6640–6696. Copyright © 2016, with permission from Wiley (https://onlinelibrary.wiley.com/doi/10.1002/smll.201600382).

capacity while photoexcited electrons on CdS have intense reduction capability. Only in Z-scheme photocatalysts does this transfer redox mediator system, a method of electron transport, take place. Z-scheme photocatalysts are so named because the charge transfer channel between these two semi-conducting systems resembles the letter Z.[57,60,61]

5.5 MATERIALS EMPLOYED IN ENVIRONMENTALLY FRIENDLY PHOTOCATALYSIS

Reverse osmosis, ion exchange, biological treatment, and adsorption are a few of the methods now used to remediate polluted water. Nevertheless, adsorption and biological treatment are thought to be more cost-effective procedures, whereas reverse osmosis and ion exchange need substantial capital inputs and operational expenses. Adsorption is a flexible method for handling hazardous pollutants and wastewater, but the largest barrier to its use is the expensive cost of adsorbents. Activated carbons, which are pricey, are now used as the adsorbents in the majority of industrial adsorption procedures for the treatment of water and wastewater. Hence, to make activated carbon more inexpensive, the budget of the production chain must be reduced.

To be considered as a replacement for activated carbons, alternative adsorbents need to possess several qualities, including high adsorption ability, low cost, availability in enormous amounts in a single location, and easy regenerability.[62] Agricultural wastes are a promising option for alternative adsorbents, according to Sen (2017).[63] Many studies—perhaps thousands—have examined the use of agricultural waste as a substitute adsorbent for water and wastewater treatment during the past 30 years. Several studies have looked into the use of biomass materials, including agricultural wastes, for the removal of hazardous substances, and have been discussed in depth in review papers.[64-66] Yet, given the proliferation of studies on this topic and the complexity of the adsorption procedure, it is crucial to present thorough and up-to-date analyses of the use of agricultural wastes as substitute adsorbents.

5.6 SUMMARY

Due to the special properties of nanomaterials, nanotechnology has enormous promise for the creation of synthetic photosynthesis systems for the removal of organic pollutants from the environment and the storage of solar energy. To solve current environmental challenges, it is crucial to use eco-friendly, sustainable production methods for these nanomaterials. Fast reaction rates, low-temperature synthesis, and little use of toxic chemicals are important features of green or biosynthetic approaches using biomass resources. We can design assemblies for energy generation, fresh methods for fuel synthesis, and tools for producing useful materials for solar cells, water-splitting devices, pollution control devices, and more by mimicking the light-harvesting capabilities of green nanomaterials found in nature.

REFERENCES

1. Al-Othman, Z.A., R. Ali, and M. Naushad, *Hexavalent chromium removal from aqueous medium by activated carbon prepared from peanut shell: Adsorption kinetics, equilibrium and thermodynamic studies.* Chemical Engineering Journal, 2012. **184**: pp. 238–247.
2. Sharma, G., et al., *Efficient removal of coomassie brilliant blue R-250 dye using starch/poly (alginic acid-cl-acrylamide) nanohydrogel.* Process Safety and Environmental Protection, 2017. **109**: pp. 301–310.
3. Ma, J., et al., *Sources of water pollution and evolution of water quality in the Wuwei basin of Shiyang River, Northwest China.* Journal of Environmental Management, 2009. **90**(2): pp. 1168–1177.
4. Bhateria, R., and D. Jain, *Water quality assessment of lake water: A review.* Sustainable Water Resources Management, 2016. **2**: pp. 161–173.
5. World Health Organization, *Water for Health: Taking Charge.* World Health Organization (WHO), 2001.
6. Durgalakshmi, D., et al., *Principles and mechanisms of green photocatalysis.* Green Photocatalysts, 2020: pp. 1–24.
7. Dittmar, M., *Development towards sustainability: How to judge past and proposed policies?* Science of the Total Environment, 2014. **472**: pp. 282–288.
8. Bina, O., *The green economy and sustainable development: An uneasy balance?* Environment and Planning C: Government and Policy, 2013. **31**(6): pp. 1023–1047.
9. Biermann, F., *Curtain down and nothing settled: Global sustainability governance after the 'Rio+20' Earth Summit.* Environment and Planning C: Government and Policy, 2013. **31**(6): pp. 1099–1114.
10. Ghasemi, M., et al., *Adsorption of Pb (II) from aqueous solution using new adsorbents prepared from agricultural waste: Adsorption isotherm and kinetic studies.* Journal of Industrial and Engineering Chemistry, 2014. **20**(4): pp. 2193–2199.
11. Naushad, M., et al., *Adsorption kinetics, isotherms, and thermodynamic studies for Hg^{2+} adsorption from aqueous medium using alizarin red-S-loaded amberlite IRA-400 resin.* Desalination and Water Treatment, 2016. **57**(39): pp. 18551–18559.
12. Pandey, S., and J. Ramontja, *Turning to nanotechnology for water pollution control: Applications of nanocomposites.* Focus on Sciences, 2016. **2**(2): pp. 1–10.
13. Rashed, M.N., *Adsorption technique for the removal of organic pollutants from water and wastewater.* Organic Pollutants-Monitoring, Risk and Treatment, 2013. **7**: pp. 167–194.

14. Sharma, G., et al., *Modification of* Hibiscus cannabinus *fiber by graft copolymerization: Application for dye removal.* Desalination and Water Treatment, 2015. **54**(11): pp. 3114–3121.

15. Gupta, V.K., et al., *Chemical treatment technologies for waste-water recycling—an overview.* RSC Advances, 2012. **2**(16): pp. 6380–6388.

16. Javadian, H., M.T. Angaji, and M. Naushad, *Synthesis and characterization of polyaniline/γ-alumina nanocomposite: A comparative study for the adsorption of three different anionic dyes.* Journal of Industrial and Engineering Chemistry, 2014. **20**(5): pp. 3890–3900.

17. Daniel, H., and B.-T. Perinaz, *What a Waste: A Global Review of Solid Waste Management.* World Bank, 2012.

18. Crini, G., *Non-conventional low-cost adsorbents for dye removal: A review.* Bioresource Technology, 2006. **97**(9): pp. 1061–1085.

19. Yagub, M.T., et al., *Dye and its removal from aqueous solution by adsorption: A review.* Advances in Colloid and Interface Science, 2014. **209**: pp. 172–184.

20. Alqadami, A.A., et al., *Novel metal-organic framework (MOF) based composite material for the sequestration of U (VI) and Th (IV) metal ions from aqueous environment.* ACS Applied Materials & Interfaces, 2017. **9**(41): pp. 36026–36037.

21. Hayashi, J., et al., *Preparation and characterization of high-specific-surface-area activated carbons from K_2CO_3-treated waste polyurethane.* Journal of Colloid and Interface Science, 2005. **281**(2): pp. 437–443.

22. Attia, A.A., B.S. Girgis, and N.A. Fathy, *Removal of methylene blue by carbons derived from peach stones by H_3PO_4 activation: Batch and column studies.* Dyes and Pigments, 2008. **76**(1): pp. 282–289.

23. Mittal, A., et al., *Fabrication of MWCNTs/ThO_2 nanocomposite and its adsorption behavior for the removal of Pb (II) metal from aqueous medium.* Desalination and Water Treatment, 2016. **57**(46): pp. 21863–21869.

24. Foster, R., M. Ghassemi, and A. Cota, *Solar Energy: Renewable Energy and the Environment.* CRC Press, 2009.

25. Nozik, A.J., *Photochemical diodes.* Applied Physics Letters, 1977. **30**(11): pp. 567–569.

26. Osterloh, F.E., *Photocatalysis versus photosynthesis: A sensitivity analysis of devices for solar energy conversion and chemical transformations.* ACS Energy Letters, 2017. **2**(2): pp. 445–453.

27. Hashimoto, K., H. Irie, and A. Fujishima, *TiO_2 photocatalysis: A historical overview and future prospects.* Japanese Journal of Applied Physics, 2005. **44**(12R): p. 8269.

28. Chowdhury, F.A., et al., *A photochemical diode artificial photosynthesis system for unassisted high efficiency overall pure water splitting.* Nature Communications, 2018. **9**(1): p. 1707.

29. Lems, S., H. Van Der Kooi, and J. De Swaan Arons, *Exergy analyses of the biochemical processes of photosynthesis.* International Journal of Exergy, 2010. **7**(3): pp. 333–351.

30. Dayan, F.E., S.O. Duke, and K. Grossmann, *Herbicides as probes in plant biology.* Weed Science, 2010. **58**(3): pp. 340–350.

31. Whitmarsh, J., *The photosynthetic process.* In *Concepts in Photobiology: Photosynthesis and Photomorphogenesis* (pp. 11–51). Springer, 1999.

32. Sayama, K., et al., *A new photocatalytic water splitting system under visible light irradiation mimicking a Z-scheme mechanism in photosynthesis.* Journal of Photochemistry and Photobiology A: Chemistry, 2002. 148(1–3): pp. 71–77.
33. Wang, X., et al., *A metal-free polymeric photocatalyst for hydrogen production from water under visible light.* Nature Materials, 2009. 8(1): pp. 76–80.
34. Reza, K.M., A. Kurny, and F. Gulshan, *Parameters affecting the photocatalytic degradation of dyes using TiO₂: A review.* Applied Water Science, 2017. 7: pp. 1569–1578.
35. Akpan, U.G., and B.H. Hameed, *Parameters affecting the photocatalytic degradation of dyes using TiO₂-based photocatalysts: A review.* Journal of Hazardous Materials, 2009. 170(2–3): pp. 520–529.
36. Davis, R.J., et al., *Photocatalytic decolorization of wastewater dyes.* Water Environment Research, 1994. 66(1): pp. 50–53.
37. Zhu, J., et al., *Hydrothermal doping method for preparation of Cr³⁺–TiO₂ photocatalysts with concentration gradient distribution of Cr³⁺.* Applied Catalysis B: Environmental, 2006. 62(3–4): pp. 329–335.
38. Kormann, C., D.W. Bahnemann, and M.R. Hoffmann, *Photocatalytic production of hydrogen peroxides and organic peroxides in aqueous suspensions of titanium dioxide, zinc oxide, and desert sand.* Environmental Science & Technology, 1988. 22(7): pp. 798–806.
39. Ameen, S., et al., *Novel graphene/polyaniline nanocomposites and its photocatalytic activity toward the degradation of rose Bengal dye.* Chemical Engineering Journal, 2012. 210: pp. 220–228.
40. Zhu, Z., *An overview of carbon nanotubes and graphene for biosensing applications.* Nano-Micro Letters, 2017. 9(3): p. 25.
41. Mamba, G., et al., *Nd, N, S-TiO₂ decorated on reduced graphene oxide for a visible light active photocatalyst for dye degradation: Comparison to its MWCNT/Nd, N, S-TiO₂ analogue.* Industrial & Engineering Chemistry Research, 2014. 53(37): pp. 14329–14338.
42. Mills, A., R.H. Davies, and D. Worsley, *Water purification by semiconductor photocatalysis.* Chemical Society Reviews, 1993. 22(6): pp. 417–425.
43. Asahi, R., et al., *Visible-light photocatalysis in nitrogen-doped titanium oxides.* Science, 2001. 293(5528): pp. 269–271.
44. Maeda, K., *Photocatalytic water splitting using semiconductor particles: History and recent developments.* Journal of Photochemistry and Photobiology C: Photochemistry Reviews, 2011. 12(4): pp. 237–268.
45. Abe, R., *Recent progress on photocatalytic and photoelectrochemical water splitting under visible light irradiation.* Journal of Photochemistry and Photobiology C: Photochemistry Reviews, 2010. 11(4): pp. 179–209.
46. Ola, O., and M.M. Maroto-Valer, *Review of material design and reactor engineering on TiO₂ photocatalysis for CO₂ reduction.* Journal of Photochemistry and Photobiology C: Photochemistry Reviews, 2015. 24: pp. 16–42.
47. Gupta, S.M., and M. Tripathi, *A review of TiO₂ nanoparticles.* Chinese Science Bulletin, 2011. 56: pp. 1639–1657.
48. Fox, M.A., and M.T. Dulay, *Heterogeneous photocatalysis.* Chemical Reviews, 1993. 93(1): pp. 341–357.
49. Bahnemann, D.W., C. Kormann, and M.R. Hoffmann, *Preparation and characterization of quantum size zinc oxide: A detailed spectroscopic study.* Journal of Physical Chemistry, 1987. 91(14): pp. 3789–3798.

50. Neamen, D.A., *Semiconductor Physics and Devices: Basic Principles*. McGraw-Hill, 2003.

51. Luo, Y., et al., *Hydrogen sensors based on noble metal doped metal-oxide semiconductor: A review*. International Journal of Hydrogen Energy, 2017. 42(31): pp. 20386–20397.

52. Siriwong, C., et al., *Doped-metal oxide nanoparticles for use as photocatalysts*. Progress in Crystal Growth and Characterization of Materials, 2012. 58(2–3): pp. 145–163.

53. Zhang, Q., et al., *Recent advancements in plasmon-enhanced visible light-driven water splitting*. Journal of Materiomics, 2017. 3(1): pp. 33–50.

54. Wang, C., and D. Astruc, *Nanogold plasmonic photocatalysis for organic synthesis and clean energy conversion*. Chemical Society Reviews, 2014. 43(20): pp. 7188–7216.

55. Morgan, B.J., and G.W. Watson, *Intrinsic n-type defect formation in TiO₂: A comparison of rutile and anatase from GGA+ U calculations*, 2010. 114(5): pp. 2321–2328.

56. Jiang, Y., et al., *A review of recent developments in graphene-enabled membranes for water treatment*. Environmental Science: Water Research & Technology, 2016. 2(6): pp. 915–922.

57. Suárez-Iglesias, O., et al., *Graphene-family nanomaterials in wastewater treatment plants*. Chemical Engineering Journal, 2017. 313: pp. 121–135.

58. Hebbar, R.S., et al., *Carbon nanotube-and graphene-based advanced membrane materials for desalination*. Environmental Chemistry Letters, 2017. 15: pp. 643–671.

59. Li, X., et al., *Graphene in photocatalysis: A review*. Small, 2016. 12(48): pp. 6640–6696.

60. Qin, J., et al., *Two-dimensional porous sheet-like carbon-doped ZnO/g-C₃N₄ nanocomposite with high visible-light photocatalytic performance*, 2017. 189: pp. 156–159.

61. Saravanan, R., et al., *Mechanothermal synthesis of Ag/TiO₂ for photocatalytic methyl orange degradation and hydrogen production*. Materials Letters, 2018. 120: pp. 339–347.

62. Febrianto, J., et al., *Equilibrium and kinetic studies in adsorption of heavy metals using biosorbent: A summary of recent studies*. Journal of Hazardous Materials, 2009. 162(2–3): pp. 616–645.

63. Sen, T.K., *Air, Gas, and Water Pollution Control Using Industrial and Agricultural Solid Wastes Adsorbents*. CRC Press, 2017.

64. Abdolali, A., et al., *Typical lignocellulosic wastes and by-products for biosorption process in water and wastewater treatment: A critical review*. Bioresource Technology, 2014. 160: pp. 57–66.

65. Bhatnagar, A., and M.J.C. Sillanpää, *Removal of natural organic matter (NOM) and its constituents from water by adsorption—A review*. Chemosphere, 2017. 166: pp. 497–510.

66. Levchuk, I., J.J.R. Màrquez, and M.J.C. Sillanpää, *Removal of natural organic matter (NOM) from water by ion exchange—A review*. Chemosphere, 2018. 192: pp. 90–104.

Part II

Applications

Chapter 6

Nanostructured Photocatalysts and Applications

ABSTRACT

Nanotechnology constitutes a multidisciplinary field having a variety of practical applications. Nanomaterials are extremely popular in many industries due to their abundance of distinctive catalytic, electrical, and optical capabilities. Nano-photocatalysts are tiny particles of semiconductors with a minimum dimension of just a few nanometers. The photocatalytic efficiency of nano-photocatalysts are quite promising. They provide the remediation of different toxic pollutants, such as dyes, heavy metals, organophosphorus compounds, volatile organic compounds, pesticides, etc. When a photocatalyst is present, a photo-induced reaction known as photocatalysis takes place. We have defined different photocatalyst such as Binary Semiconductor Photocatalysts in which different types of semiconductor materials are used and it depends upon the electron/hole excitation, Chalcogenides-Based Binary Photocatalysts were semiconductors with a tiny bandgap which have been extensively utilized as photocatalysts, Binary Photocatalysts Based on Nitrides are the photocatalyst utilized to makes the environment pollutant free, Ternary Photocatalysts which are made up of two different metallic cations as well as single anion, a Solid Solution Photocatalysts, Nanocomposite Photocatalysts and Z-Scheme Photocatalysts. Its different optical parameters including light intensity, nature and concentration of substrate and photocatalyst, its pH and the reaction temperature is discussed. It focuses on different titania systems for efficient photocatalytic activity and different doping with other metals have been discussed. Different suitable synthesis methodologies to prepare the nano-photocatalyst is introduced. Therefore, nano-photocatalyst have different applications includes hydrogen production. In order to produce hydrogen, it has to be separated from the elements. For use as fuel, hydrogen may be created from a wide variety of sources and in a variety of ways. The two of the most popular processes for generating hydrogen were electrolysis (the electrical splitting of water)

DOI: 10.1201/9781003403357-8

135

as well as steam-methane reforming. Although, nanoparticles are very efficient but face different challenges for their synthesis process. To overcome these challenges, we used different strategies.

6.1 INTRODUCTION TO NANOSTRUCTURED PHOTOCATALYSTS

In the last few years nanotechnology has brought a huge increase of interest in nanotechnology as miniaturization; nanoparticles are frequently anticipated to be close to a viable future. Nanochemistry, in its vast definition, employs the instruments of materials; chemistry brings about nanomaterials having surface, properties of size and shape that may be tailored to provoke a particular outcome with the objective of being used in a specific application or final use.

The accessibility of energy sources has a significant impact on how well human existence is lived. Within the next few decades, two major difficulties humanity has to face are a lack of energy on a global scale and environmental pollution. An estimated 80% of the world's energy supply comes from non-renewable fossil fuels like coal, petroleum, and natural gas. Therefore, by making use of the abundant and pure solar energy, photocatalysis has been shown to be a feasible and effective method for simultaneously addressing the urgent energy issue and environmental pollution.[1,2]

6.1.1 Fundamentals of Photocatalysis

In photocatalysis, photo-induced electron/hole (e/h) pairs are produced when photons with energies equivalent to or higher than the bandgap energy activate a semiconductor photocatalyst. These photo-generated e^-/h^+ pairs recombine, but some charge carriers may migrate to the uppermost layer of the photocatalyst, initiating a series of chemical reactions with the species that have been adsorbing to the photocatalyst's interface. Accordingly via thermodynamics, the photo-generated e and h can only drive surface reduction and oxidation reactions when their respective reduction and oxidation potentials are between the conduction band (CB) and valence band (VB) potentials of the semiconductor.[3] Thus, the CB/VB potentials of semiconductor photocatalysts and the redox potentials of the target reactions are tightly correlated with each other, and this correlation greatly influences the thermodynamic driving force in photocatalytic processes. The reduction reactions benefit from semiconductors with more negative CB locations, while the oxidation reactions benefit from semiconductors with more positive VB positions.[4]

Photocatalytic materials are those which can transform an incident photon into a usable or storable energy source, when an electron and hole pair are formed at the photocatalyst level. A photocatalyst, broadly speaking, is a catalyst that, when exposed to light, exhibits its catalytic properties through photon absorption. These photocatalytic systems can be found in nature, such as the photosynthesis process that takes place in green leaves. Chlorophyll serves as a photocatalyst in this process, enabling the transformation of water and CO_2 into oxygen and carbohydrates. Engineered photocatalysts frequently imitate this natural process by using photo-generated electrons and holes to produce energetic radicals, which can be employed for a variety of purposes, such as water treatment or separating water into O_2 and H_2. However, in heterogeneous photocatalysis, photocatalysts are solid substances like particles or nanoparticles supported on a substrate like a membrane or suspended in water. There is no requirement for sacrificial or any further chemicals.[5]

6.1.2 Types of Photocatalysts

Numerous synthetic techniques have been documented in the past few decades for altering photocatalysts that are active in visible wavelengths or in any other part of the electromagnetic range. Moreover, photocatalytic systems were produced as a result of these innovative efforts and divided semiconductor materials into four distinct families or categories shown in Figure 6.1, based on the progressive theme in their evolution under the regime of sensitized photoreactions.

6.1.2.1 Binary Semiconductor Photocatalysts

Through the creation of monodisperse, distinctive morphologies with a high specific surface area, the photocatalytic efficiency of binary semiconductors has been enhanced via synthesized approaches. Different binary

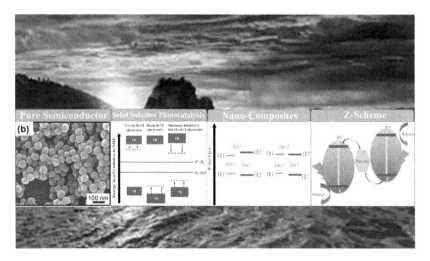

Figure 6.1 Four categories are used to group semiconductor photocatalysts.

semiconductors stated as having a variety of nanomaterials with precisely defined surface features are discussed in the following sections.

6.1.2.2 Chalcogenides-Based Binary Photocatalysts

Binary oxide semiconductors and binary chalcogenides were evaluated equal in intensity in the context of heterogeneous photocatalysts. Over the past 40 years, there has been extensive research on the use of zinc sulphide (ZnS) and cadmium sulphide (CdS) nano-photocatalysts derivatives for various purposes. These include the efficient reduction of CO_2, alkaline compounds, water splitting, and rapid dehalogenation of benzene. Mainly ZnS was given a lot of attention due to its broader energy gap of roughly about 3.2 to 4.4 eV and its easy phase control of ZnS nanocrystals.[6]

6.1.2.3 Binary Photocatalysts Based on Nitrides

Binary nitrides have just lately been investigated as photocatalysts, in contrast with binary oxides/chalcogenides. GaN examination serving as photocatalyst began following its enormous favourable outcome in the sensing sectors because this material has superior mechanical strength and chemically stable properties (at different pHs).[7] Kida et al. reported the first use of powdered GaN for the water-splitting reaction through photocatalysis.[8] Tantalum nitride (Ta_3N_5) is an intriguing and promising type of binary nitride made from transition metals. It is commonly used in the process of solar water splitting. One of the notable characteristics of Ta_3N_5 is its narrow linear bandgap, measuring 2.1 eV. This means that it has the right energy levels at its band edges, making it well suited for certain applications. In fact, Ta_3N_5 has been found to exhibit higher photocatalytic activity compared to N-doped titania nanoparticles. These nanoparticles are known for their ability to break down dye, such as methylene blue, under visible light.[9]

6.1.2.4 Ternary Photocatalysts

A single anion and two distinct metallic cations make up ternary oxide/chalcogenide semiconductors. These have received much research in the realm of heterogeneous photocatalysis due to their stability and capacity to support a variety of chemistries as compared to binary semiconductors. In this section, we explore various types of photocatalysts based on their composition and crystalline structure.

   i. *Ternary oxide photocatalysts*: Photocatalysts made of perovskites (ABO_3 kinds); perovskite oxide semiconductors, with a typical formula of a cubic crystal. The most promising photocatalysts currently being researched are ABO_3.

ii. *Ternary chalcogenide photocatalysts*: Ternary chalcogenides of the type AB_xC_y, these materials have a high absorption coefficient over a broad spectral range, making them a good choice for solar photon absorption.

6.1.2.5 Solid Solution Photocatalysts

The VB to CB transition in semiconductors' electrical structure is reflected in the material's optical characteristics. Large bandgap semiconductors have an optical gap that encompasses a large range of redox potentials, but their application is constrained by their UV-active behaviour. In the realm of heterogeneous photocatalysis, careful bandgap modulation through the fabrication of solid solution semiconductors is typically more effective than doped or loaded semiconductors (binary or ternary). According to the method used to design the bandgap, the numerous solid solution photocatalysts (Figure 6.2) evaluated can be grouped into three groups:

   i. Solid solution via ongoing CB modulation
  ii. Solid solution via ongoing VB modulation
 iii. Solid solution achieved through ongoing CB and VB modulation

6.1.2.6 Nanocomposite Photocatalysts

Different kinds of heterostructure with possible different heterojunction based on the positions of CB and VB levels: (i) straddling gap (type I) and (ii)

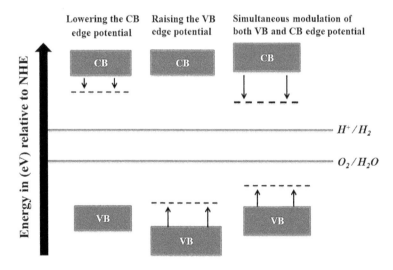

Figure 6.2 Various approaches to bandgap engineering to achieve the best possible equilibrium among visible light intensity and semiconductor band-edge redox potential.

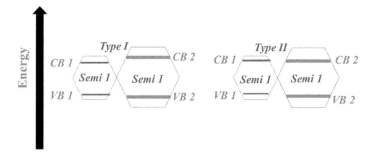

Figure 6.3 Energy gap of two types of semiconductor–semiconductor heterojunctions.

staggered gap (type II) (Figure 6.3). The type II heterocomposite system is the most ideal among the three types of heterojunctions semiconductors for photocatalysis. In a type II system, when there is a change in chemical potential, it leads to band bending at the boundary between two different semiconductors. This bending is caused by the appropriate positioning potential of a band edge. These type II systems are designed in a way that enhances the light absorption and efficient separation of charges in the photocatalyst they produce.

6.1.2.7 Z-Scheme Photocatalysts

The Z-scheme photocatalytic system is designed to mimic the natural process used by plants to produce carbohydrates and oxygen when they are exposed to sunlight.[10–12] The photoexcitation process system provides clear evidence that the reduction and oxidation areas on the semiconductor are connected to active sites created by co-catalysts that are loaded to enhance water redox activities. In order to promote water splitting, it is essential to include co-catalysts such as Pt, NiOx, or Rh on the surface of the semiconductor, although there are supercapacitors capable of performing total water splitting without the need for co-catalysts. Figure 6.4 showcases the pathway followed by charge carriers within the Z-scheme photocatalytic device.

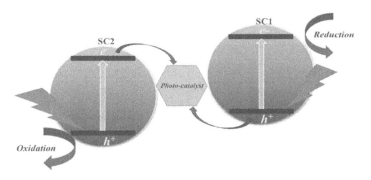

Figure 6.4 Charge-carrier path in a Z-scheme photocatalytic system.

6.2 OPERATIONAL PARAMETER'S EFFECT ON PHOTOCATALYST EFFICIENCY

Following are the operational parameters that affect the efficiency of the photocatalyst.

6.2.1 Light Intensity

The photocatalyst's ability to absorb photons significantly affects the rate of the photocatalytic reaction.[10] According to recent findings, the rate of photocatalytic degradation increased as light intensity increased.[11,12] The reaction pathway is unaffected by the type or nature of the light.[13] We can say that photocatalytic degradation is not significantly influenced by the bandgap sensitization mechanism. Unfortunately, only 5% of the entire amount of naturally irradiated sunlight contains enough energy to effectively cause photosensitization. Additionally, energy losses from light reflection, transmission, and heat generation are unavoidable during the photo process.[14] Due to this limitation, more researchers are now interested in using TiO_2 for decontamination and detoxification. Any photocatalyst or reactant's total amount of light absorption is stated by $\varnothing_{overall}$, quantum yield.

$$\phi_{overall} = Rate\,of\,reaction\,for\,absorption\,of\,radiation \qquad (6.1)$$

In this scenario, the reaction rate is determined by measuring the amount of reactant consumed or product produced over a certain period. On the other hand, the rate of absorbing radiation is measured by the quantity of photons, at specific wavelengths, that are absorbed by the photocatalyst. Light scattering plays a crucial role, especially in solid regimes. Measuring quantum yield experimentally poses a challenge due to the incomplete absorption of incident radiation by metal oxides like TiO_2, caused by refraction.[15] Another factor that can limit the efficiency of photonics is the process of thermal recombination, where electrons and holes recombine. Therefore, it is suggested that we should discourage the use of references to quantum yield or efficiencies in complex systems. The nominal comparative photonic efficiency ζ_r was defined as a useful and easy alternative for comparing process efficiencies.[15] We can determine quantum yield from ζ_r as

$$\phi = \zeta_r \phi_{compound} \qquad (6.2)$$

where $\phi_{compound}$ represents quantum yield of this chemical's photocatalyzed oxidative disappearance employing a photocatalyst (Degussa P-25 TiO_2).

6.2.2 Nature and Concentration of Substrate

Non-chemical molecules that could successfully adsorb the photocatalyst's exterior shall be further vulnerable to straight oxidation. The substituent group affects the photocatalytic degradation of aromatic compounds. According to reports, nitrophenol degrades more quickly because it is a considerably stronger adsorbing substrate than phenol.[16,17] Hugul et al. showed that monochlorinated phenol destroys chloroaromatic more quickly than di- or tri-chlorinated compounds.[17] In general, it has been found that molecules containing electron-withdrawing groups, like nitrobenzene and benzoic acid, tend to adsorb more in the absence of light compared to molecules with electron-donating groups.[18] During photocatalytic oxidation, photonic efficiency is also a factor in determining the organic substrate's concentration over time. However, at elevated substrate concentrations, its titanium dioxide layer becomes saturated and the photonic efficiency falls, which deactivates the catalyst.[19,20]

6.2.3 Nature and Concentration of Photocatalyst

The exterior of organic solvents and the amount of TiO_2 photocatalyst on the surface are directly correlated. Kogo et al. claim that the rate of the reaction is actually determined by the number of photons that hit the photocatalyst.[21] The latter suggests the reaction is only taking place in the adsorption phase of the semiconductor particle. Size of the particles and agglomerates on the surface, is a crucial factor affecting how well nanomaterials work in photocatalytic oxidation.[22] TiO_2 has been synthesized in a variety of forms using various techniques to produce materials with suitable physical characteristics, stability and effectiveness for photocatalytic use. Apparently, the possible use of the sample generated to use for certain uses and the rational development of enhanced synthesis processes are directly related to the surface features.[23, 24] For instance, it has been reported that using nanostructured titanium dioxide, smaller nanoparticle sizes can increase the activities of organic compound photo mineralization in the gaseous phase. As might be expected, the concentration of the photocatalyst has a significant impact on the rate of the photocatalytic reaction. With increased catalyst loadings, diverse photocatalytic processes are familiar to exhibit a proportionate give rise to photodegradation.[25, 26] In order to prevent using too much catalyst and guarantee complete photon absorption in any specified photocatalytic applications, the ideal catalyst concentration ought to be identified. Therefore, an excess of photocatalyst results in unfavourable light dispersed and less light entrance into the mixture.[27]

6.2.4 pH and Reaction Temperature

As the pH of the solution determines the surface charge characteristics of the photocatalyst and the size of the generated aggregates, it is a crucial factor in reactions occurring on particulate surfaces. According to the latest information, the Degussa P-25 material is composed of 80% anatase and

20% rutile. It is claimed to have a point of zero charge (pzc) of 6.9. The pzc refers to the state of a surface where there is no electric charge density.[27] At acidic or alkaline circumstances, respectively, the titania layer can be formed or broken down through the following reactions:

$$TiOH + H^+ \rightarrow TiOH^{2+} \tag{6.3}$$

$$TiOH + OH^- \rightarrow TiO^- + H_2O \tag{6.4}$$

The result is that a titania surface will continue to be negatively charged in a medium that is alkaline (pH > 6.9) while becoming positively charged in an acidic solution (pH 6.9). According to reports, TiO_2 has more oxidizing activity at lower pH levels, although excess H^+ at very low pH levels can slow down reaction rates. Based on reports, it has been observed that TiO_2 exhibits higher oxidizing activity when the pH levels are lower. However, it is worth noting that excessive amounts of H^+ at extremely low pH levels can hinder the reaction rates. Many studies have been conducted to explore the impact of pH on the photocatalytic processes and absorption of organic molecules on TiO_2 surfaces.[28,29] While TiO_2 is present, changes in pH can increase the effectiveness of photo removal of organic contaminants without changing the rate equation. Under ideal circumstances, it has been claimed that these chemicals degrade more effectively.[30] Since the 1970s, there has been experimental investigation on the connection among the rate of reaction and temperature for the oxidation of organic compounds. Numerous researchers developed experimental proof of the relationship between photocatalytic activity and temperature.[31,32] In general, temperature rises speed up charge-carrier recombination and the desorption of adsorbed reactant species, which reduces photocatalytic activity. The Arrhenius equation, which predicts that the apparent first-order rate constant lnk_{app} should increase linearly with $exp(1/T)$, is well supported by these findings.

6.3 PROPERTIES AND CHARACTERISTICS OF PHOTOCATALYSTS: TITANIA VERSUS OTHER PHOTOCATALYSTS

The following characteristics define an optimal photocatalyst for photocatalytic oxidation:

i.   Photo constancy
ii. Static nature in terms of chemistry and biology
iii. Accessibility and affordability
iv. Capacity to adsorb reactants while undergoing effective photonic activation $\left(hv \geq E_g \right)$

Even though there are other materials now being studied, titania is one of the most widely utilized materials in photocatalytic methods because it can be used as photocatalyst and/or encourages photocatalysis. These include zeolites (as supports), metal chalcogenides, and related metal oxides.[33]

6.3.1 Titania (TiO₂)

Atomic scale titania is among the most often utilized mechanisms for photocatalysis due to its substantially greater photocatalytic capacity than other materials, low toxicity, chemical stability, and extremely low cost. According to reports, titania's anatase form offers the optimum balance of photoactivity and photostability.[34] Anatase and rutile, the crystalline forms of titania, have mostly been the subject of investigations. The photon must have a minimum bandgap energy of 3.2 eV, or a wavelength of 388 nm, in order to generate charge carriers on the semiconductor TiO_2 (anatase form). Practically, TiO_2 photoactivation occurs between 300 and 388 nm. When organisms are absorbed onto photocatalysts made of semiconductors, they experience a process called photo-induced exchange of electrons. Two factors influence this process: the position of the semiconductor's band-edge and the redox potentials of the adsorbates.[35,36] A schematic showing the band locations for different semiconductors is presented in Figure 6.5.

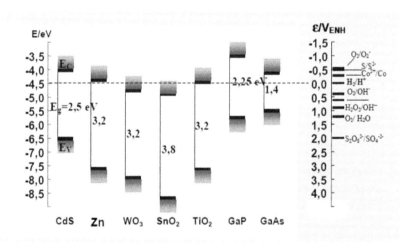

Figure 6.5 Diagrams of the valence and conduction bands for various semiconductors.

Source: Adapted with permission under the terms of the Attribution 3.0 Unported (CC BY 3.0) AT (https://creativecommons.org/licenses/by/3.0/). Colmenares, J.C., et al., *Nanostructured photocatalysts and their applications in the photocatalytic transformation of lignocellulosic biomass: An overview.* Materials, 2009. **2**(4): pp. 2228–2258.

6.3.1.1 Types of TiO_2 Catalysts

The Degussa P25 substance is the TiO_2 photocatalyst that is utilized the most frequently.[37] It has a surface area of only 50 m² per gram and a particle size of approximately 25 nm. The advantage of decreasing the particle size is that it increases the exterior surface area by up to a few nanometers. Once the particles reach the nanometric size range, they tend to aggregate together because of the strong forces between them. When the size of the particles is reduced to a few nanometers or less, something interesting happens. At this point, the effects of quantum size come into play. This leads to a widening of the bandgap in the semiconductor and a shift in the absorption of light towards the blue end of the spectrum. Titania has also gained interest for its usage in photocatalysis using materials having one dimension, such as nanotubes, nanowires as photocatalyst titania.[38–40] Numerous studies have focused in particular on titania nanotubes with micrometric lengths and diameters between 10 and 100 nm. When comparing spherical particles to one-dimensional TiO_2 nanoparticles, it is observed that the latter offer a larger surface area and a faster rate of surface charge exchange. The probability of e^-/h^+ recombination is expected to decrease because the carriers have the freedom to move across the entire length of the nanostructure. Compared to nonporous morphology of titania, titania nanotubes possess a larger surface area, and the tubular shape of the titania prevents charge recombination, as shown by time-resolved diffuse-reflectance spectrometry. These titania nanotubes are easily produced by digesting titania nanoparticles, such as P-25, for many hours at roughly 150°C in an autoclave under strong basic conditions.[41] To create nanotubes that are 100% anatase-containing, such nanotube-TiO_2 were subjected to annealing at 400°C for 3 hours. In comparison to ordinary titania nanoparticles, the visible quantum yield for separation of charges of the nanotube-TiO_2 has been calculated. For the past 20 years, titania has continued to serve as a standard for evaluating any newly discovered candidate, despite its research efforts for pursuit of fresh photocatalysts.[42] Zhang and Maggard also discovered the production of hydrated amorphous titania, which possesses a broader energy bandgap compared to acetate. Furthermore, they observed that this material exhibits noteworthy photocatalytic activity.[43]

Numerous important restrictions severely restrict the employment of TiO_2 the photocatalyst as an all-around instrument to either remove organic contaminants in the liquid or gas state or to carry out advantageous modifications of organic molecules.[44,45] The shortage of visible-light TiO_2 photocatalytic activity is one of the most significant restrictions. The reason for this is that the acetate form of TiO_2, which is a type of wide bandgap semiconductor with a bandgap of 3.2 eV in most cases, has a visible absorption band that begins around 350 nm. Since TiO_2 can only absorb around 5% of the power of solar illumination, this TiO_2 adsorption onset is insufficient to provide effective photocatalytic activity with sun energy. The aforementioned remarks clarify why there is ongoing interest in increasing TiO_2's photocatalytic performance.[46]

6.3.1.2 Modified Titania Systems for Increased Visible-Light Photocatalytic Activity

To improve the ability of TiO_2 to photocatalyze when exposed to visible light, the following techniques were recently developed: doping of TiO_2 with metallic and non-metallic components or utilizing a natural dye as a photo-sensitizer.[47,48] When oxygen and air are removed, the initial approach using a natural dye to absorb observable light has performed admirably. By using the proper electrolyte to successfully quench the dye's oxidation state, the rate of dye degradation is decreased.[49] Otherwise, the photocatalytic system loses its ability to react with the presence of oxygen, light that is visible, especially the dye, which is quickly mineralized in the absence of oxygen. Researchers have recently made a breakthrough in the field of TiO_2 dye sensitized by utilizing advanced time-resolved sub-nanosecond laser pulse photolysis techniques. The visible dye's absorption spectrum and the power of the dye's stimulated electronic state of an electron, that must be highly sufficient to transfer to the conduction band of the semiconductor, are the two most important factors in dye sensitization. Another chemical technique to boost titania photoresponse in the spectrum of visible light is the doping of TiO_2 materials using metallic and non-metallic components.[47] In this instance, the introduction of doping introduces either occupied or unoccupied orbitals into the bandgap area, resulting in negative or positive doping, respectively. Recently, a list of current developments in unique titania photocatalyst preparations that are responsive to UV and visible light has been put together.[50]

6.3.1.3 Doping with Metals

There has been a recent surge of interest in doping titania due to its promising ability to enhance the photo-oxidation rate, particularly in the gas phase. There have been reports stating that Pt-TiO_2 has the ability to accelerate the photo-oxidation process of ethanol, acetaldehyde, and acetone when present in the gaseous phase.[51] Li et al. have reported on the continued development of mesoporous titania as a photocatalyst.[52] The photocatalyst created by these authors (Au/TiO_2) is made of mesoporous titania with embedded gold nanoparticles. To prepare it, you will need to use ethanol as the solvent. In addition, it is recommended to use P-123 as the structure-directing agent. Furthermore, a combination of $TiCl_4$ and $Ti(Obu)_4$ is necessary. Finally,

$AuCl_3$ will be used as the gold source. To create a homogeneous mesostructured nanocomposite, a thin layer of the gel is applied onto a Petri dish. This layer is then aged at 100°C, allowing it to develop a uniform structure. The template is eliminated during calcination at 350°C while also causing TiO_2 to crystallize and the production of gold nanoparticles.

Besides these metals, TiO_2 has also been infused with V, Cr, Mn, Fe, and Ni. It was observed that the presence of the dopant caused a significant shift of the absorption band of TiO_2 into the visible range. There have been studies with conflicting results regarding photocatalytic activity, particularly in relation to metal doping.[53,54] One of the main reasons for this debate is the challenge of accurately comparing the photocatalytic activity of different materials. This difficulty arises from using inconsistent settings and various probe molecules, which makes it hard to draw reliable conclusions. Moreover, the characterization of the doping process and the resulting material are often inadequate due to the lack of information on how the metal was injected. As a result, the outcomes of this process are often disputed and can lead to conflicting conclusions. The final concentration of the dopant is also a determining factor. It has been commonly mentioned that there exists an optimal doping level for achieving maximum efficiency. Beyond this point, however, the photocatalytic activity starts to decrease once again. However, it is generally agreed that using metal doping is not an effective way to enhance the photocatalytic activity of TiO_2.

6.3.1.4 Doping with Non-Metallic Elements

It has been discovered recently that TiO_2 can absorb visible light by adding carbon, nitrogen, sulphur, and other non-metallic components.[54] The researchers Asahi et al. were the first to demonstrate that nitrogen doping has the effect of enhancing absorption in the visible region.[54] Researchers were able to investigate the doping of titania with non-metallic elements. However, it is important for us to explore whether the addition of non-metallic elements can be a practical and efficient method to improve the photocatalytic efficiency of titania. We must consider the potential problems of corrosion and instability that may arise from using doped materials. Although it is expected that p-type metal-ion dopants would act as acceptor centres, the complete understanding of the photophysical mechanism of doped TiO_2 is still lacking.

6.3.1.5 Binary Metal Oxides

Other conventional metal oxides, such as TiO_2, have also been thoroughly researched because of their unique benefits. Numerous metal oxide semiconductors have been studied and used as photocatalysts for organic pollutant degradation, including ZnO, ZrO_2, Fe_2O_3, and WO_3. ZnO, $-Fe_2O_3$, and WO_3 are three examples of them. Yet, they all have limitations that prevent their use in photocatalysis. Under photo-generated hole bandgap irradiation, ZnO is quickly photo-corroded. Because of its low conduction band level, WO_3 is not a good candidate to produce H_2, even though it is a consistent photocatalyst to produce O_2 when exposed to visible light. -Fe_2O_3 is not especially stable in acidic conditions, in addition to having the same difficulty as WO_3. Recently, scientists developed a new type of photocatalyst called nanocrystalline mesoporous Ta_2O_5. They used a combination of sol-gel approach and a surfactant-assisted templating mechanism to create this catalyst. Its purpose is to produce H_2, or hydrogen gas. Researchers have also conducted studies on the impact of loading and doping NiO with Fe as a co-catalyst.[53,54]

6.3.1.6 Metal Sulphides

Typically, metal sulphides are viewed as desirable candidates for photocatalysis that responds to visible light. When opposed to metal oxides, metal sulphides have a narrower bandgap and a valence band that is typically composed of 3p orbitals of S. Recent study has focused on CdS, ZnS, and their solid solutions.

CdS has a suitable bandgap (2.4 eV) and excellent band positions for visible light-aided water splitting. However, S_2 in CdS is easily oxidized by photo-generated holes, resulting in the release of Cd^{2+} into solution, similar to ZnO. The majority of metal sulphide photocatalysts experience such photocorrosion as a regular issue.

Four directions have emerged recently that attempt to improve CdS and ZnS photocatalysts:

    i. Constructing one-dimensional, spongy CdS
    ii. CdS and ZnS solid solutions can be produced via doping
    iii. Co-catalyst added to CdS
    iv. Developing the CdS matrix and support structures

The researchers used a solvothermal technique to synthesize CdS nanorods and nanowires. These were then used to create porous CdS materials.[55] The researchers, White et al., have developed a method to study the pore structure of materials that cannot be observed using traditional microscope imaging. They achieved this by using CdS quantum dots supported on porous polysaccharides. This innovative approach provided them with a better understanding of the inner workings of these materials.[56] As opposed to this, the bandgap of ZnS (3.6 eV) is too wide for it to react to visible light. ZnS has the ability to absorb visible light that can be improved via doping, the formation of solid ZnS solutions, and the use of narrow bandgap semiconductors. Similar crystal structures in ZnS and CdS make it simple for them to create solid solutions. Between ZnS and $AgInS_2$, the bandgap in solid solutions $(AgIn)_xZn_{2(1-x)}S_2$ is smaller. The absorbance of solid solutions increased monotonically when the ratio of $AgInS_2$ to ZnS increased. Because of the contributions of the Ag 4d and S 3p orbitals to the valence and conduction bands, respectively, the band position has shifted, and these nanomaterials' photophysical and photocatalytic capacity were significantly sensitive on composition.[57]

6.4 PREPARATION OF NANOSTRUCTURED PHOTOCATALYSTS

The sol-gel method is a promising technique that can be used to synthesize nano-photocatalysts made of TiO_2. The sol-gel approach is very promising due to its many advantages. These include the ability to use low sintering temperatures, flexibility in processing, and the uniformity of molecules. TiO_2-anatase can be made with this technique at a low temperature. Due to its intense activity in photocatalytic applications, this phase has undergone substantial research.[58] Recently, there has been a growing focus on the creation of porous TiO_2 films that possess a large specific surface area. On the surface, photocatalytic activities include chemical reactions. The process ought to function more efficiently as a result of the increase in surface area because it results in more contact surfaces exposed to the reagents. By adding surfactants to the traditional alkoxide sol-gel method or by utilizing templating membranes, porous inorganic TiO_2-anatase films can be produced. With the help of templates, the original polymer morphology can be kept all the way up to the completed porous structure. Because polyethylene glycol completely decomposes at very low temperatures, it is especially well suited for altering the porosity structure of coatings. In order to obtain coatings that are thick, free from cracks, and have a uniform appearance, it is necessary to carefully manage both the synthesis and deposition processes.

Using mesoporous sol-gel synthesis offers the advantage of being able to create uniform layers on a substrate. The study conducted by Sanchez and colleagues. As the glass is slowly removed from the solution and ethanol evaporates, the concentration of the surfactant gradually increases until it reaches the critical micellar concentration (cmc). At this stage, the surfactant starts to assist in the creation of thin layers of mesoporous TiO_2. In order to achieve the desired result, it is important to effectively manage the rate at which the solvent evaporates. This process enables the formation of TiO_2 polymer around the self-assembled micelles. These micelles were initially created by the surfactant while it was in its liquid crystal state. Stucky and his colleagues utilized the previously mentioned method to create meticulously arranged materials consisting of anatase nanoparticles measuring 5 to 10 nm, which were organized in a grid-like pattern.

6.4.1 Ultrasonic Preparation of Nanostructured Photocatalysts

To prevent or at least reduce crystallite expansion, it may be helpful to apply novel irradiation techniques (such as sonication and microwave radiation) throughout the synthesis. In this regard, it has been observed that using ultrasonic irradiation during the synthetic process facilitates the creation of smaller, homogenous nanoparticles and increases surface area.[59-61]

The cavitation phenomena during ultrasonic treatment are enhanced by metal supports or other particles that are solid in the liquid system and because microbubbles tend to fragment into smaller particles, there are more areas with high temperatures and pressure overall. In addition, ultrasound has the ability to improve the movement of mass towards the barrier between liquid and solid.[62-64]

Regardless of the titanium precursor used, pure-anatase tiny particles, a highly efficient photocatalytic phase in titania, were generated by aging under sonication, which also significantly increased the total surface area of this nanostructured substance. When Pt/TiO_2 was used as a photocatalyst in the oxy-dehydrogenation reaction of 2-propanol to ethanol, it led to a higher molar conversion rate. This can be attributed to the increased surface area. When platinum is added to zeolites that already contain Ti and V, it results in a notable enhancement in molar conversion for the same reaction. Additionally, there is very little or negligible deactivation observed over time, and a noticeable increase in selectivity towards acetone (ranging from 90% to 96%).[64]

6.4.2 Other Non-Conventional Synthesis Methodologies

Pulsed laser deposition (PLD), chemical vapor deposition, sputtering, reactive technique, or hydrothermal method are a few of the techniques that have been employed to create TiO_2 materials.[65,66] Mergel et al. used granular TiO_2 with a purity of 99.5% to create films using electron-beam evaporation in a BAK640 high-vacuum room that was pushed by a diffusion pump. This method produced films with thicknesses ranging from 0.9 to 2.4 mm.[67]

Yamamoto et al. conducted a study where they produced films using TiO_2 mixed with acetate and having the rutile structure. They employed the PLD technique, specifically using an Nd-YAG laser, in a carefully controlled atmosphere of O_2.[66] In addition, the researchers have achieved successful creation of epitaxial acetate films on various oxide substrates that possess different lattice properties. Some of the substrates that can be used include $LaAlO_3$, $SrTiO_3$, MgO, and yttria-stabilized zirconia. Additionally, the researchers were able to successfully create high-quality epitaxial rutile coatings on an Al_2O_3 substrate. During the deposition process, the substrates were kept at a temperature of 500°C and exposed to a gas pressure of 35 mtorr of O_2.

Epitaxial films usually had a thickness that fell within the range of 200 to 880 nm. Researchers conducted optical absorption measurements and discovered that the energy difference between the bands in acetate and rutile TiO_2 epitaxial layers is 3.22 and 3.03 eV, respectively. When it comes to these epitaxial TiO_2 films, using extra techniques to reduce crystal defects can improve the importance of photoactivated electrons and holes in the photocatalysis process.

6.5 HYDROGEN PRODUCTION

Hydrogen is considered as the most efficient clean energy carrier because of its high-energy density and potential demand to create water vapors when it is used as a fuel. It can be produced by several methods having advantages as well as limitations such as steam-methane reforming, partial oxidation, autothermal reforming, electrolysis, biological fermentation, thermochemical water-splitting, coal gasification, and photocatalytic water splitting. In this section, we discuss photocatalytic hydrogen production, the principle of photocatalytic hydrogen production, photocatalytic materials, mechanism of photocatalytic hydrogen production, factors affecting photocatalytic hydrogen production, and applications of photocatalytic hydrogen production in detail.

6.5.1 Photocatalytic Hydrogen Production

Photocatalytic water splitting is a process by which hydrogen is produced when a semiconductor-based photocatalyst harnesses the solar energy to drive the chemical reaction for water splitting and as a result, that water splits into O_2 gas and H_2 gas. The photocatalyst is subjected to solar light, it absorbs the light and heat energy of the incoming photons, and thus the electron-hole pair generation occurs. These generated holes oxidize water and produce O_2 and H_2 gas. This is the simplest and most cost-effective hydrogen production method and does not produce any extra and harmful emissions.

6.5.1.1 Principle of Photocatalytic Hydrogen Production

This process relies on semiconductors usually known as photocatalysts that can absorb light and activate redox reactions. When the light radiates, it provides energy to the photocatalyst, which then absorbs the light and creates electron-hole pairs. When water molecules are adsorbed on the surface of a photocatalyst, they undergo a reaction with electron-hole pairs, and this chain of reactions continues without stopping. These reactions reduce the water into O_2 gas and H^+ ions. Thus, the H^+ ions combine to generate hydrogen gas.

6.5.1.2 Photocatalytic Materials

There are several semiconductors that have been explored as efficient photocatalysts for the hydrogen production mechanism. The most prominent among them include TiO_2, cadmium sulphide, zinc sulphide, tungsten trioxide, and dibismuth trioxide. TiO_2 has been widely studied due to greater stability and ultraviolet light absorption properties. These materials, when doped and co-doped with suitable materials, show promising applications for photocatalytic hydrogen production lying in visible regions.

6.5.1.3 Mechanism of Photocatalytic Hydrogen Production

Photocatalytic hydrogen production can be done by direct water splitting when electrons reduce the water molecules and holes oxidize the water molecules on the photocatalyst's surface. In the indirect mechanism, electrons that are photo-generated react with redox peacekeepers in the solution which then take part in water splitting.

6.5.1.4 Factors Affecting Photocatalytic Hydrogen Production

Several factors are reported that can affect photocatalytic hydrogen production such as the following:

i. *Properties of photocatalyst:* Electron-hole generation and recombination rates, bandgap energy, crystallinity, surface area, and the dopants affect the reactivity of the photocatalyst.
ii. *Radiation source:* The wavelength of light used, and the intensity of incident light strongly affect the photocatalytic activity as well as hydrogen production rate. For example, ultraviolet and visible light.
iii. *Conditions of reaction:* Water concentration, pH value, temperature, and photocatalyst concentration affect the rate of hydrogen production and water splitting.

Therefore, the development of novel candidate materials as well as the correction of currently used photocatalysts via hydrogen production have been accelerated by nanotechnology. The significance of this popular subject is demonstrated by the sharp increase in the number of research papers regarding nanophotocatalytic H_2 generation (1.5 times annually since 2004). Numerous papers investigated how various nanostructures and nanomaterials affected the functionality of photocatalysts because the efficiency of their energy conversion is mostly driven by nanoscale characteristics. Hydrogen may now be produced sustainably using biomass sources.[68,69] Several procedures, including steam gasification, quick pyrolysis, and supercritical conversion, have been developed for this purpose.[70–73] However, these procedures necessitate harsh reaction environments, such as high temperatures and/or pressures, which implies expensive costs.

Unlike these energy-consuming thermochemical processes, photocatalytic reformation offers a promising alternative. This technique has the advantage of being powered by sunlight and can be carried out at room temperature. Using renewable biomass for photocatalytic reformation to produce hydrogen might be a more practical and feasible option compared to photocatalytic water splitting, mainly because of its potential for higher efficiency. The effectiveness of water-splitting operations is mostly hindered by the recombination reaction between the electrons and holes generated by light, which limits their efficiency.[74]

However, to our knowledge, there are not many publications in the literature regarding the photocatalytic reformation converting biomass into hydrogen. In 1980, pioneering research was carried out. According to Kawai and Sakata, a photocatalyst made of $RuO_2/TiO_2/Pt$ may produce hydrogen from sugars when exposed to a 500W Xe lamp.[75] Equation 6.5, which describes the procedure, is combined with Equation 6.6, which describes how green plants produce carbohydrates by photosynthesis:

$$\left(C_6H_{12}O_6\right)_n + 6n\,H_2O\,light, \overline{\frac{RuO_2}{TiO_2}} / pt\, 6n\,CO_2 + 12n\,H_2 \tag{6.5}$$

$$6n\,H_2O + 6n\,CO_2\,\overline{light, green\,plant}\,\left(C_6H_{12}O_6\right)_n + 6n\,O_2 \tag{6.6}$$

$$12n\,H_2O \rightarrow 12n + 6n\,O_2 \tag{6.7}$$

where $(C_6H_{12}O_6)_n$ denotes hydrolyzed cellulose (n = 1000–5000), starch (n = 100), or saccharose (n = 2), respectively. This technique may also be used to break down excretions that contain cellulose, protein, and fat while simultaneously producing H_2 as a biofuel. Later, the same researchers discovered that various biomass sources, like cellulose, insect carcasses, and waste materials, could create hydrogen under the same conditions.[76–78] This research shows that it is possible to produce hydrogen using photocatalysis from biomass.

In recent studies, researchers have been exploring a new method for producing H_2 gas. They are specifically looking at the photocatalytic reformation of glucose, which is a component found in cellulose. In order to perform this process, they utilize catalysts composed of M/titania, where M can refer to Pt, Rh, Ru, Au, or Ir. The titania used is an industrial-grade TiO_2 known as Degussa P25.[78] The best catalytic system, Rh/TiO_2, was discovered to produce around 3500 mol of H_2 for 0.5 g of photocatalyst and 5 hours of radiation.

Alcohols are model molecules for biomass structure, and Verykios et al. discovered that utilizing them as hole scavengers can increase quantum yields and photocatalytic hydrogen production rates.[79] This is an excellent example of how solar energy may be converted to chemical energy utilizing H_2 as the energy carrier. Both a "chemical storage of light energy" and a modest 100% specific oxidation process are involved in the reaction. Additionally, photocatalytic hydrogen production via chemicals and biomass is possible using Pt/TiO_2 photocatalyst.[80,81] The reason for the selectivity of this procedure is because it involves the oxidation of a specific type of oxygen species (O [ads]) using photoactive atomic oxygen organisms. These organisms are neutral in nature and can be detected through photoconductivity. The neutralization of the O (ads) species occurs when it interacts with positively photo-generated holes (h^+).

$$O^-_{(ads)} + h^+ \rightarrow O^*_{(ads)} \tag{6.8}$$

6.6 FUTURE CHALLENGES AND PROSPECTS

6.6.1 Difficulties in Synthesis of Nano-photocatalysts

Successful continuous synthetic procedures have been described for four distinct groups of photocatalytic materials. These materials include traditional photocatalysts composed solely of semiconductors such as TiO_2 and ZnO. Several efforts are now being conducted to increase the yield of the disclosed procedure, while at the same time, novel synthetic approaches employing supercritical medium are being reported by several research groups for synthesizing atypical photocatalysts (GaN and CeO_2). For the continuous synthesis of ternary semiconductors, a one-source precursor is used, and there are investigations into operational flow systems. The cost of single-source precursors for solid solution photocatalysts is being reduced by extensive investigation into existing or developing new synthetic methods. Because there are not any single-source, reliable, and ready-to-use predecessors that are free of oxidation, continuous synthetic processes must switch from batch to flow since solid solutions are fully inert. The present focus is on using a double reactor configuration for type II plus directing the Z-scheme photocatalysts to continuously sequentially synthesize the core (base semiconductor) and shell.

6.6.2 Photocatalysts Employed in Heterogeneous Photocatalysis

In a multistage system, heterogeneous photocatalysis refers to the process of connecting different phases through mass transfer (adsorption) and charge transport when exposed to light. The arrangement of a reactor using heterogeneous photocatalysis is greatly affected by the chemical reactions occurring within the system and the type of electromagnetic energy used as a source. The primary objective in the design of a photoreactor is to attain a high level of photocatalytic activity. To achieve this, it is important to make sure that the solidified photocatalytic semiconductor is effectively exposed to light and that the rate at which molecules move from the photocatalyst surfaces to the reaction vessel is enhanced. The choice of materials for constructing the photoreactor depends on the composition and chemistry of the different phases in the system. In heterogeneous photocatalysis, three primary multiphasic systems are employed: (i) the solid-liquid system, (ii) the solid-gas system (organic/inorganic), and (iii) the solid-liquid-gas system. Researchers are studying two main types of photoreactors: batch reactors and flow reactors. In multiphasic systems, the batch reaction vessel is usually made of materials like quartz, pyrex, and sometimes stainless steel. The source of radiation can be located either indoors or outdoors.[82] Suspension and immobilization are two alternative ways to effectively utilize photocatalysts in batch and flow reactors. In the previous approach, a photocatalytic material is kept suspended in the fluid by utilizing magnetic stirring techniques such as stirring with magnets, toroidal or non-toroidal agitation with the help of gas molecules, and magnetic stirring. The slurry system is the term used to describe the suspended state of the photocatalytic components, regardless of whether the liquid phase is aqueous or non-aqueous. The photocatalysts in the immobilized system are bonded to inactive surfaces either chemically or physically.

Additionally, several potential future developments are now being researched. Although many of these study areas are still relatively new, they are anticipated to hold immense potential in the coming years. These include the following:

i. Developing photocatalytic nanoparticles that can selectively break down organic pollutants

ii. Developing unique titanium oxo-photocatalyst formulations when future family members are available

iii. Creating more trustworthy photocatalysts that can be photoactivated by solar or visible light

iv. Looking into the possibility of using materials other than titania, like sulphide metals

v. Platinizing TiO_2 to make it photosensitize in the visible, as well as continuing to research anionic doping

vi. Utilizing photocatalysis in advance for fine chemical preparation

vii. Using photocatalysis as a novel medicinal instrument, such as in the treatment of cancer

viii. Using emerging photocatalysts for energy production including bio-hydrogen, which can be produced from biomass or by photocatalytical splitting water

REFERENCES

1. Grätzel, M., *Mesoscopic solar cells for electricity and hydrogen production from sunlight*. Chemistry Letters, 2005. 34(1): pp. 8–13.
2. Grätzel, M., *Photoelectrochemical cells*. Nature, 2001. 414: pp. 338–344.
3. Li, X., et al., *Design and fabrication of semiconductor photocatalyst for photocatalytic reduction of CO_2 to solar fuel*. Science China Materials, 2014. 57: pp. 70–100.
4. Li, X., J. Yu, and C. Jiang, *Principle and surface science of photocatalysis*. Interface Science and Technology, 2020. 31: pp. 1–38.
5. Ameta, S.C., and R. Ameta, *Advanced Oxidation Processes for Wastewater Treatment: Emerging Green Chemical Technology*. Academic Press, 2008.
6. Banerjee, I.A., L. Yu, and H. Matsui, *Room-temperature wurtzite ZnS nanocrystal growth on Zn finger-like peptide nanotubes by controlling their unfolding peptide structures*. Journal of the American Chemical Society, 2005. 127(46): pp. 16002–16003.
7. Yonenaga, I., *Thermo-mechanical stability of wide-bandgap semiconductors: High temperature hardness of SiC, AlN, GaN, ZnO and ZnSe*. Physica B: Condensed Matter, 2001. 308: pp. 1150–1152.
8. Kida, T., et al., *Photocatalytic activity of gallium nitride for producing hydrogen from water under light irradiation*. Journal of Materials Science, 2006. 41: pp. 3527–3534.
9. Zhang, Q., and L. Gao, *Ta_3N_5 nanoparticles with enhanced photocatalytic efficiency under visible light irradiation*. Langmuir, 2004. 20(22): pp. 9821–9827.
10. Ustinovich, E.A., D.G. Shchukin, and D.V. Sviridov, *Heterogeneous photocatalysis in titania-stabilized perfluorocarbon-in-water emulsions: Urea photosynthesis and chloroform photodegradation*. Journal of Photochemistry and Photobiology A: Chemistry, 2005. 175(2–3): pp. 249–252.
11. Qamar, M., M. Muneer, and D. Bahnemann, *Heterogeneous photocatalysed degradation of two selected pesticide derivatives, triclopyr and daminozid in aqueous suspensions of titanium dioxide*. Journal of Environmental Management, 2006. 80(2): pp. 99–106.
12. Karunakaran, C., and S. Senthilvelan, *Photooxidation of aniline on alumina with sunlight and artificial UV light*. Catalysis Communications, 2005. 6(2): pp. 159–165.
13. Stylidi, M., D.I. Kondarides, and X.E. Verykios, *Visible light-induced photocatalytic degradation of Acid Orange 7 in aqueous TiO_2 suspensions*. Applied Catalysis B: Environmental, 2004. 47(3): pp. 189–201.
14. Yang, L., and Z. Liu, *Study on light intensity in the process of photocatalytic degradation of indoor gaseous formaldehyde for saving energy*. Energy Conversion and Management, 2007. 48(3): pp. 882–889.
15. Serpone, N., *Relative photonic efficiencies and quantum yields in heterogeneous photocatalysis*. Journal of Photochemistry and Photobiology A: Chemistry, 1997. 104(1–3): pp. 1–12.
16. Tariq, M.A., et al., *Photochemical reactions of a few selected pesticide derivatives and other priority organic pollutants in aqueous suspensions of titanium dioxide*. Journal of Molecular Catalysis A: Chemical, 2007. 265(1–2): pp. 231–236.

17. Bhatkhande, D.S., et al., *Photocatalytic and photochemical degradation of nitrobenzene using artificial ultraviolet light.* Chemical Engineering Journal, 2004. **102**(3): pp. 283–290.

18. Palmisano, G., et al., *Selectivity of hydroxyl radical in the partial oxidation of aromatic compounds in heterogeneous photocatalysis.* Catalysis Today, 2007. **122**(1–2): pp. 118–127.

19. Friesen, D.A., et al., *Factors influencing relative efficiency in photo-oxidations of organic molecules by $Cs_3PW_{12}O_{40}$ and TiO_2 colloidal photocatalysts.* Journal of Photochemistry and Photobiology A: Chemistry, 2000. **133**(3): pp. 213–220.

20. Arana, J., et al., *Photocatalytic degradation of formaldehyde containing wastewater from veterinarian laboratories.* Chemosphere, 2004. **55**(6): pp. 893–904.

21. Kogo, K., H. Yoneyama, and H. Tamura, *Photocatalytic oxidation of cyanide on platinized titanium dioxide.* Journal of Physical Chemistry, 1980. **84**(13): pp. 1705–1710.

22. Ding, H., H. Sun, and Y. Shan, *Preparation and characterization of mesoporous SBA-15 supported dye-sensitized TiO_2 photocatalyst.* Journal of Photochemistry and Photobiology A: Chemistry, 2005. **169**(1): pp. 101–107.

23. Mohammadi, M., et al., *Preparation of high surface area titania (TiO_2) films and powders using particulate sol-gel route aided by polymeric fugitive agents.* Sensors and Actuators B: Chemical, 2006. **120**(1): pp. 86–95.

24. Diebold, U., *The surface science of titanium dioxide.* Surface Science Reports, 2003. **48**(5–8): pp. 53–229.

25. Maira, A.J., et al., *Gas-phase photo-oxidation of toluene using nanometer-size TiO_2 catalysts.* Applied Catalysis B: Environmental, 2001. **29**(4): pp. 327–336.

26. Krýsa, J., et al., *The effect of thermal treatment on the properties of TiO_2 photocatalyst.* Materials Chemistry and Physics, 2004. **86**(2–3): pp. 333–339.

27. Chun, H., W. Yizhong, and T. Hongxiao, *Destruction of phenol aqueous solution by photocatalysis or direct photolysis.* Chemosphere, 2000. **41**(8): pp. 1205–1209.

28. Wang, W.-Y., and Y. Ku, *Effect of solution pH on the adsorption and photocatalytic reaction behaviors of dyes using TiO_2 and Nafion-coated TiO_2.* Colloids and Surfaces A: Physicochemical and Engineering Aspects, 2007. **302**(1–3): pp. 261–268.

29. Mrowetz, M., and E. Selli, *Photocatalytic degradation of formic and benzoic acids and hydrogen peroxide evolution in TiO_2 and ZnO water suspensions.* Journal of Photochemistry and Photobiology A: Chemistry, 2006. **180**(1–2): pp. 15–22.

30. Mansilla, H., et al., *Photocatalytic EDTA degradation on suspended and immobilized TiO_2.* Journal of Photochemistry and Photobiology A: Chemistry, 2006. **181**(2–3): pp. 188–194.

31. Evgenidou, E., K. Fytianos, and I. Poulios, *Semiconductor-sensitized photodegradation of dichlorvos in water using TiO_2 and ZnO as catalysts.* Applied Catalysis B: Environmental, 2005. **59**(1–2): pp. 81–89.

32. Muradov, N.Z., et al., *Selective photocatalytic destruction of airborne VOCs.* Solar Energy, 1996. **56**(5): pp. 445–453.

33. Carp, O., C.L. Huisman, and A. Reller, *Photoinduced reactivity of titanium dioxide.* Progress in Solid State Chemistry, 2004. **32**(1–2): pp. 33–177.

34. Zeltner, W.A., and D.T. Tompkins, *Shedding light on photocatalysis.* Ashrae Transactions, 2005. **111**: p. 523.
35. Perkowski, J., et al., *Decomposition of detergents present in car-wash sewage by titania photo-assisted oxidation.* Polish Journal of Environmental Studies, 2006. **15**(3).
36. Fujishima, A., T.N. Rao, and D.A. Tryk, *Titanium dioxide photocatalysis.* Journal of Photochemistry and Photobiology C: Photochemistry Reviews, 2000. **1**(1): pp. 1–21.
37. Hoffmann, M.R., et al., *Environmental applications of semiconductor photocatalysis.* Chemical Reviews, 1995. **95**(1): pp. 69–96.
38. Zhong, Z., et al., *Synthesis of one-dimensional and porous TiO$_2$ nanostructures by controlled hydrolysis of titanium alkoxide via coupling with an esterification reaction.* Chemistry of Materials, 2005. **17**(26): pp. 6814–6818.
39. Xiong, C., and K.J. Balkus, *Fabrication of TiO$_2$ nanofibers from a mesoporous silica film.* Chemistry of Materials, 2005. **17**(20): pp. 5136–5140.
40. Yuan, Z.-Y., and B.-L. Su, *Titanium oxide nanotubes, nanofibers and nanowires.* Colloids and Surfaces A: Physicochemical and Engineering Aspects, 2004. **241**(1–3): pp. 173–183.
41. Tachikawa, T., M. Fujitsuka, and T. Majima, *Mechanistic insight into the TiO$_2$ photocatalytic reactions: Design of new photocatalysts.* Journal of Physical Chemistry C, 2007. **111**(14): pp. 5259–5275.
42. Rajeshwar, K., et al., *Titania-based heterogeneous photocatalysis. Materials, mechanistic issues, and implications for environmental remediation.* Pure and Applied Chemistry, 2001. **73**(12): pp. 1849–1860.
43. Zhang, Z., and P.A. Maggard, *Investigation of photocatalytically-active hydrated forms of amorphous titania, TiO$_2$ nH$_2$O.* Journal of Photochemistry and Photobiology A: Chemistry, 2007. **186**(1): pp. 8–13.
44. Kitano, M., et al., *Recent developments in titanium oxide-based photocatalysts.* Applied Catalysis A: General, 2007. **325**(1): pp. 1–14.
45. Legrini, O., E. Oliveros, and A. Braun, *Photochemical processes for water treatment.* Chemical Reviews, 1993. **93**(2): pp. 671–698.
46. Kisch, H., and W. Macyk, *Visible-light photocatalysis by modified titania.* ChemPhysChem, 2002. **3**(5): pp. 399–400.
47. Bacsa, R., et al., *Preparation, testing and characterization of doped TiO$_2$ active in the peroxidation of biomolecules under visible light.* Journal of Physical Chemistry B, 2005. **109**(12): pp. 5994–6003.
48. Jin, Z., et al., *5.1% Apparent quantum efficiency for stable hydrogen generation over eosin-sensitized CuO/TiO$_2$ photocatalyst under visible light irradiation.* Catalysis Communications, 2007. **8**(8): pp. 1267–1273.
49. Bauer, C., et al., *Interfacial electron-transfer dynamics in Ru (tcterpy)(NCS) 3-sensitized TiO$_2$ nanocrystalline solar cells.* Journal of Physical Chemistry B, 2002. **106**(49): pp. 12693–12704.
50. Gaya, U.I., and A.H. Abdullah, *Heterogeneous photocatalytic degradation of organic contaminants over titanium dioxide: A review of fundamentals, progress and problems.* Journal of Photochemistry and Photobiology C: Photochemistry Reviews, 2008. **9**(1): pp. 1–12.
51. Kryukova, G.N., et al., *Structural peculiarities of TiO$_2$ and Pt/TiO$_2$ catalysts for the photocatalytic oxidation of aqueous solution of Acid Orange 7 dye upon ultraviolet light.* Applied Catalysis B: Environmental, 2007. **71**(3–4): pp. 169–176.

52. Li, H., et al., *Mesoporous Au/TiO₂ nanocomposites with enhanced photocatalytic activity.* Journal of the American Chemical Society, 2007. **129**(15): pp. 4538–4539.

53. Sreethawong, T., et al., *Nanocrystalline mesoporous Ta₂O₅-based photocatalysts prepared by surfactant-assisted templating sol-gel process for photocatalytic H₂ evolution.* Journal of Molecular Catalysis A: Chemical, 2005. **235**(1–2): pp. 1–11.

54. Jing, D., and L. Guo, *Hydrogen production over Fe-doped tantalum oxide from an aqueous methanol solution under the light irradiation.* Journal of Physics and Chemistry of Solids, 2007. **68**(12): pp. 2363–2369.

55. Colmenares, J.C., et al., *Nanostructured photocatalysts and their applications in the photocatalytic transformation of lignocellulosic biomass: An overview.* Materials, 2009. **2**(4): pp. 2228–2258.

56. Imai, H., and H. Hirashima, *Preparation of porous anatase coating from sol-gel-derived titanium dioxide and titanium dioxide-silica by water-vapor exposure.* Journal of the American Ceramic Society, 1999. **82**(9): pp. 2301–2304.

57. Tsuji, I., et al., *Photocatalytic H₂ evolution reaction from aqueous solutions over band structure-controlled (AgIn)ₓ Zn₂₍₁₋ₓ₎ S₂ solid solution photocatalysts with visible-light response and their surface nanostructures.* Journal of the American Chemical Society, 2004. **126**(41): pp. 13406–13413.

58. Kato, K., et al., *Crystal structures of TiO₂ thin coatings prepared from the alkoxide solution via the dip-coating technique affecting the photocatalytic decomposition of aqueous acetic acid.* Journal of Materials Science, 1994. **29**: pp. 5911–5915.

59. Jimmy, C.Y., et al., *Preparation of highly photocatalytic active nano-sized TiO₂ particles via ultrasonic irradiation.* Chemical Communications, 2001(19): pp. 1942–1943.

60. Oh, C.W., et al., *Synthesis of nanosized TiO₂ particles via ultrasonic irradiation and their photocatalytic activity.* Reaction Kinetics and Catalysis Letters, 2005. **85**: pp. 261–268.

61. Colmenares, J., et al., *Synthesis, characterization and photocatalytic activity of different metal-doped titania systems.* Applied Catalysis A: General, 2006. **306**: pp. 120–127.

62. Suslick, K.S., *Mechanochemistry and sonochemistry: Concluding remarks.* Faraday Discussions, 2014. **170**: pp. 411–422.

63. Mason, T., and J. Luche, *Ultrasound as a new tool for synthetic chemists.* In van Eldk, R., and C.D. Hubbard (eds.), *Chemistry Under Extreme or Non Classical Conditions* (pp. 317–380). John Wiley and Sons, Inc., Spektrum Akademischer Verlag, 1996.

64. Aramendía, M., et al., *Screening of different zeolite-based catalysts for gas-phase selective photooxidation of propan-2-ol.* Catalysis Today, 2007. **129**(1–2): pp. 102–109.

65. Yamamoto, S., et al., *Preparation of epitaxial TiO₂ films by pulsed laser deposition technique.* Thin Solid Films, 2001. **401**(1–2): pp. 88–93.

66. Stillings, R.A., and R.J. Nostrand, *The action of ultraviolet light upon cellulose. I. Irradiation effects. II. Post-irradiation effects.* Journal of the American Chemical Society, 1944. **66**(5): pp. 753–760.

67. Mergel, D., et al., *Density and refractive index of TiO₂ films prepared by reactive evaporation.* Thin Solid Films, 2000. **371**(1–2): pp. 218–224.

68. Ni, M., et al., *An overview of hydrogen production from biomass.* Fuel Processing Technology, 2006. **87**(5): pp. 461–472.
69. Iwasaki, W., *A consideration of the economic efficiency of hydrogen production from biomass.* International Journal of Hydrogen Energy, 2003. **28**(9): pp. 939–944.
70. Rapagnà, S., N. Jand, and P.U. Foscolo, *Catalytic gasification of biomass to produce hydrogen rich gas.* International Journal of Hydrogen Energy, 1998. **23**(7): pp. 551–557.
71. Li, S., et al., *Fast pyrolysis of biomass in free-fall reactor for hydrogen-rich gas.* Fuel Processing Technology, 2004. **85**(8–10): pp. 1201–1211.
72. Watanabe, M., H. Inomata, and K. Arai, *Catalytic hydrogen generation from biomass (glucose and cellulose) with ZrO$_2$ in supercritical water.* Biomass and Bioenergy, 2002. **22**(5): pp. 405–410.
73. Hao, X., et al., *Hydrogen production from glucose used as a model compound of biomass gasified in supercritical water.* International Journal of Hydrogen Energy, 2003. **28**(1): pp. 55–64.
74. Ni, M., et al., *A review and recent developments in photocatalytic water-splitting using TiO$_2$ for hydrogen production.* Renewable and Sustainable Energy Reviews, 2007. **11**(3): pp. 401–425.
75. Kawai, T., and T. Sakata, *Conversion of carbohydrate into hydrogen fuel by a photocatalytic process.* Nature, 1980. **286**(5772): pp. 474–476.
76. Kawai, M., T. Kawai, and K. Tamaru, *Production of hydrogen and hydrocarbon from cellulose and water.* Chemistry Letters, 1981. **10**(8): pp. 1185–1188.
77. Kawai, T., and T. Sakata, *Photocatalytic hydrogen production from water by the decomposition of poly-vinylchloride, protein, algae, dead insects, and excrement.* Chemistry Letters, 1981. **10**(1): pp. 81–84.
78. Sakata, T., *Hydrogen production from biomass and water by photocatalytic processes.* New Journal of Chemistry, 1981. **5**: p. 279.
79. Patsoura, A., D.I. Kondarides, and X.E. Verykios, *Photocatalytic degradation of organic pollutants with simultaneous production of hydrogen.* Catalysis Today, 2007. **124**(3–4): pp. 94–102.
80. Herrmann, J., *From catalysis by metals to bifunctional photocatalysis.* Topics in Catalysis, 2006. **39**(1–2): pp. 3–10.
81. Pichat, P., et al., *Photocatalytic hydrogen production from aliphatic alcohols over a bifunctional platinum on titanium dioxide catalyst.* New Journal of Chemistry, 1981. **5**(12): pp. 627–636.
82. Umena, Y., et al., *Crystal structure of oxygen-evolving photosystem II at a resolution of 1.9 Å.* Nature, 2011. **473**(7345): pp. 55–60.

Chapter 7

Wastewater Treatment

ABSTRACT

It is important to have a good knowledge of the different types of water contamination, like industrial waste, home sewage, municipal sewage, and accidental spills. However, there is a practical and sustainable solution available with photocatalysis. This versatile method effectively removes pollutants, has low energy consumption, and can fully mineralize contaminants. It is essential to remove pollutants and impurities before releasing water into the environment for the sake of public health and the environment. The different types of wastewater treatments are discussed in detail, along with their individual purposes and drawbacks. Photocatalysts are highly effective for removing pathogens, heavy metals, pharmaceutical residues, organic pollutants, and dyes from water. Nanomaterials are of considerable importance in the advancement of wastewater treatment methodologies, as they provide enhanced efficacy, cost-effectiveness, and ecological compatibility in the pursuit of solutions. This chapter provides comprehensive knowledge about photocatalytic reactors utilizing photocatalysts, storage, utilization, mechanism, and future implications, making the photocatalytic reactor a great option for treating wastewater.

7.1 WATER POLLUTION AND SOURCES OF WATER POLLUTION

Water pollution occurs when commercial and industrial waste, agricultural operations, human activity, and various contaminants affect water sources such as rivers, lakes, oceans, groundwater, and drinking water supplies. These pollutants enter the water, affect aquatic life, and harm human health.[1] Metals are unsafe to marine life and may serve as primary or secondary contaminants. The dye effluent contains a variety of

DOI: 10.1201/9781003403357-9

toxic metals, including Cr, Cd, Ni, Zn, Cu, Pb, and Fe, among others. Metals combine with other compounds in dye effluent to produce complex metal salts, which are extremely difficult to remove using conventional methods; therefore, it is necessary to develop suitable methods for water purification.[2]

Generally, the pollutants come from three prominent sources:

i.   Wastewater from domestic discharge into rivers
ii. Untreated industrial wastewater
iii. Surface runoff from agricultural areas treated with pesticides, herbicides, and chemical fertilizers

The main causes of water pollution are both direct and indirect sources, as well as other sources.

7.1.1 Direct Pollution

Direct water pollution is the deliberate or unintentional discharge of pollutants into rivers, lakes, seas, or groundwater. This type of water pollution is more dangerous to the environment and human health when toxic chemicals add to water bodies without being purified.[3] As an example, municipal wastewater discharge, industrial discharge, and agricultural runoff are a few examples of direct pollution. By consuming this contaminated water, both humans and animals end their lives.

7.1.2 Indirect Water Pollution

Indirect water pollution is occurring through the addition of toxic chemicals into the water via rainwater. Chemical fertilizers and pesticides slowly permeate the soil and enter the groundwater before reaching various streams and rivers.[4] Water contamination can cause disease and even death to any living organism, whether it occurs directly or indirectly e.g. fertilizers and pesticides discharged into the environment.

7.2 SOURCES OF NATURAL RUNOFF

Pollutants can be recognized in water reservoirs through a variety of actions, some of which are as follows:

i.   Rainwater
ii. Domestic sewage
iii. Industrial wastage

7.2.1 Rainwater

Rainwater is a significant natural source of water contamination in which hazardous air pollutants and harmful substances reach into the groundwater. Toxic gases such as nitrogen and sulphur oxides were found in rainwater, which led to acid rain.[5] Waters may become contaminated by the falling of leaves, branches, and other plant materials and a variety of elements, including dust, pollen, contaminants, and gases, may alter its composition and purity.

7.2.2 Domestic Sewage

Domestic sewage is any waste or wastewater produced by humans or domestic activities discharged into the environment.[6] It includes water used for cleaning, such as in domestic washing machines or laundromats as well as for other cleaning obligations, such as washing automobiles at home or at a car wash. Wastewater treatment primarily tries to remove nutrients and organic contaminants etc.

7.2.3 Industrial Wastage

Industrial waste can be in the form of a solid, liquid, or gas and is the waste generated by industrial activities during a production process. It may or may not be dangerous. Hazardous wastes may contaminate the air, soil, and water supplies and are toxic, inflammable, corrosive, reactive, or radioactive.[7]

7.3 PHOTOCATALYTIC DEGRADATION OF ORGANIC POLLUTANTS IN WATER

A semiconductor photocatalyst is employed to facilitate the degradation of contaminants that exhibit high concentrations and low biodegradability when exposed to visible light irradiation. Photocatalytic degradation represents a highly efficient and secure approach for the elimination of pollutants from various surfaces, encompassing both aqueous and gaseous environments. During this particular phenomenon, the semiconductor material possesses the capability to assimilate radiant energy and generate electron-hole pairs, thereby functioning as a photocatalyst.[8] As a result, a redox reaction takes place that eventually decomposes hazardous contaminants into non-toxic by-products There are various factors that affect photocatalytic degradation as shown in Figure 7.1.

7.3.1 Photocatalyst Preparation

A semiconductor-based photocatalyst is typically used for the removal of specific pollutants that need to be degraded under ultraviolet (UV) or visible light irradiation.[9] These photocatalysts are prepared in the form of

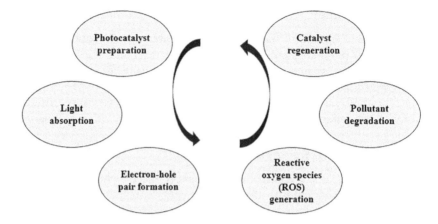

Figure 7.1 Photocatalytic degradation of organic pollutants.

nanoparticles, thin films, and other nanostructured materials by using physical or chemical methods. For example, ZnO- and WO_3-based nanoparticles were utilized for the degradation of organic dyes i.e. methylene blue.

7.3.2 Light Absorption

The photocatalyst is often exposed to light using a UV-visible light source. The bandgap of a semiconductor material determines by the wavelength of light that it can absorb, and this absorbed light produces electron-hole pairs.[10] For example, the bandgap of WO_3 is 2.8 eV which lies under visible light irradiation. This material is used for purification of contaminated water.

7.3.3 Electron-Hole Pair Formation

The efficient photocatalyst decreases the recombination electron-hole pair by absorbing light. The holes remain in the valence band while the electrons travel from the valence band to the conduction band.[11] Electrons in the semiconductor's valance band are stimulated into the conduction band when photons of light with energy equal to or greater than the semiconductor's bandgap energy irradiate the semiconductor. For example, the photodegrading efficiency is increased by the doping of Mg into ZnO because it lowers the rate of electron-hole pair recombination.

7.3.4 Reactive Oxygen Species Generation

The holes and electrons generate highly reactive oxygen species (ROS), such as hydroxyl radicals ($\bullet OH$) and superoxide ions ($O_2 \bullet -$), and they can interact with oxygen and water molecules in the environment to form strong oxidizing agents.[12] The ROS will break the organic pollutant molecules that adsorb on the surface of the photocatalyst. These ROS oxidize the pollutant and convert it into less harmful molecules like water, carbon dioxide, and other non-toxic by-products.

7.3.5 Catalyst Regeneration

The photocatalyst is regenerated and can be reused for further photocatalytic degradation cycles after the degradation reaction. Radicals can then interact with the molecules of contaminants to completely or partially degrade organic pollutants. In all aspects, the regeneration process's enhancements— which include raising the photocatalysis's kinetics and optimizing the catalyst recovery and regeneration processes' efficacies—will make it valuable for water remediation at any scale.[13,14]

There are various limitations regarding the most common photocatalysts:

- Few semiconductor materials only exhibit photocatalysis in the UV region due to the larger bandgap values i.e. TiO_2. Higher energy light sources are consequently necessary for these materials.
- The kinetics of the photocatalytic degradation process and the recombination of photo-generated electron-hole pairs are the limitations of a semiconductor-based photocatalyst. Thus, the incorporation of carbon nanotubes (CNTs) is an exciting approach for eliminating these problems and enhancing the degradation efficiency of the photocatalyst.[15]

7.4 FACTORS INFLUENCING PHOTOCATALYTIC DEGRADATION

The efficiency of photocatalysts is significantly influenced by a number of operational parameters that affect the photodegradation of organic contaminants. Numerous studies have discussed the importance of operational parameters.[16]

7.4.1 Effect of Dye Concentration

The concentration of the dye at the beginning of the experiment is a crucial variable that impacts the adsorption process. This is due to its indirect influence on the efficacy of dye removal, which is determined by the alteration of binding sites on the adsorbent's surface. In a general context, it has been observed that as the concentration of dye is increased, there is a corresponding decrease in the efficiency of degradation. It is important to note that the number of catalysts employed in the process remains constant. By increasing dye concentration, more organic molecules adsorb on the surface of the photocatalyst (TiO$_2$), and thus fewer •OH decreases the percentage of degradation.[17]

7.4.2 Size and Structure of the Photocatalyst

The particle size and agglomerate size of a photocatalyst are crucial in organic dye degradation by photocatalysis and pollutants etc. In comparison to larger particles, smaller photocatalyst particles typically have a higher surface area per unit volume. More active sites for catalytic processes are made available by the increased surface area, increasing the catalytic activity. The breakdown of contaminants is thus influenced by morphology of photocatalyst.

7.4.2.1 For Example

TiO$_2$ nanoparticles with Cu doping show the highest photodegradation, whereas pure TiO$_2$ catalysts exhibit the least photodegradation. This is due to the fact that doping changes the morphology as well as particle size of photocatalyst. The rough and zigzag surface of the Cu-doped TiO$_2$ enhances the adsorption of organic molecules on the photocatalytic surface. Thus, Cu-doped TiO$_2$ has higher photocatalytic degradation efficiency than pristine TiO$_2$.[18]

7.4.3 Effect of Light Intensity and Irradiation Time

The intensity of light and the duration of irradiation have a significant impact on photocatalytic degradation efficiency. The rate of degradation increased steadily as the light intensity increased. It seems like there was a direct correlation between the two, where the degradation was proportional to the amount of light.[19] The rate of degradation exhibits a positive correlation with the passage of time, ultimately culminating in the attainment of the maximum level, owing to the progressive generation of a greater quantity of hydroxyl radicals. As temporal progression ensues, the percentage of photodegradation concurrently escalates until it attains an optimal threshold owing to the augmented generation of hydroxyl radicals. As a result, the photodegradation on the photocatalytic reaction increases with increasing time of irradiation.

7.4.4 Effect of pH

The photodegradation of dye is greatly influenced by the pH level of the solution. It seems that the degradation process increases in an acidic medium and decreases when the pH of the solution is raised. This information is important to keep in mind when monitoring photodegradation to achieve optimal outcomes and prevent the formation of hydroxyl radicals through the highest level of degradation. The degradation of dye increased in acidic medium and decreased as the pH of the solution increased. In contrast, the surface charge and dispersion of nanoparticles can be influenced by the acidity of solutions.

7.4.4.1 For Example

It is important to keep in mind that the surface charge of TiO_2 particles and the potentials of catalytic processes can be varied by the pH level of the solution. Therefore, the pH level is crucial for achieving the desired results in these processes. The presence of dye on the surface affects the adsorption process, leading to a change in the rate at which the reaction occurs.[20] When the initial pH of the reaction mixture was kept at 11, there was a decline in degradation. It seems that there is a change in the breakdown of dye percentage at higher pH levels. This change may change the degradation percentage of dye molecules themselves.

7.4.5 Effect of Catalyst Amount

The overall rate of the photocatalytic process is directly dependent on the amount of catalyst. This implies that if you increase the amount of catalyst used, it can result in a higher rate of the process, and vice versa, and it is important to keep this in mind when performing experiments that involve photocatalysis. The speed of the photocatalytic process increases in direct proportion to the quantity of catalyst used.

7.4.5.1 For Example

The amount of TiO_2 particles affects the photocatalytic rate of reaction in a heterogeneous catalytic regime. By increasing the amount of catalysts, the number of active sites on the photocatalyst surface is enhanced. This leads to the production of a greater number of •OH radicals, which can effectively participate in the degradation of the dye.[21]

7.5 WATER TREATMENT METHODS

7.5.1 Sonolysis

Sonolysis is a relatively advanced method for degradation. It is also known as sono-chemical degradation, and it uses acoustic cavitation and the cyclical growth of gas bubbles to break down materials. It is a great way to remove dangerous water pollutants from water without requiring adding additional chemicals to the water supply. This is because the method uses acoustic cavitation and the cyclical growth of gas bubbles to break down materials, making it an effective and environmentally friendly solution.[22]

7.5.1.1 Working Principle

Ultrasound waves propagate in an aqueous solution which forms cavitation bubbles. By the cavitation process, pre-existing gas radii start to oscillate in the cavity due to changing the pressure field of the ultrasonic waves periodically. The collapse of gas bubbles produces very high temperature and pressure which dissociates H_2O and converts it into its radical. These radicals are used for the destruction of organic substrates. This process required high ultrasonic frequency (400 kHz) for the production of radicals, and ultrasonic energy generation can be formed by electrochemical or liquid-driven methods which are also expansive and difficult to use at a large scale. Organic waste materials can be decomposed by indirect radiolysis which is interacting with an aqueous solution of water and generates hydrogen peroxide. But these radiations generated 100 eV absorbed energy.

7.5.2 Ozonolysis

The ozone/UV process is also useful for dye degradation from wastewater because it has a strong oxidant (redox pH 1 and potential is 2.07 V). Hydroxyl radicals are generated when UV photolysis reacts with hydrogen. Hydroxyl radicals are produced when these hydroxyl radicals make a reaction with hydrogen peroxide which is also known as an oxidation reagent. When radicals of hydroxyl form, it further reacts with organic substrates by using electron transfer processes. Removal of inorganic materials from water by the HO/UV process is widely researched. The best advantage to utilizing this process is that this method is easy to use, and hydrogen peroxide is very low in cost and very easily reacts with water and there is no need for separating the water treatment. However, the adsorption of hydrogen peroxide and generation of quantum efficiency is very low at A > 250 nm.

$$3O_3 + H_2O \rightarrow 2OH^{\cdot} + 4O_2 \tag{7.1}$$

When other oxidants are present, the decomposition of H_2O_2 and O_3 leads to the production of hydroperoxide (HO_2^-) and the generation of OH^- ions.

$$H_2O_2 \rightarrow HO_2^- + H^+ \tag{7.2}$$

$$HO_2^- + O_3 \rightarrow O_2^- + OH^- + O_2 \tag{7.3}$$

In the process of ozone/UV treatment, hydrogen peroxide (H_2O_2) is primarily produced as an additional oxidant by the photolysis of O_3, as in Equation 7.4.

$$O_3 + H_2O + hv \rightarrow + H_2O_2 + O_2 \tag{7.4}$$

There are three ways in which OH· can be generated. The first pathway is through ozonation (Equation 7.1). The second pathway involves O_3/H_2O_2 (Equations 7.2 and 7.3). Last, OH· can be generated through the photolysis of H_2O_2, as shown in Equation 7.5.

$$H_2O_2 + hv \rightarrow 2OH \quad (7.5)$$

7.5.3 UV-Based Advanced Oxidation Processes

It is important to note that when there are catalysts or oxidants, photons can start the creation of hydroxyl radicals. One of the most commonly employed catalysts is TiO_2, which is triggered to generate electrons in the conduction band and holes in the valence band.

$$TiO_2 + hv \rightarrow e^- \left(c + hv + vb\right) \quad (7.6)$$

When OH^-, H_2O, and O_2^- react at the surface of TiO_2, the resulting holes and electrons have the ability to combine and produce hydroxyl radicals.

$$hv + v + OH - \left(surface\right) \rightarrow OH^· \quad (7.7)$$

$$hv + vb + H_2O\left(absorbed\right) \rightarrow OH^· + H^+ \quad (7.8)$$

$$e - cb + O_2 \left(absorbed\right) \rightarrow O_2^{-} \quad (7.9)$$

In the presence of H_2O_2 or O_3 oxidants, additional OH· may undergo the UV irradiation, for example, an H_2O_2 molecule produced by two OH· under UV light irradiation.

$$H_2O_2 + hv \rightarrow 2OH^· \quad (7.10)$$

$$H_2O + hv \rightarrow OH^· + H^· \quad (7.11)$$

7.5.4 Photo-Fento Reactions

Fenton reagent forms hydroxyl radicals, and these reactions are useful for the removal of metal oxides from water. Additionally, it has been seen that the fundamental Fenton reaction under UV light reduction is a much faster and more beneficial procedure for oxidative degradation. The reaction of

H_2O_2 and Fe^{2+} produces a powerful reactive species and a so-called Fenton reaction. While there have been suggestions for hypothetical alternative molecules such as ferri ions, the resulting reactive species are commonly known as hydroxyl radicals.[23] The Fenton reaction is used to purify wastewater, and it has been extensively discussed. The classic Fenton radical processes' primary reactions are as follows:

$$Fe_2^+ + H_2O_2 \rightarrow Fe_3^+ + OH^{\cdot} + OH^- \tag{7.12}$$

$$Fe_3^+ + H_2O_2 \rightarrow Fe_2^+ + HO^{\cdot}_2 + H^+ \tag{7.13}$$

$$OH^{\cdot} + H_2O_2 \rightarrow + HO^{\cdot}_2 + H_2O \tag{7.14}$$

$$OH^{\cdot} + Fe^{2+} \rightarrow + Fe^{3+} + OH^- \tag{7.15}$$

$$Fe_3^+ + HO^{\cdot}_2 \rightarrow Fe_2^+ + O_2H^+ \tag{7.16}$$

$$Fe_3^+ + HO^{\cdot}_2 + H^+ \rightarrow Fe_3^+ + H_2O_2 \tag{7.17}$$

$$2HO^{\cdot}_2 \rightarrow H_2O_2 + O_2 \tag{7.18}$$

Although Equation 7.13 indicates that Fe_3^+ can be reduced to Fe_2^+, the iron cannot be a catalyst in the Fenton system. It can be quite challenging to dispose of the sludge separately, which adds to the treatment's complexity and operational expenses. It requires careful planning and execution to ensure that the process is done safely and thoroughly. It should be emphasized that the pH level must be acidic for the Fenton reaction to generate hydroxyl radicals. Therefore, the practical use of the Fenton reaction to wastewater treatment is limited.

7.5.5 Other Advanced Oxidation Processes

It is interesting to know that advanced oxidation processes, like ultrasonic irradiation and electronic beam irradiation, have been explored as potential methods for wastewater treatment. This could potentially simplify the process of disposing of sludge separately and make the treatment process more efficient and cost-effective. However, thorough planning and execution are still necessary to ensure safety and optimal results. High pressure (200–500 atm) and temperatures (4200–5000 K) may be encountered in a millisecond when a microbubble bursts. Under these harsh conditions, gaseous water molecules within microbubbles dissolve and generate hydroxyl radicals.[24,25]

7.5.6 Photocatalysis

A photocatalyst is a substance that absorbs light and acts as a catalyst for chemical reactions. It is interesting that the word "photocatalyst" is made up of two words, "photo" and "catalyst," which both play important roles in its function. The "photo" part relates to photons and their absorption, while the "catalyst" part refers to how it affects the rate of reaction. When catalyst is exposed to light, it alters the rate of a chemical reaction. These substances are called photocatalysts and they are really interesting. This phenomenon is known as photocatalysis.[26] The photocatalytic reaction generates an electron-hole pair when a semiconducting material (WO_3) is exposed to light, then the photocatalytic reaction occurs and generates an electron-hole pair. The process of oxidation-reduction occurs, and less hazardous by-products are obtained as shown in Figure 7.2.

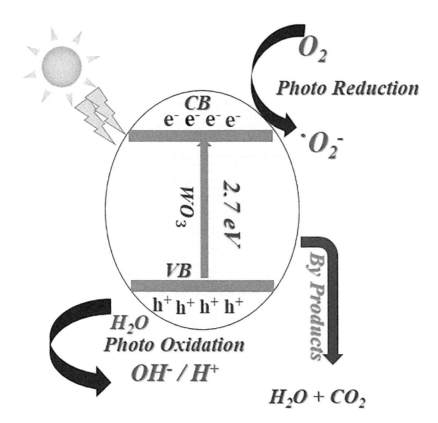

Figure 7.2 Photocatalytic degradation of organic pollutants utilizing WO_3 photocatalyst.

Two different types of photocatalytic reactions can be distinguished based on the physical properties of the reactants.

7.5.6.1 Homogeneous Photocatalysis

When the reactant and semiconductor photocatalysts exist in the same phase, it is called homogeneous photocatalysis. This can occur in gases, solids, or liquids and is an important phenomenon in chemical reactions.

7.5.6.2 Heterogeneous Photocatalysis

Heterogeneous photocatalysis refers to the photocatalytic reactions that occur when the reactant and semiconductor photocatalysts are in separate phases. This phenomenon is often observed in various chemical reactions and can have significant implications for their outcomes.

The electronic structure and bandgap of materials are classified into three types: conductor, semiconductor, and insulator. Semiconductor-based photocatalysis is a technique employed to break down pollutants, such as organic dyes i.e. methylene blue, rhodamine B etc. A sufficient bandgap energy is required for photocatalytic degradation, and all materials have their definite bandgap values. The bandgap energy of semiconductor material lies in the visible region; thus, it is used for photocatalysis and the values of all materials shown in Figure 7.3. TiO_2 is a widely employed semiconductor material in the field of photocatalysis due to its notable attributes of elevated photocatalytic efficacy and robustness.

7.5.6.3 Enhancement of Degradation Efficiency

There are numerous methods for engineering TiO_2 catalyst to improve its photoresponse for the solar spectrum. There are some ways to improve photocatalysis efficiency, formation of composites using CNTs, which seems to be quite effective. Another process involves the doping of noble

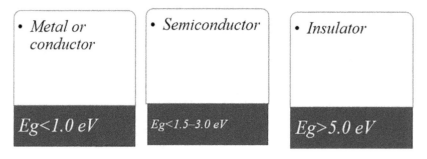

Figure 7.3 The bandgap energy of metal, semiconductor, and insulator.

and transition metals into semiconductor materials, which also appears to enhance degradation efficiency. These methods include the formation of CNTs-based composites, doping of noble metals and transition metals into semiconductor-based photocatalysis that enhances the degradation efficiency. It seems that modified TiO_2 catalysts exhibit higher degradation efficiency when exposed to sunlight. It has been observed that CNTs connected to TiO_2 have the capability to increase the number of electron-hole pairs by effectively trapping the electron within its structure.[27]

7.6 DEVELOPMENT OF NANOMATERIALS AS ADSORBENT FOR WASTEWATER TREATMENT

The adsorption technique is used for the removal of inorganic pollutants from wastewater. The nanoparticles such as oxides of aluminum, iron, and titanium are most frequently used to remove inorganic ions. Certain studies showed that iron oxide nanoparticles are excellent adsorbents for effectively removing inorganic compounds due to their larger surface area. For example, As, Cd, Cr, Cu, and other heavy metal contaminants were removed by using different metal and metal oxide nanoparticles. The size of nanoparticles is crucial, and it lies within 1 to 100 nm. Water can contain three types of contaminants: organic, inorganic, and biological pollutants. It is essential to detect these pollutants because they can harm our health and the environment.

7.6.1 Nano-Adsorbents

Nano-adsorbents play a crucial role for the purification of water because of their large surface area, which enables them to effectively adsorb and remove a variety of pollutants from water. Wastewater treatment uses a wide variety of nano-adsorbents. The nanomaterials that have been widely employed for the purpose of wastewater treatment encompass graphene, Fe_3O_4, MnO_2, Co_3O_4, TiO_2, MgO, and ZnO. They can be derived in a variety of morphological shapes, including sheets, tubes, and particles. Following are a few common nano-adsorbents for water purification.

7.6.2 Activated Carbon

Activated carbon is a remarkably efficient material utilized for the purpose of water purification. The compound exhibits a characteristic of weak acidic ion exchange, rendering it highly suitable for the elimination of metallic impurities and contaminants from aqueous waste streams. This makes it a popular choice for industries to improve water quality. With its ability to absorb and remove harmful substances, activated carbon is a safe and reliable solution for clean water.

For example, activated carbon made from coconut tree fibers is the optimal adsorbent for extracting Cr(VI) metal from aqueous solutions. It excels in removing metal impurities and pollutants from water, making it the ideal choice for water purification. Its acidic ion exchange character is weak, but it still effectively absorbs and removes harmful substances, providing a safe and reliable solution for clean purification.[28]

7.6.3 Carbon Nanotubes

CNTs are a highly effective adsorbent for removing impurities from wastewater, making them a promising choice for wastewater treatment. Multi-walled carbon nanotubes (MWCNTs) are often used without further processing due to their exceptional synthetic dye adsorption capabilities, providing a safe and reliable method for water purification. Single-wall carbon nanotubes (SWCNTs) have the ability to adsorb organic pollutants more effectively compared to hybrid carbon nanotubes (HCNTs) and MWCNTs. This is primarily because SWCNTs possess a higher specific surface area. It was observed that methylene red and methylene blue could be effectively eliminated from aqueous solutions using oxidized MWCNTs. SWCNTs are able to better adsorb organic pollutants than HCNTs and MWCNTs due to their enhanced specific surface area.

Various types of CNTs have been employed in scientific investigations pertaining to the elimination of divalent metal ions such as Cd(II), Cu(II), Ni(I), Pb(II), and Zn(II) from aqueous solutions.[29] CNTs have proven to be particularly desirable adsorbents for environmental pollution.

7.6.4 Metal Oxide

The oxides of zinc, zirconium, lead, manganese, iron, titanium, magnesium, cerium and aluminum are known to be some of the most famous and promising adsorbents for various applications, including environmental protection. Metal oxides have the ability to purify water by removing different harmful metal impurities from the water. This is because metal oxides have a significant surface area, which allows them to effectively eliminate these contaminants. One major drawback of metal oxides is that their stability is significantly threatened when they are scaled down from micro to nano size due to the increase in surface energy.

7.6.5 Nano Aluminum Oxides

For removing heavy metals, alumina has traditionally been used as an adsorbent. In contrast, it is expected that gamma-alumina (γ-Al_2O_3) will be a more effective adsorbent than alpha-alumina (α-Al_2O_3). Sol-gel preparation is used to create nanoscale γAl_2O_3. It is anticipated that physical or chemical modification of γ-Al_2O_3 by combining with Ag, CNTs, CeO_2, and γ-Mn_2O_3, ZnO, CdO, and TiO_2, can be utilized to create nanocomposites, and their enhanced photocatalytic activities are employed to remove industrial textile effluents from wastewater.

7.7 PHOTOCATALYTIC REACTOR

A photoreactor is a device that brings together the necessary light and reactants, ensuring they come into effective and proper contact with each other. When it comes to photocatalytic reactors for water treatment, they can typically be divided into two main configurations based on how the photocatalysts are used:

i. Reactors with photocatalyst particles suspended
ii. Reactors with a continuous inert carrier that is immobilized with a photocatalyst

The utilization of various reactor types, including but not limited to the annular slurry photoreactor, cascade photoreactor, and downflow contactor reactor, has been observed in the field of photocatalytic water treatment. The primary disparity observed between these two configurations resides in the requirement of an additional downstream separation unit in the former, for the purpose of recuperating the photocatalyst particles. Conversely, the latter configuration facilitates uninterrupted operation.

Advanced oxidation processes, heterogeneous photocatalysis using semiconductor photocatalysts (ZnO, TiO_2, CdS, Fe_2O_3, ZnS, and GaP), has shown its effectiveness in the breakdown of a wide range of organic materials into immediately biodegradable mixtures and ultimately mineralized them to safe water and carbon dioxide. The metal-oxide-semiconductor photocatalysts have garnered significant interest due to their capacity for remediation of waste and pollutants in aqueous and atmospheric environments. The augmentation of the degradation rate is contingent upon the attainment of a substantial surface area-to-volume ratio, as the phenomenon of photocatalysis transpires exclusively on the surface of the catalyst.[30]

7.7.1 Fundamental Constituents in the Design of Photocatalytic Reactors

Photocatalytic reactors can be classified into three primary constituents: the luminous emitter, the catalyst, and the reactor apparatus. Let's take a closer look at each of these components. Light possesses various characteristics that make it a fascinating phenomenon. One of its fundamental properties is its ability to travel in straight lines, allowing it to propagate through space.

Furthermore, the parameters pertaining to light encompass the specific wavelength required to initiate the catalytic process, such as UV or visible radiation, alongside the light source employed for this purpose.

7.7.1.1 The Wavelength of Light

The light source's wavelength should be proportional to the bandgap energy of the catalyst, and it plays a key role in photocatalysis. As the size of the photocatalyst decreases, more of it comes into contact with the reaction media and gets lighted. Nanotechnology represents a highly sophisticated method for the purification of wastewater. The decolorization and mineralization of textile dyes were explored using photocatalytic reactors with immobilized nanostructured TiO_2.

7.7.1.2 Light Source

The light source is a crucial element in photocatalytic reactors. Photocatalytic reactors can make use of either natural light, such as sunlight, or artificial light from lamps. These lamps can be conventional or light-emitting diodes (LEDs). The prevailing UV light sources employed in photocatalytic processes encompass low- and medium-pressure mercury lamps. While continuous photon flux is advantageous, it is important to note that these lamps also have some drawbacks. For instance, they have a relatively short lifespan of around 9000 to 12,000 hours. Additionally, their energy efficiency is quite low, and they can have negative environmental consequences due to the presence of mercury. In order to mitigate these constraints, scientists resorted to employing high-pressure mercury lamps as the illuminating apparatus in subsequent inquiries.

Photocatalytic reactors, which harness natural light sources for their operation, commonly referred to as solar photoreactors, can be classified into two distinct categories according to the type of radiation they are exposed to: concentrating and non-concentrating light. Despite being less expensive than LEDs, traditional lights have a short lifespan and contain hazardous mercury. Since then, photocatalysis research has focused on creating LEDs that are stronger, more energy efficient, and non-toxic.[30] Due to the greater degree of freedom provided by a photocatalytic reactor, the LEDs can also be positioned in different directions.

UV-LEDs are another type of artificial light source that can be an excellent substitute for traditional mercury lamps. This light source offers several advantages, including taking up less space, being highly durable, starting up quickly, consuming less energy, and having a longer lifespan of 35,000 to 50,000 hours. One distinguishing characteristic of photocatalytic reactors lies in the spatial arrangement of the lamp or radiation source. Based on the spatial orientation of the light source, photocatalytic reactor configurations can be classified into three over-arching categories:

i. Reactors equipped with an internal radiation source, specifically a bulb that is placed within the confines of the reactor.
ii. Reactors that employ an extrinsic radiation source (a luminary positioned external to the reactor).
iii. Reactors featuring light sources that are dispersed, wherein radiation is transmitted from the source to the reactor through the utilization of reflectors and other optical apparatus.

7.7.1.3 Catalyst

Various catalysts, such as TiO_2, ZnO, WO_3, ZnS, Fe_2O_3, CdS, and numerous others, have been employed in the process of photocatalytic degradation of diverse contaminants. The utilization of the catalyst is a fundamental part of designing photocatalytic reactors. Photocatalyst materials must be chosen based on the available light source. Titanium oxide is a highly efficient photocatalyst that exhibits a wide range of UV light absorption. It has been extensively studied due to its exceptional photoactivity efficiency, non-toxic nature, and remarkable photochemical stability.[30] A photocatalyst can be made more active utilizing a variety of techniques such as doping, bandgap, and shape of crystal.

7.8 TYPES OF PHOTOCATALYTIC REACTORS

There are various types of reactors that can be classified according to their light source and the specific wavelength of light they utilize to activate the catalyst. There are two options for the light source: traditional lamps or LEDs. Reactors can be classified into two categories: TiO_2-based and non-TiO_2–based, depending on the catalyst employed. There are several types of reactors, including the *slurry type*, *immersion type*, *external type*, and *distributive type*. In the slurry type, catalyst particles are suspended, while in the immersion type, one or more lamps are submerged in the reactor. The external type is designed to be used for lamps that are placed outside the reactor. On the other hand, the distributive type is used to distribute light by utilizing optical components such as reflectors, light conductors, or optical fibers. When formulating a photocatalytic reactor, it is imperative to take into account two fundamental variables: the aggregate surface area of the catalyst in relation to the volume it occupies, and the extent to which light is dispersed throughout the reactor. The aforementioned elements are of paramount importance in determining the efficacy of the reactor.[31]

7.8.1 Slurry Photocatalytic Membrane Reactor

A hybrid technique that combines a photocatalysis and membrane filtration is a photocatalytic membrane reactor (PMR). Strong oxidizing radicals produced by photocatalysis cause the contaminant to be degraded. The reactor uses a membrane module that serves as a support layer or an intermediate step in the separation process. Organic molecules can be broken down by the system via photocatalytic activities that take place on the catalyst surface. An efficient photocatalytic membrane reactor has a catalyst placed on its membrane surface. In order to refill the catalyst, the photocatalytic membrane must be completely replaced due to its limited catalyst active surface area.[32] A second type of PMR was developed by using suspension catalytic particles to overcome these limitations. For the recovery of the suspended catalyst using this innovative approach, an additional filtration step is required.

7.8.1.1 Working Principle of Photocatalytic Membrane Reactor

When light falls on a material, the electrons within the catalyst material undergo a transition from the valence band to the conduction band, resulting in the creation of a free electron-hole pair. The mechanism of slurry PMR to degrade organic pollutants is illustrated in Figure 7.4. These free charges come into contact with species that are absorbed on the catalyst surface, leading to the formation of active radicals like hydroxyl

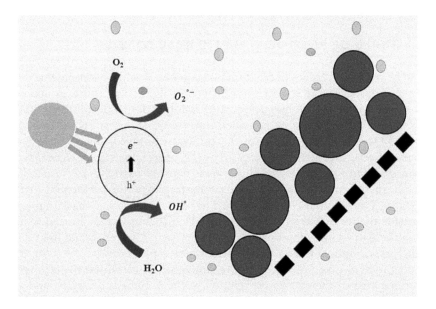

Figure 7.4 Scheme of slurry PMR operation with catalyst particles in green and pollutant in sky blue.

radical and superoxide radicals (OH$\bullet$ and $O_2\bullet$, respectively), which possess strong oxidizing potentials to degrade both organic and inorganic materials. Eliminating organic pollutants like dyes and waste from drugs can be achieved by taking this action in contaminated water. The system also includes a membrane module that simultaneously controls the catalyst particles and produces a treated stream that is free from catalyst particles on the penetrated side.

7.8.1.2 Photocatalyst in Photocatalytic Membrane Reactor

Photocatalyst particles are essential to slurry PMR systems because they improve degradation kinetics. The following sections give an idea of the different light sources being used today, as well as the most significant catalyst characteristics that affect its photocatalytic activity.

7.8.1.3 Light Source

Did you know that LEDs can be used in place of mercury lamps as a light source? LEDs have many benefits, including increased energy efficiency and the ability to avoid the problem of mercury disposal. Additionally, using LEDs extends the lifespan of mercury lamps from 9000 to 12,000 hours to 35,000 to 50,000 hours. If you are interested in the technical details, the semiconductor industry currently offers diodes with emission peak wavelengths between 255 and 405 nm. Mercury lamps are a popular type of lamp that are available in two variations: low pressure (LP Hg) and medium pressure (MP Hg). It has been found that LP lamps have a faster rate of destroying cytotoxic drugs compared to MP lamps. This is due to the fact that the photons generated by LP lamps possess higher energy levels, which means they have shorter wavelengths, typically around 254 nm.

7.8.1.4 Photocatalytic Material

The photocatalyst material must be selected depending on the available light source. One of the most studied photocatalysts is titanium oxide, which has high photoactivity efficiency, is non-toxic, and has strong photochemical stability. Other photocatalyst materials such as WO_3 and ZnO were also used for the purification of contaminated water.

It has been discovered that cytostatic medicines can be broken down by cyclophosphamide using a slurry UV/TiO_2 system. In the case of TiO_2, anatase and rutile are two different crystalline forms that have bandgaps of 3.2 and 3.02 eV, respectively. Their absorption edge values are 416 nm and 280 to 400 nm used as catalysts. Thus, titanium-based catalysts must be stimulated by UV wavelengths up to 387 nm, or around 3% of the solar spectrum received by the earth. A photocatalyst can be made more active using a variety of techniques, such as doping, bandgap nanoengineering, and crystal shape tuning.

7.8.2 Methods to Enhance Catalyst Efficiency

There are various photocatalysts used for breaking down organic contaminants, but their efficiency is not as high as doped and composite-based photocatalysts. Dopants can improve electron-hole separation and introduce intermediate energy levels, which can improve the surface absorption types. Transition metals or noble metals can be doped in titanium oxide photocatalysts to decrease their bandgap values. For instance, platinum doping (0.15%) using a straightforward sol-gel method in TiO_2 has been shown to increase cyclophosphamide removal from 66% to 99% when exposed to artificial visible light.

Another way to increase the photocatalytic efficiency is by altering the geometry of nanoparticles. In the process of manufacturing photocatalytic particles, it is possible to manipulate operational parameters in order to generate nanoparticles exhibiting distinct crystal sizes and morphologies. The aforementioned alterations have the potential to exert an influence on the recombination rate of electrons and holes. For instance, the morphology of nanoparticles can be altered by modifying the calcination temperature.[33]

7.8.3 Comparison of Photocatalytic Reactors

A photocatalytic slurry reactor with suspended TiO_2 particles is the most common type of reactor utilized in scientific research. The high specific surface area of the suspended catalytic particles in these reactors is their primary advantage. However, for practical applications, immobilized TiO_2 reactors are preferred because they permit continuous operation, do not necessitate the separation of catalyst particles, and permit the recycling of catalytic supports over a number of cycles. Low area-to-volume ratios in immobilized systems can result in limited mass transfer and slow reaction rates. This is their fundamental flaw. A suitable reactor design, which should aim to enhance catalyst illumination and prevent mass transfer limits, could solve this drawback. By contrasting various photocatalytic reactors, the optimal photocatalytic reactor design for industrial applications will be discovered. Figure 7.5 represents the main parameters affecting the comparative study of photocatalytic reactors.

7.8.3.1 Space-Time Yield

A standardized benchmark called space-time yield (STY) has been proposed to evaluate the efficiency of reactors. The benchmark we are discussing here focuses on reactors that have a capacity of 1 m³. Its purpose is to determine the extent to which pollutants can be reduced, specifically from an initial concentration of 100 mmol/L down to 0.1 mmol/L. This calculation is done

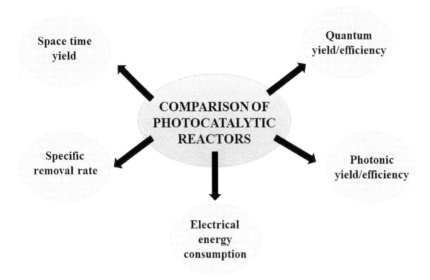

Figure 7.5 Parameters affecting photocatalytic reactor.

under the assumption that a constant STY value is applied. The SI unit for STY is cubic metres of pollutant per cubic metre of reactor per day. It is important to mention that reactors with plug flow specific area generally have a larger STY value compared to continuously stirred tank reactors.

7.8.3.2 Specific Removal Rate

In order to ascertain the specific removal rate (SRR), it is necessary to perform a division operation, whereby the mass (mg) of the pollutant eliminated within a given hour is divided by the mass (g) of the catalyst dosage. The SRR, denoted as SR, is quantified in the International System of Units (SI) as milligrams (mg) of chemical substance removed per gram (g) of catalyst per hour (h). The utilization of this particular metric is frequently employed for the quantification of SRR. The rate of pollutant removal is contingent upon a multitude of factors, encompassing the specific nature of the pollutant, its concentration, the catalyst employed, the quantity of catalyst, and the duration of the reaction.

7.8.3.3 Electrical Energy Consumption

In general, approximately 70% of the overall expenditure can be ascribed to the financial outlays associated with the luminaire apparatus and the expenditure incurred for electrical power consumption. Therefore, the quantification of electrical energy consumption (EEC) assumes a pivotal role in evaluating the efficacy of the photochemical oxidation process. The performance of the EEC is subject to the influence of various factors, including the power output of the lamp, the degree of degradation experienced, and the optimized flow rate. The parametre is operationally defined as the quantification of electrical energy, expressed in kilowatt-hours (kWh), required to induce a reduction in the concentration of a pollutant by a factor of 10 in a volume of 1 m³ of water that has been contaminated.

Typically, the initial cost of the light along with the cost of electricity make up 70% of the total cost. The efficiency of the photochemical oxidation process is greatly influenced by EEC. Light power, degradation percentage, and ideal flow rate all have an impact on EEC. The value represents the quantity of kilowatt-hours (kWh) of electrical energy required to effectively reduce the concentration of a contaminant in 1 m³ of polluted water. The value is determined by calculating the amount of electrical energy, measured in kilowatt-hours (kWh), needed to decrease the concentration of a pollutant.

7.8.3.4 Photonic Efficiency/Yield

Photonic efficiency pertains to the correlation between the rate of a photocatalytic reaction occurring within a designated timeframe and the rate of incident photons that reside within a specified wavelength spectrum within the irradiation of the reactor. To ascertain the photonic yield, it is imperative to juxtapose the temporal rate of the photocatalytic reaction with the rate of monochromatic light incident within the confines of the irradiation chamber. This enables us to quantify the photonic yield. By employing light sensor technology, the incident photon flux within the reactor can be ascertained. Light sensor technology possesses the inherent capacity to make estimations regarding the quantity of photons that penetrate the reactor.

7.8.3.5 Quantum Yield/Efficiency

Quantum yield is a crucial concept that reveals the number of specific events that occur for each photon absorbed by a system. It provides valuable insights into reactor designs and their efficiency. In order to attain comprehensive success, it is imperative to take into account multiple factors, encompassing electrical efficiency, outer geometrical incidence efficiency, inner incidence efficiency, absorption efficiency, and reaction efficiency. Each of these components exhibits a pivotal function in ascertaining the efficacy of the system. In the realm of heterogeneous photocatalysis, the concept of "quantum efficiency" pertains to the utilization of a specific range of wavelengths, in contrast to monochromatic illumination, for the absorption of radiation.[34] In the context of a heterogeneous catalytic system, it is important to note that the entirety of incident photons entering the reactor does not undergo absorption. Certain phenomena are propagated through transmission, whereas others undergo scattering or reflection.

$$\varphi(i) = \varphi(a) + \varphi(t) + \varphi(s) \qquad (7.19)$$

where $\varphi(i)$, $\varphi(a)$, and $\varphi(s)$ represent the flux of incident photon, absorbed photon flux, and the flux of reflected photon, respectively. Experimental studies have been conducted using radiometer devices to determine the absorbed photons for calculating quantum yield/efficiency.

7.9 SUMMARY

This chapter provides details of various types of water pollution and various factors that degrade contaminants. Energy supply is crucial for various chemical processes to occur smoothly. Without it, reactions like the oxidation of pollutants and the separation of water would be difficult or even impossible to carry out under normal circumstances. The use of nanomaterials in the photocatalysis process offers a reliable and eco-friendly solution for treating contaminated water. The harmful effects of hazardous dyes released by the industry on the environment have been well studied. Many types of nanomaterials have been employed for degrading these dyes, and various photocatalytic reactors have been studied for this purpose. It is concluded that semiconductor-based photocatalysis methods were used for degradation of dyes, drugs, and pollutants etc.

REFERENCES

1. Owa, F., *Water pollution: Sources, effects, control and management*. Mediterranean Journal of Social Sciences, 2013. 4(8): p. 65.
2. Gosavi, V.D., and S. Sharma, *A general review on various treatment methods for textile wastewater*. Journal of Environmental Science, Computer Science and Engineering & Technology, 2014. 3(1): pp. 29–39.
3. Speight, J.G., *Sources of water pollution*. Natural Water Remediation, 2020: pp. 165–198.
4. Heath, A.G., *Water Pollution and Fish Physiology*. CRC Press, 2018.
5. Gould, J. *Is rainwater safe to drink? A review of recent findings*. In *9th International Rainwater Catchment Systems Conference*. Citeseer, 1999.
6. Cui, Q., et al., *Diversity and abundance of bacterial pathogens in urban rivers impacted by domestic sewage*. Environmental Pollution, 2019. 249: pp. 24–35.
7. Singh, J., et al., *Water pollutants: Origin and status*. Sensors in Water Pollutants Monitoring: Role of Material, 2020: pp. 5–20.
8. Kumar, S., et al., *Photocatalytic degradation of organic pollutants in water using graphene oxide composite*. A New Generation Material Graphene: Applications in Water Technology, 2019: pp. 413–438.
9. Humayun, M., et al., *Perovskite-type lanthanum ferrite based photocatalysts: Preparation, properties, and applications*. Journal of Energy Chemistry, 2022. 66: pp. 314–338.
10. Szilágyi, I.M., et al., WO_3 *photocatalysts: Influence of structure and composition*. Journal of Catalysis, 2012. 294: pp. 119–127.

11. Zhou, W., and H. Fu, *Defect-mediated electron-hole separation in semiconductor photocatalysis.* Inorganic Chemistry Frontiers, 2018. 5(6): pp. 1240–1254.
12. Nosaka, Y., and A.Y. Nosaka, *Generation and detection of reactive oxygen species in photocatalysis.* Chemical Reviews, 2017. 117(17): pp. 11302–11336.
13. Abbasi, S., and M. Hasanpour, *The effect of pH on the photocatalytic degradation of methyl orange using decorated ZnO nanoparticles with SnO$_2$ nanoparticles.* Journal of Materials Science: Materials in Electronics, 2017. 28: pp. 1307–1314.
14. Kaur, K., et al., *Photodegradation of organic pollutants using heterojunctions: A review.* Journal of Environmental Chemical Engineering, 2020. 8(2): p. 103666.
15. Valian, M., et al., *Sol-gel auto-combustion synthesis of a novel chitosan/ HO$_2$Ti$_2$O$_7$ nanocomposite and its characterization for photocatalytic degradation of organic pollutant in wastewater under visible illumination.* International Journal of Hydrogen Energy, 2022. 47(49): pp. 21146–21159.
16. Gusain, R., N. Kumar, and S.S. Ray, *Factors influencing the photocatalytic activity of photocatalysts in wastewater treatment.* Photocatalysts in Advanced Oxidation Processes for Wastewater Treatment, 2020: pp. 229–270.
17. Krishnan, J., et al., *Effect of pH, inoculum dose and initial dye concentration on the removal of azo dye mixture under aerobic conditions.* International Biodeterioration & Biodegradation, 2017. 119: pp. 16–27.
18. Farzaneh, A., et al., *Optical and photocatalytic characteristics of Al and Cu doped TiO$_2$: Experimental assessments and DFT calculations.* Journal of Physics and Chemistry of Solids, 2022. 161: p. 110404.
19. Bell, S., G. Will, and J. Bell, *Light intensity effects on photocatalytic water splitting with a titania catalyst.* International Journal of Hydrogen Energy, 2013. 38(17): pp. 6938–6947.
20. Alkaim, A., et al., *Effect of pH on adsorption and photocatalytic degradation efficiency of different catalysts on removal of methylene blue.* Asian Journal of Chemistry, 2014. 26(24): p. 8445.
21. Yunus, N., et al. *Effect of catalyst loading on photocatalytic degradation of phenol by using N, S Co-doped TiO$_2$.* In *IOP Conference Series: Materials Science and Engineering.* IOP Publishing, 2017.
22. Torres-Palma, R.A., and E.A. Serna-Galvis, *Sonolysis.* In *Advanced Oxidation Processes for Waste Water Treatment* (pp. 177–213). Elsevier, 2018.
23. Lamkhanter, H., et al., *Photocatalytic degradation of fungicide difenoconazole via photo-Fento process using α-Fe$_2$O$_3$.* Materials Chemistry and Physics, 2021. 267: p. 124713.
24. Deng, Y., and R. Zhao, *Advanced oxidation processes (AOPs) in wastewater treatment.* Current Pollution Reports, 2015. 1: pp. 167–176.
25. Saleh, I.A., N. Zouari, and M.A. Al-Ghouti, *Removal of pesticides from water and wastewater: Chemical, physical and biological treatment approaches.* Environmental Technology & Innovation, 2020. 19: p. 101026.
26. Ameta, R., et al., *Photocatalysis.* In *Advanced Oxidation Processes for Waste Water Treatment* (pp. 135–175). Elsevier, 2018.
27. Nur, A.S., et al., *A review on the development of elemental and codoped TiO$_2$ photocatalysts for enhanced dye degradation under UV-vis irradiation.* Journal of Water Process Engineering, 2022. 47: p. 102728.

28. Liu, Z., et al., *Synthesis of carbon-based nanomaterials and their application in pollution management*. Nanoscale Advances, 2022. 4(5): pp. 1246–1262.
29. Zhao, H., et al., *Wse_2-loaded co-catalysts Cu_3P and CNTs: Improving photocatalytic hydrogen precipitation and photocatalytic memory performance*. Journal of Colloid and Interface Science, 2023. **629**: pp. 937–947.
30. Sundar, K.P., and S. Kanmani, *Progression of photocatalytic reactors and its comparison: A review*. Chemical Engineering Research and Design, 2020. **154**: pp. 135–150.
31. Ray, A.K., *Design, modelling and experimentation of a new large-scale photocatalytic reactor for water treatment*. Chemical Engineering Science, 1999. 54(15–16): pp. 3113–3125.
32. Janssens, R., et al., *Slurry photocatalytic membrane reactor technology for removal of pharmaceutical compounds from wastewater: Towards cytostatic drug elimination*. Science of the Total Environment, 2017. **599**: pp. 612–626.
33. Erdei, L., N. Arecrachakul, and S. Vigneswaran, *A combined photocatalytic slurry reactor–immersed membrane module system for advanced wastewater treatment*. Separation and Purification Technology, 2008. 62(2): pp. 382–388.
34. Brandi, R.J., et al., *Absolute quantum yields in photocatalytic slurry reactors*. Chemical Engineering Science, 2003. 58(3–6): pp. 979–985.

Chapter 8

Photocatalytic Carbon Dioxide Reduction

ABSTRACT

This chapter is noteworthy because it addresses the world's most pressing CO_2 reduction challenge. CO_2 levels in the atmosphere are rising on a daily basis, wreaking havoc on the environment and human existence. The chapter begins with a discussion of the significance of CO_2, since it is known as a life gas and is essential for photosynthesis. Following this, types of photocatalytic CO_2 reduction, reaction conditions, role of semiconductors and novel photocatalysts in CO_2 reduction, the idea of photocatalytic reactor for CO_2 conversion, several key photocatalytic reactors including hybrid reactors, benefits and problems are described. Finally, the use of photocatalysts in real-world CO_2 reduction scenarios has been addressed. It has also been described how it can be integrated with carbon collection and utilization technologies, as well as how it may be employed in artificial synthesis of plants to convert CO_2 into electricity. This chapter will assist readers in developing an understanding ranging from very basic information to advanced research in the topic of photocatalytic CO_2 reduction.

8.1 INTRODUCTION

CO_2, CH_4, and chlorofluorocarbons (CFCs) are the principal factors contributing to global warming. CO_2 is mostly emitted due to the combustion of fossil fuels, and during the last several years, the atmospheric concentration of CO_2 has increased due to increased human activity that has intensified the greenhouse effect. Because of individual's high need for energy, global use of fossil fuels continues to rise year after year. CO_2 extraction and storage, whether physical or chemical, can only provide a temporary solution. Furthermore, the traditional chemical technique for CO_2 reduction necessitates the use of energy. As a result, turning CO_2 to lucrative hydrocarbons is one of the greatest solutions to both global warming and energy scarcity.

DOI: 10.1201/9781003403357-10

Approximately 80% of the supply of energy in the world is dependent on fossil fuels, and humanity is anticipated to confront a catastrophic catastrophe due to the rapid reduction of fossil fuels and rising CO_2 intensities in the atmosphere. The current global energy consumption rate of 16.3 TW (2012) is expected to increase to over 40 TW by 2050 and nearly 60 TW by 2100. According to the United Nations Framework Convention on Climate Change, CO_2 levels might reach 590 ppm by 2100, and the average world temperature could climb by 1.9°C. Greenhouse gas (GHG) levels are rising dramatically, according to the Intergovernmental Panel on Climate Change's fifth assessment report (2014). The steady increase of 1.3% per year (1970–2000) has been replaced by a sharp increase of 2.2% per year (since 2000). Increasing CO_2 levels in the atmosphere can have a negative impact on the planet, producing a rise in average sea levels and average global temperature. CO_2 levels in the atmosphere might be balanced by producing other fuels or turning CO_2 back to fuel forms. Also, the efficient use of solar energy by photosynthesis sustains all life, and fossil fuels also serve as reservoirs for solar energy obtained in the past. Photosynthesis removes CO_2 from the atmosphere in the natural world. The energy from the sun is utilized to transform CO_2 into glucose, a sugar molecule that stores solar energy as chemical energy. Except for geothermal and nuclear energy, most energy types, such as fossil fuels, biomaterials, hydropower, wind, and so on, are just past or current transformations of solar energy. As a result, the sun is the ultimate source of energy for the earth. One interesting use is the use of artificial photosynthesis to create hydrocarbons by the photoreduction of CO_2. In simple terms, solar energy is converted and stored directly as chemical energy. As a result, photoreduction of CO_2 with H_2O to create compounds such as methane or methanol (CH_3OH) is particularly appealing, and there is a strong desire to improve the efficiency of this process. Methanol is simple to carry, store, and utilize as a fuel additive in cars. Furthermore, using existing chemical technology, methanol, methane, and ethylene may be easily converted into other valuable compounds.

$$CO_2 + 2H_2O \rightarrow CH_3OH + \frac{3}{2}O_2 \qquad (8.1)$$

Because CO_2 has a poor energy grade from a thermodynamic standpoint, any change to hydrocarbons needs the addition of energy. Equation 8.1 depicts a full photoreduction of CO_2 to generate methanol. According to thermodynamics, at 298 K, turning 1 mole of CO_2 into methanol needs 228 kJ (DH) of energy. That is, the procedure is both time-consuming and complicated. Energy for photocatalytic CO_2 reduction should be delivered without creating more CO_2. Plants use solar energy to produce photosynthesis in nature; however, the energy transition is inefficient since some of the energy is required to maintain their survival. According to Gust et al. (2008), the development of photosynthesis was not prompted by the demand for the most effective energy storage.[1] Natural photosynthesis has a relatively poor energy conversion efficiency, with a maximum of roughly 6% but an average of 0.8%. Aside from energy dissipation through leaves, energy loss is mostly caused by the energy required for plant development and upkeep. Even under ideal artificial conditions, energy efficiency in macroalga can only reach around 7% in full sunshine.[2]

Photocatalysis initiates processes in the exposure of light radiation by using semiconductor materials. Semiconductors have a bandgap, which causes an electron-hole pair to form when exposed to light. Many studies have demonstrated that photocatalysts can decrease CO_2 in water vapor or a solvent. TiO_2 is a well-known photocatalyst that has been used to reduce CO_2 with H_2O to generate CH_4 and CH_3OH.[3-8]

8.2 TYPES OF PHOTOCATALYTIC CO₂ REDUCTION

The H_2O molecule is an excellent reducing agent for CO_2, giving protons for using holes while also creating oxygen and hydrogen throughout the process. Nearly all photocatalytic CO_2 reduction processes require the construction of reaction mechanisms based on H_2O molecules. The reaction system is divided into two kinds: (i) gas-solid and (ii) liquid-solid systems. Figure 8.1 depicts schematics of gas-solid and liquid-solid photocatalytic CO_2 reduction processes.

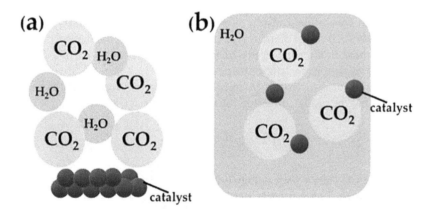

Figure 8.1 Represents (A) gas-solid and (B) liquid-solid systems for photocatalytic CO_2 reduction.

Source: Adapted with permission from Cui, Y., et al., *Research semiconductor material innovations with applications in CO₂ photocatalytic reduction*. Catalysts, 2022. 12(4): p. 372, under the terms of the Creative Commons Attribution 4.0 International (CC BY 4.0) AT (https://creativecommons.org/licenses/by/4.0/at/mdpi.com).

8.2.1 Gas-Solid Systems

The H_2O molecule appears as a vaporized form in a gas-solid system, and a catalyst is evenly disseminated at the bottom surface of the reactor or evenly layered on the substrate and deposited at the bottom of the reactor. CO_2 gas flows continuously into the reactor or exists as a saturated gas in the closed off-line system. Zhang et al. discovered that when TiO_2 nanotubes supported by Pt were utilized as a photocatalyst, there was a lot of fluid between the catalyst including the outermost -OH group.[9] The CO_2/H_2O ratio had no influence on methane generation when Pt-TiO_2 nanoparticles were utilized as a catalyst. The energy produced by water molecule adsorption on the catalyst surface affects the activity of the gas-phase photocatalytic system, showing that it modulates the system's activity. Due to the microporous shape and surface properties of the materials, the adsorption of CO_2 and H_2O on the sites that are active leads in a variety of photocatalytic reactions and the creation of numerous products.[10]

8.2.2 Liquid-Solid Systems

Catalyst is still suspended in aqueous solution in a liquid-solid system. To achieve the saturation point, CO_2 gas is pumped into the water prior to the reaction.[11] When H_2O is used as a reducing agent, the product yield is extremely poor. It was revealed that incorporating triethanolamine or isopropanol in the resulting solution uses holes, hence boosting the efficiency of photocatalytic reduction of CO_2. TiO_2 was used by Kaneco et al. to photocatalytically decrease CO_2 in isopropanol solution. CO_2 was converted to methane, while isopropanol was converted to acetone.[12] The pH of the suspension is another important element in the reaction system. Alkaline chemicals (NaOH, $NaHCO_3$ etc.) are commonly added to solutions to boost CO_2 solubility and promote CO_2 reduction.[13] Adding NaOH into the photocatalytic system solution greatly enhanced the production of CH_3OH.[14] Furthermore, the addition of electrolytes increased total production in addition to C_2 products such as ethanol as well as acetaldehyde.

The primary distinction among gas-solid as well as liquid-solid systems exists in the changing quantities of carbon dioxide and water on the catalyst's surface; the outcomes are also varied at various concentrations. The main drawback of gas-solid systems is that there is minimal contact space among reactants and a catalyst; however, liquid-solid systems include many H_2O molecules that occupy CO_2 adsorption sites. As a result, different catalyst optimization designs are required for distinct reaction systems in order to address reaction system flaws and enhance photocatalytic CO_2 reduction efficiency.

8.2.3 Photocatalytic Reaction Conditions of CO_2 Reduction

Ambient environmental and operational considerations, the range of wavelengths and intensity of light, the type of catalyst, the pH value of the medium, CO_2 pressure, the agent that decreases it, and temperature each have affect photocatalytic processes. Hou et al. studied the influence of different irradiation sources on photocatalytic CO_2 reduction employing an Au nanoparticle/ TiO_2 catalyst, yielding a range of products such as CH_4, CO, CH_3OH, and HCHO.[15] At different excitation wavelengths, the photocatalytic decrease product of liquid CO_2 by Au nanoparticle/TiO_2 catalysts was shown to be CH_4 (= 532 nm, 254 nm, and 365 nm). Under ultraviolet (UV) light (254 nm mercury lamp), nevertheless, other reaction products (which involve C_2H_6, CH_3OH, and HCHO) were identified. By raising the concentration of CO_2, it is possible to considerably boost the selectivity and activity of photocatalysis.

According to Mizuno et al.,[16] when CO_2 pressure increased, the process of adsorption of hydrogen on the TiO_2 surface gradually outpaced that of carbon dioxide. This hydrogen adsorption resulted in the formation of low-mass hydrocarbons like CH_4 and CH_2CH_2. In gas-phase networks, increasing water and carbon dioxide vapor pressure may improve reactant binding to the catalyst's active sites, resulting in an increase in the photocatalytic CO_2 reduction activity.[9] The system temperature is another component that must be addressed in photocatalytic reduction processes. Fox and Dulay came to the conclusion that the photocatalytic process was largely unaffected by a little temperature shift,[17] however, a high temperature also accelerates the pace of thermal activation stages, in addition to increasing the collision frequency and diffusivity. Under conditions of high temperature and sunlight, Guan et al. used a Pt-loaded $K_2Ti_6O_{13}$ photocatalyst in conjunction with Fe-based catalysts to reduce water and carbon dioxide vapor.[18] The yield of HCOOH, CH_3OH, and C_2H_5OH rose dramatically as the reaction temperature increased.

The catalyst's crystal phase composition must also be examined. The photocatalytic CO_2 reduction performance will change if the surface of the crystal plane is altered but the structure as a whole remains the same. Yamashita et al. investigated photocatalytic CO_2 reduction by TiO_2 single crystals with rutile phases exposed to surfaces (100) and (110). The activity of the (100) surface was found to be much greater than that of the (110) surface, and CH_4/CH_3OH products were discovered in the exposure of the TiO_2 system to the (100) surface. It follows that enhancing the internal structure of catalytic materials can significantly increase photocatalytic reduction of CO_2 activity.[19]

8.3 PHOTOCATALYSIS AND PHOTOCATALYTIC REDUCTION OF CO_2 WITH H_2O

The term "photocatalysis" is made up of two elements: "photo" and "catalysis". "Photo" refers to light, while "catalysis" refers to the ability of medium to influence the reaction rate while retaining the reactants.[20] Photocatalysis, in practice, refers to the initiation of a photoreaction while the catalyst is present.[21] Under solar irradiation with a photocatalytic CO_2 reduction mechanism containing water, photoreduction of CO_2 as well as oxidation by sunlight of H_2O take place simultaneously. The distribution of this reaction's by-products is heavily influenced by a number of reaction circumstances, incorporating illumination kind, catalyst type, reactor shape, and sacrificial reagents. As a result, forecasting a particular photocatalytic technique is a very challenging distribution of products. Photocatalytic CO_2 reduction is a highly efficient technology since it uses no extra power and has no adverse effects on the environment. Because of its low cost, another appealing idea is to employ inexpensive, abundant sunshine to transform this significant greenhouse gas into other products that include carbon. Figure 8.2 represents the most common CO_2-reduced organic pollutants. Solar radiation provides the significant activation energy required to split the extremely stable CO_2 molecule.[22]

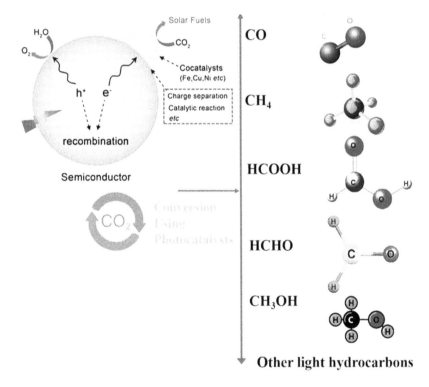

Figure 8.2 Principle of photocatalytic CO_2 reduction and most common reduced organic compounds.

The fact that the majority of CO_2-reducing photocatalysts do not respond to visible light is one of the biggest obstacles to the advancement of this research.[23] Numerous photocatalysts have been produced in this field. A couple of these catalysts had excellent conversion rates and selectivity when exposed to visible light, but others were poorly sensitive and demonstrated a relatively low rate of reaction yield.[24] In the visible area, it has been demonstrated that plasmonic metal can be used for semiconductor materials to increase photocatalytic activity.

8.3.1 Photocatalytic Reactions and Mechanisms

Photocatalysis is the use of sunshine or artificial light to activate a semiconductor. When a semiconductor absorbs enough photons, it produces electron-hole pairs when the electrons are induced to move between the valence band (VB) towards the conduction band (CB). The most basic energy band, CB, is completely devoid of electrons at the ground state, while VB is the optimum energy band in which electrons are present. Such photo-generated electrons may go to the surface of a semiconductor and interact with the species that have been adsorbed there. Electron-hole recombination is possible in the meantime.[24] The photocatalytic reaction's ability is determined by the conflict between these two mechanisms.[25] The fundamental photocatalytic process is given as (i) energy-appropriate photon absorption and production of electron-hole pairs; (ii) division and movement of pairs of electrons and holes (charge carriers); and (iii) interaction between surface species and charge carriers.[26,27]

Figure 8.3 shows an illustration of this technique. Because the procedure for charge recombination (10^{-9} s) is often significantly quicker than a reaction time (10^{-3}–10^{-8} s), accelerating the electron-hole distance stage has a significant impact on reaction yield.[28]

In addition to the immediate photon-excited carrier charges generation method in semiconductors seen in Figure 8.3, collisions between photon-electron interaction[29–31] or all the electron-hole pairs can be produced by the transfer of electrons from the metal nanoparticles energized by surface plasmon resonance.[32,33] Nevertheless, every photoexcited electron that reaches the surface is incapable of reducing the thermodynamically neutral and highly stable CO_2 molecule. This endergonic reduction process needs simultaneously hydrogen and energy.[10] As a result, sunlight-based photocatalytic CO_2 reduction and water might turn out to be the most practical method of removing atmospheric CO_2.

Table 8.1 shows the potential for multiple different CO_2 reduction products at pH 7. On the other hand, a single-electron CO_2 reduction process needs a potential that is 1.9 eV highly negative, making the procedure undesirable. The proton-aided multi-electron CO_2 reduction process, on the other hand, needs a lower redox potential (Table 8.1) and is therefore

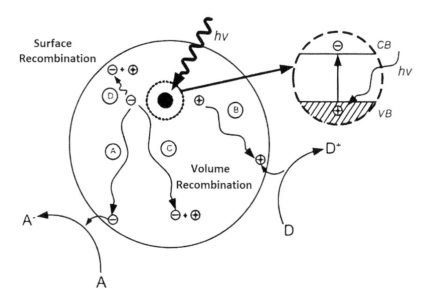

Figure 8.3 The photoexcitation process in photocatalytic CO_2 reduction.

Source: Reprinted with permission from Usubharatana, P., et al., *Photocatalytic process for CO_2 emission reduction from industrial flue gas streams.* Industrial & Engineering Chemistry Research, 2006. **45**(8): pp. 2558–2568. Copyright © 2006 American Physical Society.

Table 8.1 Reduction potential (E^0) for CO_2 reduction

Reactions	E^0/eV
$CO_2 + e^- \rightarrow CO_2$	≥ -1.9
$CO_2 + 2e^- + 2H^+ \rightarrow HCOOH$	-0.61
$CO_2 + 2e^- + 2H^+ \rightarrow CO + H_2O$	-0.53
$CO_2 + 4e^- + 4H^+ \rightarrow HCHO + H_2O$	-0.48
$CO_2 + 6e^- + 6H^+ \rightarrow CH_3OH + H_2O$	-0.38
$CO_2 + 8e^- + 8H^+ \rightarrow CH_4 + 2H_2O$	-0.24

more advantageous. Photocatalysts have a reduced potential for facilitating these reduction reactions. An excellent photocatalyst for this purpose should have two characteristics: (i) photocatalyst should be excited-by-photon for a hole to function as an acceptor of electrons, and the redox potential of the VB hole needs to be adequately positive; and (ii) the redox potential of the photoexcited CB electron must be lower than that of the redox pair of CO_2/reduced product. Furthermore, when reducing CO_2, HCOOH, HCHO, CH_3OH, and CH_4 are among the substances that its conduction band produces as a result of photo-generated electrons. The relationship between the photocatalyst and redox agent's energy levels here determines the type of reaction that takes place.

The bulk of research on photocatalytic CO_2 reduction techniques continues to rely on artificial UV rays from a powerful lamp.[34–36] Visible light makes up 43% of solar energy, whereas UV radiation makes up just around 4%. As a result, there is a high need for photocatalysts with narrow band-gaps that can utilize visible light.[37,38] A large number of researchers are focusing on the direct use of visible light, including from man-made and organic sources. Utilizing visible light instead of UV light is preferable since visible light is easily available from sunshine. However, as compared to UV light, the energy level of visible light is not as efficient. As a result, in photocatalytic reduction, visible light might not provide enough energy for the catalyst to be excited. As a result, photocatalysis using sunlight and visible light faces numerous difficulties.[39]

8.3.2 Measurement of Photocatalytic Efficiency

The yield for the outcome is frequently used to determine how effectively CO_2 is reduced using photocatalysis. R is commonly expressed as mol. H or intensity units (ppm), whereas the product is measured in grams (g) of catalyst.

$$R = \frac{n(Product)}{Time \times m(Catalysts)} \qquad (8.2)$$

The efficiency of the photocatalyst in catalyst-based evaluations generally under light irradiation, the total quantity of product produced per gram of photocatalyst over a particular period of time, may be calculated using its apparent quantum yield. This yield depends on the weight of the photocatalyst, the intensity of the irradiated light, the exposure area, and other factors. As shown within the equations, it is evaluated employing the resultant quantity and the incoming photon number.[10,40] When the photocatalytic reduction reaction produces complicated products, the amount of reacted electrons in the equation represents the total of the reacted electrons required to make each product.[41,42] Thus, by using the following equations in light-based experiments, it is possible to determine the quantum yield, or the amount of CO_2 photoreduction in different compounds:

$$Overall\ Quantum\ Yield(\%) = \frac{Number\ of\ Reacted\ Electrons}{Number\ of\ Absorbed\ Photons} \times 100 \qquad (8.3)$$

$$Apparent\ Quantum\ Yield(\%) = \frac{Number\ of\ Reacted\ Electrons}{Number\ of\ Incident\ Photons} \times 100 \qquad (8.4)$$

$$Apparent\ Quantum\ Yield\ of\ CO(\%) = \frac{2 \times Number\ of\ CO\ Molecules}{Number\ of\ Incident\ Photons} \times 100 \quad (8.5)$$

$$Apparent\ Quantum\ Yield\ of\ HCOOH(\%) = \frac{2 \times Number\ of\ HCOOH\ Molecules}{Number\ of\ Incident\ Photons} \times 100 \quad (8.6)$$

$$Apparent\ Quantum\ Yield\ of\ HCHO(\%) = \frac{4 \times Number\ of\ HCHO\ Molecules}{Number\ of\ Incident\ Photons} \times 100 \quad (8.7)$$

$$Apparent\ Quantum\ Yield\ of\ CH_3OH(\%) = \frac{6 \times Number\ of\ CH_3OH\ Molecules}{Number\ of\ Incident\ Photons} \times 100 \quad (8.8)$$

$$Apparent\ Quantum\ Yield\ of\ CH_4(\%) = \frac{8 \times Number\ of\ CH_4\ Molecules}{Number\ of\ Incident\ Photons} \times 100 \quad (8.9)$$

8.4 PHOTOCATALYTIC MATERIALS FOR CO_2 REDUCTION

Selecting the proper photocatalyst is the initial step in enhancing photocatalytic activity. It is a critical topic for both practical applications of photocatalysts as well as knowledge of their mechanisms. Based on their structure, photocatalysts may be divided into two fundamental groups: photocatalysts that are homogenous and heterogeneous. Lehn et al. conducted a groundbreaking study that demonstrated selective CO_2 reduction into CO using Re(I) diimine complexes[43]; subsequently then, the employing metal complexes in photocatalysis for both CO_2 reduction[44-47] and H_2O oxidation[48] has been extensively researched. When triethanolamine and homogeneous photocatalysts like Re complexes are used in close proximity to electron donors, CO_2 is effectively reduced to create CO. CO_2 reduction and H_2O oxidation, on the other hand, need separate reaction conditions.

8.4.1 Semiconductor Materials for CO_2 Photocatalytic Reduction

CO_2 reduction via photocatalysis is an exterior process. As previously stated, finding and using components that are both good for the environment and work well as catalysts is essential. Inoue et al. employed semiconductor materials for example CdS, TiO_2, and WO_3 to convert CO_2 to CO, CH_4, and other chemicals in 1979.[49] Even though the photocatalysts previously discussed have been shown to decrease CO_2, they have several drawbacks, for example, having a significant oxidation capacity, while conduction band is rather positive at bottom, and also reduction efficiency is poor.[50,51] Certain photocatalysts have uncertain chemical characteristics and are disposed to photocorrosion throughout photocatalytic reactions, which produce hazardous by-products.

8.4.1.1 TiO_2 Photocatalyst

TiO_2 is recognized as an attractive semiconductor material for environmental pollution management by means of n-type semiconductor material by its systainability, strong chemical stability, and non-toxicity. It shows promise for use in pitches of environment as well as energy.[52] Fujishima and Honda reported in 1972 that TiO_2 could degrade water molecules into hydrogen and oxygen when exposed to UV light, generating widespread interest in TiO_2 as a photocatalyst material.[53] TiO_2 has been widely employed in the photocatalytic breakdown of hydrogen in water products,[54-56] the degradation of pollutants,[57-59] CO_2 reduction,[60-63] and other research results. TiO_2 photocatalyst compounds are also widely employed in everyday life, for example, solar cubicles, coverings, cosmetics, antiseptic polymers, and air purifiers.

TiO_2 survives in three crystal polymorphs: anatase, rutile, and brookite.[64] Six oxygen atoms are encircled around a titanium atom in the three crystals' twisted octahedral shape. Tetragonal crystal formations are seen in anatase and rutile. The varied electrical structures of the three TiO_2 crystal types result in significant variances in photocatalytic activity. Tang et al. investigated the impact of various mineral forms continuously on TiO_2 photocatalytic performance. Once anatase or mixtures of rutile and anatase types were utilized by way of photocatalyst, the breakdown proportion of contaminants was about 100%. When pure rutile TiO_2 was utilized as a photocatalyst, the deprivation degree was a reduced amount of 15%.[65] Figure 8.4 shows how Jin et al. generated PbO-decorated TiO_2 composites in a one-pot approach with highly photoactive CO_2 conversion. The heterojunction created by the catalyst could efficiently limit photo-generated charge recombination, though the PbO could increase CO_2 adsorption on the catalyst. As a result, the heterojunction complex's photocatalytic activity for CO_2 reduction was greatly enhanced.[66]

Morphology is another major element influencing TiO_2 photocatalytic activity. TiO_2 catalysts with varied shapes have drastically varying superficial areas, energetic locations, charge transfer proportions, and uncovered

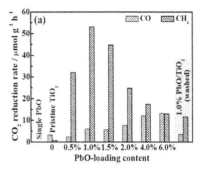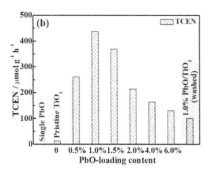

Figure 8.4 The average photocatalytic production activities of CO/CH₄ for single PbO, pristine TiO₂, and PbO/TiO₂.

Source: Reprinted with permission from Jin, J., et al., *One-pot hydrothermal preparation of PbO-decorated brookite/anatase TiO₂ composites with remarkably enhanced CO₂ photoreduction activity.*Applied Catalysis B: Environmental, 2020. **263**: p. 118353. Copyright © 2019 Elsevier.

crystal surfaces, resulting in major changes in presentation. Cao et al. created a TiO_2 photocatalyst that included nanorods and nanorod-hierarchical nanostructures. The substance outperformed commercial TiO_2 (P25) in photocatalytic CO_2 reduction. The increased charge transfer show, precise surface area, and immersion of light efficiency of the catalyst provided through the nanorod-hierarchical nanostructures were primarily responsible for the high catalytic activity.[67] Kar et al. created a TiO_2 nanotube photocatalyst that demonstrated very effective photocatalytic CO_2 reduction to CH_4. The strong photocatalytic activity was attributable mostly to the nanotube structure's increase of visible light absorption. The intensity of the external light source, in accumulation to the crystal magnitude and morphology of the TiO_2 substance, has a consequence on the photocatalytic movement of TiO_2; normally, by increasing the strength of light, a greater number of photogenerated electrons are produced by the catalyst's excitation, and the photocatalytic response is encouraged as a result.[68]

8.4.1.2 Metal-Organic Frameworks

Metal-organic frameworks (MOFs) are materials which are organic-inorganic hybrid having intramolecular holes generated by ready-made coordination bonds amid inorganic metal ions, clusters, and organic ligands. MOFs' enormous definite surface area, great porosity, and customizable assembly make them promising energy storing materials. MOF structure may be altered by varying the core metallic atoms and the interface of various carbon-based ligands. Transition metals like Fe, Co, and Ni are frequently used by means of key metal foundations. Such transition metals are frequent, widely dispersed, and extensively accessible on the earth as MOF raw materials, which helps to keep MOF raw material costs low. MOFs are generally employed in catalytic liveliness transformation and various additional applications as a result.[69] MOFs can be utilized as photocatalysts or transformers in photo-catalytic processes to boost the photocatalytic reaction.

MOFs are an emerging form of frame material with evenly scattered metal nodes that promote gas molecule adsorption and activation. Combining semiconductors with MOFs to make inorganic-organic nanocomposites allows semiconductor materials to absorb photons to produce carriers; however, it also allows the materials to adsorb and trigger very steady CO_2 molecules. Xiong et al. created a technique for producing core-shell $Cu_3(BTC)_2$@ TiO_2 structures.[70] The flexible shell structure of TiO_2 allows CO_2 molecules to flow through the shell, but the high CO_2 captivation of Cu_3 $(BTC)_2$ in the centre promotes the successful CO_2 reduction of nanocomposites.

8.4.1.3 Metal Halide Perovskites

Metal halide perovskite (MHP) resources have gained popularity in opto-electronics and energy adaptation in recent years.[71] These materials exhibit a high extinction coefficient, narrow band emission, high carrier diffusion length, and excellent defect tolerance when compared to typical semiconductor nanocrystals. Furthermore, the variation of perovskite assemblies allows the bandgap to be altered to improve light detention.[72] MHPs have a crystal structure comparable to oxide perovskite. The chemical formulation is ABX_3, where A represents a monovalent cation, B represents a divalent metallic cation (the most mutual being Pb^{2+} and Sn^{2+}), and X represents a halogen ion. The surface structure of MHP nanocrystals is halogen rich. Rendering to the consideration of cations in their chemical construction, perovskites can be categorized as either inorganic or organic-inorganic hybrid halogenated perovskites. The discovery of MHPs with distinct photoelectric properties opens up new avenues for effective photocatalytic CO_2 reduction.

Although the decreased interest in MHPs for CO_2 decrease varies with nanocrystal size, so does the catalytic movement of MHP nanocrystals. Sun et al. created $CsPbBr_3$ quantum dots (QDs) of various sizes to investigate the influence of QD size on CO_2 reduction. $CsPbBr_3$ with a diameter of 8.5 nm was discovered to have the longest carrier lifespan, the most undesirable band bottom prospective, and maximum catalytic movement.[73]

8.4.1.4 Other Semiconductor Photocatalysts

Other suitable photocatalytic materials for CO_2 reduction must be developed and explored in comparison to currently utilized photocatalysts (MHPs, MOFs, TiO_2 etc.). SiC's conduction band location has a significantly higher negative potential, which can yield photo-generated electrons through greater lessening capacity for photocatalytic CO_2 lessening. However, synthesis of SiC in a high-temperature defensive environment is not favourable to nanostructure control.[74,75] For the photocatalytic reduction of CO_2, coated double hydroxides (LDhs) such as Zn-Al LDH,[76] Mg-Al LDH,[77] and Zn-Cu-Ga LDH[78] have been utilized. Teramura et al. created a number of LDhs through surface alkaline locations for photocatalytic CO_2 exchange. Their activities are often greater than those of pure hydroxide.[79] Graphite carbon nitride (g-C_3N_4) is a metal-free polymer that has shown promise as a visible-light catalyst.[80,81] Hsu et al. employed graphene oxide (GO) as a catalyst for effective photocatalytic CO_2 to methanol conversion and manufactured a graphene catalyst with a novel technique to increase activity.[82] The harvest of CH_3OH was six times that of uncontaminated TiO_2. Without the use of additives, the photocatalytic activity of very porous Ga_2O_3 for CO_2 reduction was reported to be over four times that of commercially available Ga_2O_3.[83] By boosting the superficial area twice and trebling the adsorption capacity of porous Ga_2O_3, the performance was increased. Tanaka et al. photocatalytically reduced CO_2 on Ga_2O_3 using H_2 instead of H_2O as a plummeting agent, and the result was CO rather than CH_4. Specifically, roughly 7.3% of the CO_2 absorbed on the surface was transformed.[84]

An existing technique is to create novel semiconductor photocatalysts through visible-light responsiveness to increase sunlight usage and hence photocatalytic activity. Zhou et al. used a hydrothermal approach to create Bi_2WO_6 square nanoplates, and the product obtained by reducing CO_2 was CH_4.[84] Figure 8.5 depicts the production of CH_4 on Bi_2WO_6 photocatalysts.

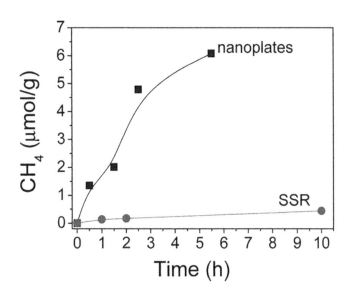

Figure 8.5 Creation of CH_4 on nanoplates and on solid-state reaction with respect to time irradiation.

Source: Reprinted with permission from Zhou, Y., et al., *High-yield synthesis of ultrathin and uniform Bi_2WO_6 square nanoplates benefitting from photocatalytic reduction of CO_2 into renewable hydrocarbon fuel under visible light.* ACS Applied Materials & Interfaces, 2011. **3**(9): pp. 3594–3601. Copyright © 2011 American Physical Society.

Cheng et al. created hollow Bi_2WO_6 microspheres and got methanol by decreasing CO_2.[86] By means of a one-step liquid-phase approach, Xi et al. created $W_{18}O_{49}$ nanowires. CO_2 was photocatalytically decreased in water vapor by visible light to produce CH_4. Using Pt and Au as co-catalysts considerably boosted the average rate of CH_4 production.[87]

8.4.1.5 Regeneration of Photocatalyst

After the reaction, the photocatalyst returns to its original state, and the pairs of electrons and holes recombine or transfer their charges to specific electron receivers, completing the catalytic cycle. As a result, the photocatalyst can be reused for subsequent CO_2 breakdown processes. Photocatalysis occurs when a catalyst uses light to accelerate a chemical process. Photocatalysis is a technique for converting CO_2 into beneficial chemicals such as hydrocarbons. Renewable energy sources such as sunshine can be used to fuel this process. This approach has received a lot of attention as a possible way to cut GHG production and combat climate change. The basic mechanism for catalyst regeneration in electrochemical CO_2 reduction (ECR) is illustrated, which is dependent on current density and electrolysis.

Catalyst regeneration is critical in the process of using light to break down CO_2. Because of several factors, such as having too much waste or other objects on its surface, the catalyst might cease operating effectively or get unclean over time. Catalyst regeneration (Figure 8.6) can be accomplished

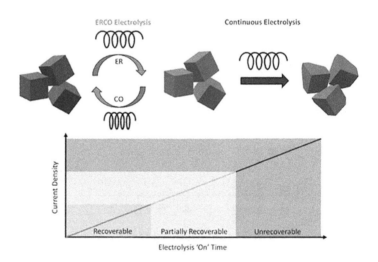

Figure 8.6 Representation of catalyst regeneration factors.

Source: Reprinted with permission from Nguyen, T.N., et al., *Catalyst regeneration via chemical oxidation enables long-term electrochemical carbon dioxide reduction.* Journal of the American Chemical Society, 2022. 144(29): pp. 13254–13265. Copyright © 2022 American Physical Society.

in a variety of methods, depending on the kind of catalyst and the reaction circumstances.

8.5 RECENT DEVELOPMENTS OF NOVEL PHOTOCATALYSTS FOR CO_2 REDUCTION

There has been a lot of attention on using electrochemical CO_2 reduction reaction (CO_2RR) to create value-added yields. With global investigative efforts, considerable progress has been made, including improved discrimination for decreased products, the achievement of well-organized reduction over binary electrons, and the supply of technologically significant present densities. In this evaluation, we discuss the most current developments in CO_2RR nanomaterials, such as zero-dimensional graphene QDs, two-dimensional materials such as metal chalcogenides, and nanostructured metal catalysts.

8.5.1 Quantum Dots

Copper has been found as the active metal to generate multicarbon hydrocarbons from CO_2RR, but other catalysts capable of converting CO_2 into useful products are required. Over the years, QDs have benefited from their tiny size, greater monodispersement, and better surface-to-volume ratio, which aid in the creation of effective CO_2RR catalysts. Metal-based QDs include Pb/Au/Ag/Cu vacancy-rich QD-derived catalysts (QDDCs),[88] Sn QDs,[89] InP colloidal quantum dots (CQDs),[90] and metal-free QDs like GQDs or other carbon nanodots.[91] GQD has been demonstrated to have copper-like activity in converting CO_2 to high-order hydrocarbons. Carbon in a honeycomb assembly with a size of 100 nm is referred to as GQDs.[92,93] GQDs have substantially higher superiority, abundance, and flaws in physical and chemical characteristics than graphene.[94,95]

GQDs produced by a hydrothermal reaction of GO were employed in the first publication.[83] The GO was created through intense oxidative exfoliation of graphite, resulting in a significant degree of variability in the final structures.[96] Wu et al. used a bottom-up technique to produce a more homogenous GQD structure by connecting 1,3,6-trinitropyrene, resulting in GQDs with a scarcely dispersed scope of 2 to 3 nm. The CO_2RR performance may be carefully managed by modifying the surface functional group of the pure GQDs by oxidation or reduction.[97]

8.5.2 Metal Chalcogenides

Metal chalcogenides are made up of a minimum of one metallic component and one chalcogen anion. These may comprise monochalcogenides, dichalcogenides, trichalcogenides, tetrachalcogenides, or analogous structures depending on the number of chalcogen anions.[98] Metallic chalcogenides and their by-products have been extensively studied for hydrogen development during the last several decades. Because of the famous hydrogen evolution reaction activities, their submissions in CO_2RR received little consideration until Salehi-Khojin et al. claimed that molybdenum disulfide (MoS_2) might serve as a robust catalyst for decreasing CO_2 into CO in a diluted solution of EMIM-BF_4 ionic liquid.[99] The utilization of an ionic liquid could steady CO_2 in an acidic situation (pH 4) as [EMIM-CO_2]+; this construction was pH-sensitive and might transition into [EMIM-HCO_3] or [EMIM-CO_3] in impartial or basic circumstances, correspondingly. Furthermore, because [EMIM-CO_2]+ may physisorb at damagingly charged MoS_2, the [EMIM]+ cation can be an effective intermediary for electron transfer to CO_2. The promising synthesis of [EMIM-CO_2]+ and adsorption of [EMIM]+ at the electrode surface led to the CO_2RR of metallic chalcogenides in a mutually beneficial way.

8.5.3 Nanostructured Metals

Transition metals and principal group metals have been widely researched as CO_2RR electrocatalysts. Metal catalysts may be classified into four classes based on CO_2RR selectivity (Figure 8.7).[100] The first category comprises primarily metals of P-block such as Sn, Pb, Bi, and In.[101,102] The great selectivity of these metal catalysts and derivatives of them toward formic acid is well

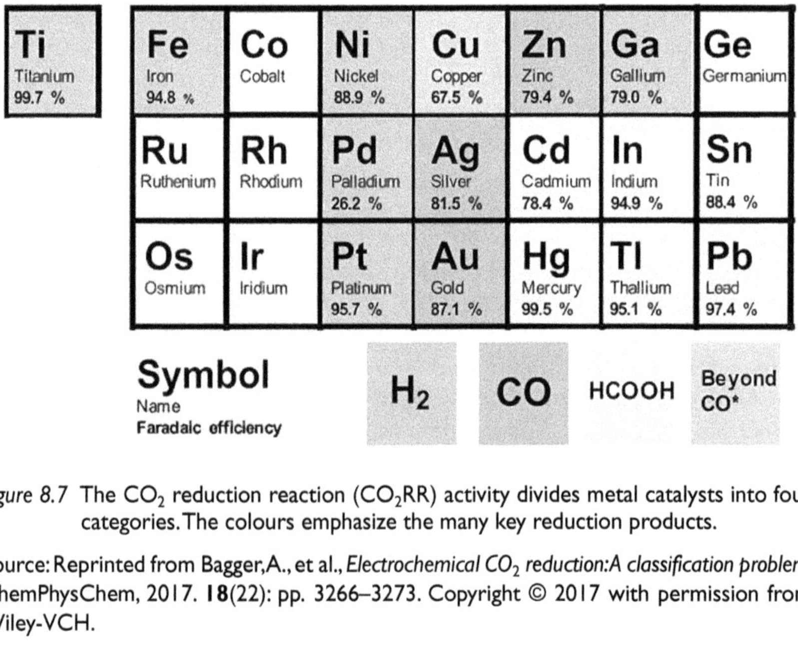

Figure 8.7 The CO_2 reduction reaction (CO_2RR) activity divides metal catalysts into four categories. The colours emphasize the many key reduction products.

Source: Reprinted from Bagger, A., et al., Electrochemical CO_2 reduction: A classification problem. ChemPhysChem, 2017. 18(22): pp. 3266–3273. Copyright © 2017 with permission from Wiley-VCH.

recognized. The second category includes metals such as Pd, Zn, Au, and Ag, which promote CO_2 reduction. Both formic acid and CO are produced by means of a two-electron reduction process, and the discrimination is mostly influenced through the energy of the very primary step, the adsorption of CO_2 onto the metallic substrate.

Pure metal catalytic activity may be able to be influenced by their dimensions, morphologies, and surfaces. Because of the differences in superficial group number and binding capacity to the adsorbates, the crystal facets will have distinct activity.[103] Optimizing the sizes of nanoparticles is an effective method for forming under-coordinated sites. Smaller nanoparticles, overall, will contain additional low-coordinated spots at their edges, which can be used to adjust the activities of metal catalysts.[104,105] Strasser et al. and Cuenya et al., for example, investigated the size-dependent CO_2RR doings of Au nanoparticles and discovered that lesser sizes were supplementary advantageous for CO_2RR.[106,107]

8.5.4 Artificial Photosynthesis

Humans are confronted with three significant concerns that are interconnected: a lack of energy resources, a lack of carbon means, and worldwide warming.[108,109] The main source of carbon in the biosphere is CO_2, which is propagatively converted into highly energetic carbohydrates using solar light as an energy source and water as a reductant. Even so, owing to both chemical and renewable energy sources, people have been consuming a significant amount of the fossilized materials produced by photosynthesis and stored behind. The majority of them were eventually torched, releasing massive amounts of CO_2 into the environment. Artificial photosynthesis mechanisms, which are capable of converting CO_2 to beneficial and energy-rich molecules employing solar light as the energy basis and water the reductant, are arguably the perfect method for addressing these significant apprehensions at the same period[110,111] (Figure 8.8).

Figure 8.9 presents the fundamental operational criteria for artificial photosynthesis systems. These criteria encompass the following aspects: efficient light absorption capabilities, considering the low photon density of sunlight and the small size of molecular photocatalysts; the generation of robust reduction and oxidation reactions utilizing low-energy visible light, which necessitates a stepwise two-photon preoccupation mechanism known as the man-made Z-scheme; the presence of a catalyst for water oxidation; and a catalyst to facilitate the reduction of CO_2.

Researchers have been focusing on artificial photosynthesis systems that replicate natural photosynthesis in order to convert CO_2 into successful supplies like fuels and chemicals.[112–116] Using light to transform CO_2 into usable molecules is a promising chemical method that might aid in the resolution of resource, energy, and environmental issues. One advantage of artificial photosynthesis is that it can convert solar energy into chemicals by utilizing

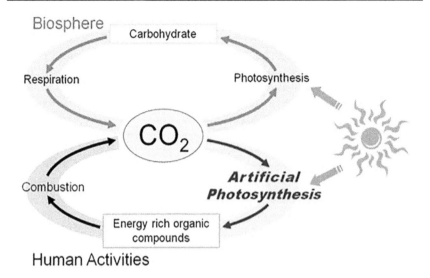

Figure 8.8 Natural and artificial photosynthesis processes.

Source: Adapted with permission from Sahara, G., and O. Ishitani, *Efficient photocatalysts for CO$_2$ reduction.* Inorganic Chemistry, 2015. **54**(11): pp. 5096–5104. Under the terms of the Creative Commons Attribution Licence (CC BY) AT (https://creativecommons.org/licenses/by/). Copyright © 2022 American Chemical Society.

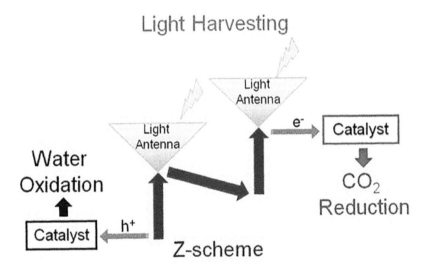

Figure 8.9 Components required for a "practical artificial photosynthesis process".

Source: Adapted with permission from Sahara, G., and O. Ishitani, *Efficient photocatalysts for CO$_2$ reduction.* Inorganic Chemistry, 2015. **54**(11): pp. 5096–5104, under the terms of the Creative Commons Attribution Licence (CC BY) AT (https://creativecommons.org/licenses/by/). Copyright © 2022 American Chemical Society.

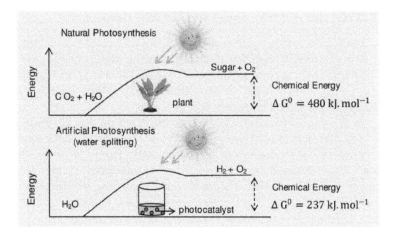

Figure 8.10 Represents the difference between natural and artificial photosynthesis.

Source: Adapted with permission under the terms of the Creative Commons Attribution (CC BY) AT (https://creativecommons.org/licenses/by/repo.uni). Hamid, S., *Stoichiometry of the photocatalytic fuel production by the reformation of aqueous acetic acid*. 2018. (uni-hannover.de)

abundant water as electron and proton sources. Figure 8.10 depicts the distinction between natural and artificial photosynthesis.

8.6 PHOTOCATALYTIC CO_2 REACTORS

Photocatalytic CO_2 reactors are technologies that employ photocatalysis to convert CO_2 into useful substances such as fuels or chemicals. They accomplish this by using light as an energy source. This method uses light to cause a particular material to react with CO_2 and transform it into valuable compounds. Due to its potential to reduce GHG emissions and expand the accessibility of sources of clean energy, this kind of technology is quite exciting. Figure 8.11 shows some of the most famous reactors.

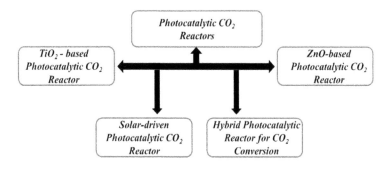

Figure 8.11 Different types of photocatalytic CO_2 reactors.

8.6.1 TiO₂-Based Photocatalytic CO₂ Reactor

Using renewable energy, photocatalytic CO_2 conversion is a promising method for reducing CO_2 emissions and producing useful chemicals or fuels. The substance TiO_2 is commonly used in studies to look at how light might be used to speed up chemical processes. It is popular since it is reliable, safe for people, and reasonably priced. A common arrangement in the literature is a photoreactor, which consists of a container that holds a chemical called a photocatalyst, typically TiO_2, and a gas that contains CO_2. As a result of the photocatalyst's absorption of light, frequently from UV or visible light sources, CO_2 is changed into a variety of other molecules, including hydrocarbons or carbon monoxide. The characteristics associated with the catalyst and the type of reaction that occurs determine this. The type of TiO_2 used (anatase or rutile); the particle size, shape, as well as manufacturing process; the reaction conditions (which include pressure, temperature, and gas configuration); as well as the intensity of the light source can all have an impact on how efficient a TiO_2-based CO_2 reactor is. TiO_2's capability to photocatalyze the reduction of CO_2 is shown in Figure 8.12. Researchers have investigated many methods for enlightening the efficiency of CO_2 alteration by modifying it. They have experimented with adding metals or non-metals to TiO_2, as well as coating it with other compounds such as Pt or Au and combining it with other materials.

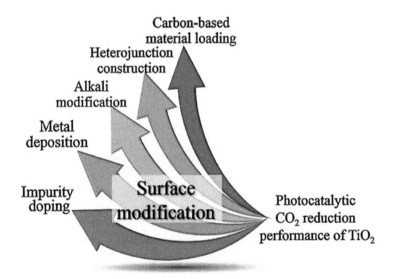

Figure 8.12 The photocatalytic CO₂ reduction performance of TiO₂.

Source: Reprinted from Low, J., B. Cheng, and J. Yu, *Surface modification and enhanced photocatalytic CO₂ reduction performance of TiO₂: A review.* Applied Surface Science, 2017. **392**: pp. 658–686. Copyright © 2017 with permission from Elsevier.

8.6.2 ZnO-Based Photocatalytic CO_2 Reactor

ZnO is a unique substance that may assist in the conversion of sunlight into energy for the production of useful compounds such as hydrocarbons and alcohols from CO_2. Light is absorbed by a material known as a ZnO catalyst during the photocatalytic process. This produces pairs of electrons and holes that collaborate to convert CO_2 into more beneficial molecules. When CO_2 is taken in during photocatalytic reduction, it serves to make the entire procedure more efficient. The surfaces of various carbon compounds, such as triggered carbon, carbon nanotubes, graphene-based materials, and carbon particles, vary. Because of these surface changes, they interact differently with CO_2 molecules. Because activated carbons have a large surface area, they may absorb a wide range of substances. They also contain microscopic perforations that make them much more absorbent. Different portions of graphene nanostructures can perform different things. Carbon dots and nanotubes can both store and transport electrical power. All of these characteristics make it simpler for CO_2 to connect to them and be converted into something else by light. As a result, adding carbon materials to ZnO improves the process of lowering CO_2 with light. This issue is gaining popularity in the field of photocatalysis with various materials. So, in general, learning a lot about these materials known as nanocomposites that contain ZnO is beneficial.[117] ZnO is a widely utilized photocatalyst that plays a role in light-related processes. These nanocomposites also contain several forms of carbon materials, and by combining them, they may make ZnO operate better and persist longer in these CO_2 degrading processes.

8.6.3 Solar-Driven Photocatalytic CO_2 Reactor

Solar-driven photocatalytic CO_2 reactors are a type of technology that uses sunlight to convert CO_2 into usable chemicals or fuels. The general mechanism is shown in Figure 8.13. These reactors often employ a specific material that can absorb sunlight and initiate a chemical reaction with CO_2 to produce the desired results. The photocatalyst is activated by sunlight, which subsequently converts CO_2 into useful chemicals. For this reason, many materials that can aid in photocatalysis have been investigated.

Metal oxides such as TiO_2, metal sulphides, and MOFs are examples of these materials. The sort of photocatalyst used is determined by how effectively it works, how long it lasts, and how well it can produce the desired goods. The products of the reaction of decreasing CO_2 via photocatalysis might be diverse. They might be as simple as methane and methanol or as complex as hydrocarbons. The goal is to convert CO_2 into usable chemicals or renewable fuels in order to minimize GHG emissions and contribute to the development of sustainable energy alternatives. Researchers are still researching this topic and attempting to improve photocatalysts' ability to convert CO_2. They are also exploring for novel materials and techniques to improve the efficiency of the process.[118–120]

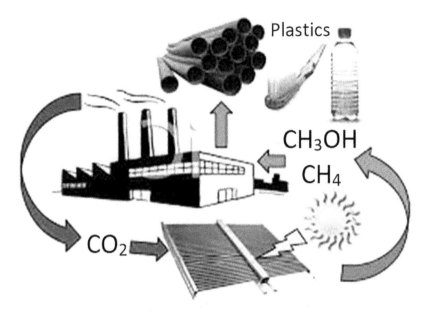

Plastics

CH_3OH
CH_4

CO_2

Photocatalytic reactor

Figure 8.13 CO_2 reduction through photocatalytic reactor.

Source: Adapted with permission from Chen, D., X. Zhang, and A.F. Lee, *Synthetic strategies to nanostructured photocatalysts for CO_2 reduction to solar fuels and chemicals.* Journal of Materials Chemistry A, 2015. **3**(28): pp. 14487–14516, under the terms of the Creative Commons Attribution 3.0 Unported Licence (CC BY 3.0) AT (https://creativecommons.org/licenses/by/3.0/).

8.6.4 Hybrid Photocatalytic Reactor for CO_2 Conversion

The hybrid photocatalytic reactor works by combining two different effects together so that efficient outcomes are achieved. Here we mainly discuss the photocatalytic membrane reactor (PMR) in hybrid photocatalytic reactors to convert CO_2.

8.6.4.1 Photocatalytic Membrane Reactor

The photocatalytic membrane was used to reduce CO_2 consuming H_2O as a decreasing agent. The membrane was irradiated through a medium-high mercury suspension pressure lamp (Zs lamp, Helios Ital quartz, Milan, Italy) with an emittance variety of 360 nm (UV-visible) to 600 nm (UV-visible). The experimental setup of PMR is shown in Figure 8.14. The flat area membrane was installed in a stainless strengthened module that had a quartz frame that permitted UV-visible irradiance of the catalytic tissue interface.

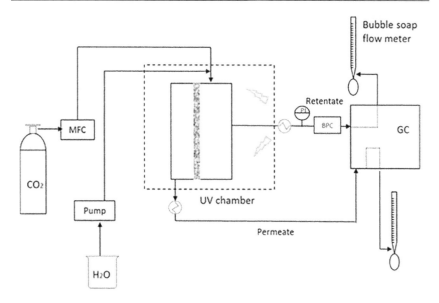

Figure 8.14 The PMR for CO_2 reduction.

Source: Adapted with permission from Pomilla, F.R., et al., *CO_2 to liquid fuels: Photocatalytic conversion in a continuous membrane reactor*. Hybrid Processes, 2018. **6**(7): pp. 8743–8753. Copyright © 2018 American Chemical Society.

The membrane reactor is made up of three fragments: the food/retentate and infuse chambers, as well as the catalyst-loaded tissue. The two apparatus chambers can be thought of as systems with aggregated parameters. Because of the poor conversion, the reaction occurs at the sheath coating, and no variation in the attention of slightly chemical class is predicted. As a result, the retentate and infuse streams have similar composition as the two measurements. The device was positioned vertically to ease permeate and retentate samples and to prevent the creation of stagnant zones. The researchers discovered that the photocatalytic tissue reactor converted at least 10 times as much carbon as the batch method as the enhanced dispersion of the photocatalyst was integrated into the Nafion matrix. It is a device that uses a special type of material to help speed up chemical reactions.

8.7 SUMMARY

In conclusion, this chapter emphasizes the critical significance of using light to minimize CO_2 emissions and tackle the CO_2 problem. This will contribute to a more environmentally friendly future. We can lessen climate change and usher in a time of clean, renewable energy by utilizing solar energy and enhancing photocatalytic technologies.

REFERENCES

1. Gust, D., et al., *Engineered and artificial photosynthesis: Human ingenuity enters the game.* MRS Bulletin, 2008. **33**: pp. 383–387.
2. Laws, E., and J.L. Berning, *Photosynthetic efficiency optimization studies with the macroalga Gracilaria tikvihae: Implications for CO_2 emission control from power plants.* Bioresource Technology, 1991. **37**(1): pp. 25–33.
3. Anpo, M., et al., *Photocatalytic reduction of CO_2 with H_2O on titanium oxides anchored within micropores of zeolites: Effects of the structure of the active sites and the addition of Pt.* Journal of Physical Chemistry B, 1997. **101**(14): pp. 2632–2636.
4. Dey, G.R., et al., *Photo-catalytic reduction of carbon dioxide to methane using TiO_2 as suspension in water.* Journal of Photochemistry and Photobiology A: Chemistry, 2004. **163**(3): pp. 503–508.
5. Lo, C.-C., et al., *Photoreduction of carbon dioxide with H_2 and H_2O over TiO_2 and ZrO_2 in a circulated photocatalytic reactor.* Solar Energy Materials and Solar Cells, 2007. **91**(19): pp. 1765–1774.
6. Pathak, P., et al., *Improving photoreduction of CO_2 with homogeneously dispersed nanoscale TiO_2 catalysts.* Chemical Communications, 2004(10): pp. 1234–1235.
7. Yamashita, H., et al., *Selective formation of CH_3OH in the photocatalytic reduction of CO_2 with H_2O on titanium oxides highly dispersed within zeolites and mesoporous molecular sieves.* Catalysis Today, 1998. **45**(1–4): pp. 221–227.
8. Anpo, M., et al., *Photocatalytic reduction of CO_2 with H_2O on titanium oxides anchored within zeolites.* Studies in Surface Science and Catalysis, 1998. **114**: pp. 177–182.
9. Zhang, Q.-H., et al., *Photocatalytic reduction of CO_2 with H_2O on Pt-loaded TiO_2 catalyst.* Catalysis Today, 2009. **148**(3–4): pp. 335–340.
10. Mao, J., et al., *Recent advances in the photocatalytic CO_2 reduction over semiconductors.* Catalysis Science & Technology, 2013. **3**(10): pp. 2481–2498.
11. Mori, K., H. Yamashita, and M. Anpo, *Photocatalytic reduction of CO_2 with H_2O on various titanium oxide photocatalysts.* RSC Advances, 2012. **2**(8): pp. 3165–3172.
12. Kaneco, S., et al., *Photocatalytic reduction of high pressure carbon dioxide using TiO_2 powders with a positive hole scavenger.* Journal of Photochemistry and Photobiology A: Chemistry, 1998. **115**(3): 223–226.
13. Li, H., et al., *Intercorrelated superhybrid of AgBr supported on graphitic-C_3N_4-decorated nitrogen-doped graphene: High engineering photocatalytic activities for water purification and CO_2 reduction.* Advanced Materials, 2015. **27**(43): pp. 6906–6913.
14. Tseng, I.-H., W.-C. Chang, and J. Wu, *Photoreduction of CO_2 using sol-gel derived titania and titania-supported copper catalysts.* Applied Catalysis B Environmental, 2002. **37**(1): pp. 37–48.
15. Hou, W., et al., *Photocatalytic conversion of CO_2 to hydrocarbon fuels via plasmon-enhanced absorption and metallic interband transitions.* ACS Catalysis, 2011. **1**(8): pp. 929–936.

16. Mizuno, T., et al., *Effect of CO_2 pressure on photocatalytic reduction of CO_2 using TiO_2 in aqueous solutions.* Journal of Photochemistry and Photobiology A: Chemistry, 1996. **98**(1–2): pp. 87–90.

17. Fox, M.A., and M.T. Dulay, *Heterogeneous photocatalysis.* Chemical Reviews, 1993. **93**(1): pp. 341–357.

18. Guan, G., et al., *Photoreduction of carbon dioxide with water over $K_2Ti_6O_{13}$ photocatalyst combined with Cu/ZnO catalyst under concentrated sunlight.* Applied Catalysis A: General, 2003. **249**(1): pp. 11–18.

19. Ikeue, K., et al., *Photocatalytic reduction of CO_2 with H_2O on Ti–β zeolite photocatalysts: Effect of the hydrophobic and hydrophilic properties.* Journal of Physical Chemistry B, 2001. **105**(35): pp. 8350–8355.

20. Shrestha, S.R., et al., *Analysis of the vehicle fleet in the Kathmandu Valley for estimation of environment and climate co-benefits of technology intrusions.* Atmospheric Environment, 2013. **81**: pp. 579–590.

21. Parmon, V.N., *Photocatalysis as a phenomenon: Aspects of terminology.* Catalysis Today, 1997. **39**(3): pp. 137–144.

22. Usubharatana, P., et al., *Photocatalytic process for CO_2 emission reduction from industrial flue gas streams.* Industrial & Engineering Chemistry Research, 2006. **45**(8): pp. 2558–2568.

23. Das, S., and W.W. Daud, *Retracted: Photocatalytic CO_2 transformation into fuel: A review on advances in photocatalyst and photoreactor.* Renewable and Sustainable Energy Reviews, 2014. Pp. 765–805.

24. Kamat, P.V., *Manipulation of charge transfer across semiconductor interface. A criterion that cannot be ignored in photocatalyst design.* Journal of Physical Chemistry Letters, 2012. **3**(5): pp. 663–672.

25. Ângelo, J., et al., *An overview of photocatalysis phenomena applied to NO_x abatement.* Journal of Environmental Management, 2013. **129**: pp. 522–539.

26. Chen, H., C.E. Nanayakkara, and V.H. Grassian, *Titanium dioxide photocatalysis in atmospheric chemistry.* Chemical Reviews, 2012. **112**(11): pp. 5919–5948.

27. Chen, X., et al., *Semiconductor-based photocatalytic hydrogen generation.* Chemical Reviews, 2010. **110**(11): pp. 6503–6570.

28. Xie, S., et al., *Photocatalytic and photoelectrocatalytic reduction of CO_2 using heterogeneous catalysts with controlled nanostructures.* Chemical Communications, 2016. **52**(1): pp. 35–59.

29. Torimoto, T., et al., *Plasmon-enhanced photocatalytic activity of cadmium sulfide nanoparticle immobilized on silica-coated gold particles.* Journal of Physical Chemistry Letters, 2011. **2**(16): pp. 2057–2062.

30. Langhammer, C., et al., *Plasmonic properties of supported Pt and Pd nanostructures.* Journal of Physical Chemistry Letters, 2006. **6**(4): pp. 833–838.

31. Zhdanov, V.P., C. Hägglund, and B. Kasemo, *Relaxation of plasmons in nm-sized metal particles located on or embedded in an amorphous semiconductor.* Surface Science, 2005. **599**(1–3): pp. L372–L375.

32. Nishijima, Y., et al., *Plasmon-assisted photocurrent generation from visible to near-infrared wavelength using a Au-nanorods/TiO_2 electrode.* Journal of Physical Chemistry Letters, 2010. **1**(13): pp. 2031–2036.

33. Mubeen, S., et al., *Plasmonic photosensitization of a wide bandgap semiconductor: Converting plasmons to charge carriers*. Nano Letters, 2011. **11**(12): pp. 5548–5552.

34. Zhao, Z., et al., *Photo-catalytic reduction of carbon dioxide with in-situ synthesized CoPc/TiO₂ under visible light irradiation*. Journal of Cleaner Production, 2009. **17**(11): pp. 1025–1029.

35. Zhao, Z., et al., *Optimal design and preparation of titania-supported CoPc using sol-gel for the photo-reduction of CO₂*. Chemical Engineering Journal, 2009. **151**(1–3): pp. 134–140.

36. Zhao, Z.-H., J.-M. Fan, and Z.-Z. Wang, *Photo-catalytic CO₂ reduction using sol-gel derived titania-supported zinc-phthalocyanine*. Journal of Cleaner Production, 2007. **15**(18): pp. 1894–1897.

37. Shen, S., et al., *Visible-light-driven photocatalytic water splitting on nanostructured semiconducting materials*. International Journal of Nanotechnology, 2011. **8**(6–7): pp. 523–591.

38. Fan, W., Q. Zhang, and Y. Wang, *Semiconductor-based nanocomposites for photocatalytic H₂ production and CO₂ conversion*. Physical Chemistry Chemical Physics, 2013. **15**(8): pp. 2632–2649.

39. Jia, L., J. Li, and W. Fang, *Enhanced visible-light active C and Fe co-doped LaCoO₃ for reduction of carbon dioxide*. Catalysis Communications, 2009. **11**(2): pp. 87–90.

40. Yuan, L., and Y.-J. Xu, *Photocatalytic conversion of CO₂ into value-added and renewable fuels*. Applied Surface Science, 2015. **342**: pp. 154–167.

41. Takeda, H., et al., *Development of an efficient photocatalytic system for CO₂ reduction using rhenium (I) complexes based on mechanistic studies*. Journal of the American Chemical Society, 2008. **130**(6): pp. 2023–2031.

42. Morris, A.J., G.J. Meyer, and E. Fujita, *Molecular approaches to the photocatalytic reduction of carbon dioxide for solar fuels*. Accounts of Chemical Research, 2009. **42**(12): pp. 1983–1994.

43. Hawecker, J., J.-M. Lehn, and R. Ziessel, *Efficient photochemical reduction of CO₂ to CO by visible light irradiation of systems containing Re (bipy)(CO)₃ X or Ru (bipy)3 2+–Co 2+ combinations as homogeneous catalysts*. Journal of the Chemical Society, Chemical Communications, 1983. (9): pp. 536–538.

44. Takeda, H., et al., *Photocatalytic CO₂ reduction using a Mn complex as a catalyst*. Chemical Communications, 2014. **50**(12): pp. 1491–1493.

45. Kuramochi, Y., M. Kamiya, and H. Ishida, *Photocatalytic CO₂ reduction in N, N-dimethylacetamide/water as an alternative solvent system*. Inorganic Chemistry, 2014. **53**(7): pp. 3326–3332.

46. Tamaki, Y., et al., *Photocatalytic CO₂ reduction with high turnover frequency and selectivity of formic acid formation using Ru (II) multinuclear complexes*. Proceedings of the National Academy of Sciences, 2012. **109**(39): pp. 15673–15678.

47. Bruckmeier, C., et al., *Binuclear rhenium (I) complexes for the photocatalytic reduction of CO₂*. Dalton Transactions, 2012. **41**(16): pp. 5026–5037.

48. Lv, H., et al., *An exceptionally fast homogeneous carbon-free cobalt-based water oxidation catalyst*. Journal of the American Chemical Society, 2014. **136**(26): pp. 9268–9271.

49. Inoue, T., et al., *Photoelectrocatalytic reduction of carbon dioxide in aqueous suspensions of semiconductor powders*. Nature, 1979. **277**(5698): pp. 637–638.

50. Möbs, J., M. Gerhard, and J. Heine, *(Hpy)₂ (Py) CuBi₃I₁₂, a low band-gap metal halide photoconductor.* Dalton Transactions, 2020. **49**(41): pp. 14397–14400.

51. Park, B.W., et al., *Bismuth based hybrid perovskites A₃Bi₂I₉ (A): Methylammonium or cesium) for solar cell application.* Advanced Materials, 2015. **27**(43): pp. 6806–6813.

52. Shen, Y., F. Ren, and H. Liu, *Progress in research on photocatalytic performance of doped-TiO (2).* Xiyou Jinshu Cailiao yu Gongcheng (Rare Metal Materials and Engineering), 2006. **35**(11): pp. 1841–1844.

53. Fujishima, A., and K. Honda, *Electrochemical photolysis of water at a semiconductor electrode.* Nature, 1972. **238**(5358): pp. 37–38.

54. Miodyńska, M., et al., *Urchin-like TiO₂ structures decorated with lanthanide-doped Bi₂S₃ quantum dots to boost hydrogen photogeneration performance.* Applied Catalysis B: Environmental, 2020. **272**: p. 118962.

55. Wang, J., et al., *Single 2D Mxene precursor-derived TiO₂ nanosheets with a uniform decoration of amorphous carbon for enhancing photocatalytic water splitting.* Applied Catalysis B: Environmental, 2020. **270**: p. 118885.

56. Li, F., et al., *TiO₂-on-C₃N₄ double-shell microtubes: In-situ fabricated heterostructures toward enhanced photocatalytic hydrogen evolution.* Journal of Colloid and Interface Science, 2020. **572**: pp. 22–30.

57. Rao, Z., et al., *Photocatalytic degradation of gaseous VOCs over Tm³⁺-TiO₂: Revealing the activity enhancement mechanism and different reaction paths.* Chemical Engineering Journal, 2020. **395**: p. 125078.

58. Feizpoor, S., et al., *Combining carbon dots and Ag₆Si₂O₇ nanoparticles with TiO₂: Visible-light-driven photocatalysts with efficient performance for removal of pollutants.* Separation and Purification Technology, 2020. **248**: p. 116928.

59. Porcar-Santos, O., et al., *Photocatalytic degradation of sulfamethoxazole using TiO₂ in simulated seawater: Evidence for direct formation of reactive halogen species and halogenated by-products.* Science of the Total Environment, 2020. **736**: p. 139605.

60. Li, K., et al., *Recent advances in TiO₂-based heterojunctions for photocatalytic CO₂ reduction with water oxidation: A review.* Frontiers in Chemistry, 2021. **9**: p. 637501.

61. Nguyen, T.P., et al., *Recent advances in TiO₂-based photocatalysts for reduction of CO₂ to fuels.* Nanomaterials (Basel), 2020. **10**(2): p. 337.

62. Shehzad, N., et al., *A critical review on TiO₂ based photocatalytic CO₂ reduction system: Strategies to improve efficiency.* Journal of CO₂ Utilization, 2018. **26**: pp. 98–122.

63. Tahir, M., and N.S. Amin, *Advances in visible light responsive titanium oxide-based photocatalysts for CO₂ conversion to hydrocarbon fuels.* Energy Conversion and Management, 2013. **76**: pp. 194–214.

64. Wu, L., C. Fu, and W. Huang, *Surface chemistry of TiO₂ connecting thermal catalysis and photocatalysis.* Physical Chemistry Chemical Physics, 2020. **22**(18): pp. 9875–9909.

65. Tang, J., Y. Chen, and Z. Dong, *Effect of crystalline structure on terbuthylazine degradation by H₂O₂-assisted TiO₂ photocatalysis under visible irradiation.* Journal of Environmental Sciences, 2019. **79**: pp. 153–160.

66. Jin, J., et al., *One-pot hydrothermal preparation of PbO-decorated brookite/anatase* TiO_2 *composites with remarkably enhanced* CO_2 *photoreduction activity.* Applied Catalysis B: Environmental, 2020. **263**: p. 118353.
67. Cao, M.-Q., et al., *Hierarchical* TiO_2 *nanorods with a highly active surface for photocatalytic* CO_2 *reduction.* Journal of Central South University, 2019. **26**(6): pp. 1503–1509.
68. Kar, P., et al., *High rate* CO_2 *photoreduction using flame annealed* TiO_2 *nanotubes.* Applied Catalysis B: Environmental, 2019. **243**: pp. 522–536.
69. Zhu, L., et al., *Metal-organic frameworks for heterogeneous basic catalysis.* Chemical Reviews, 2017. **117**(12): pp. 8129–8176.
70. Li, R., et al., *Integration of an inorganic semiconductor with a metal-organic framework: A platform for enhanced gaseous photocatalytic reactions.* Advanced Materials, 2014. **26**(28): pp. 4783–4788.
71. Chen, J., et al., *Metal halide perovskites for solar-to-chemical fuel conversion.* Advanced Energy Materials, 2020. **10**(13): p. 1902433.
72. Liang, S., et al., *Recent advances in synthesis, properties, and applications of metal halide perovskite nanocrystals/polymer nanocomposites.* Advanced Materials, 2021. **33**(50): p. 2005888.
73. Hou, J., et al., *Inorganic colloidal perovskite quantum dots for robust solar* CO_2 *reduction.* Chemistry, 2017. **23**(40): pp. 9481–9485.
74. Han, C., et al., *The functionality of surface hydroxyls on selective* CH_4 *generation from photoreduction of* CO_2 *over SiC nanosheets.* Chemical Communications, 2019. **55**(11): pp. 1572–1575.
75. Yang, T.-C., et al., *Photocatalytic reduction of* CO_2 *with SiC recovered from silicon sludge wastes.* Environmental Technology, 2015. **36**(23): pp. 2987–2990.
76. Ahmed, N., et al., *Photocatalytic conversion of carbon dioxide into methanol using zinc–copper–M (III) (M= aluminum, gallium) layered double hydroxides.* Journal of Catalysis, 2011. **279**(1): pp. 123–135.
77. Gao, G., et al., *Ultrathin magnetic Mg-Al LDH photocatalyst for enhanced* CO_2 *reduction: Fabrication and mechanism.* Journal of Colloid and Interface Science, 2019. **555**: pp. 1–10.
78. Kawamura, S., et al., *Photocatalytic conversion of carbon dioxide using Zn-Cu-Ga layered double hydroxides assembled with Cu phthalocyanine: Cu in contact with gaseous reactant is needed for methanol generation.* Oil & Gas Science and Technology–Revue de l IFP, 2015. **70**(5): pp. 841–852.
79. Teramura, K., et al., *Photocatalytic conversion of* CO_2 *in water over layered double hydroxides.* Angewandte Chemie International Edition in English, 2012. **51**(32): pp. 8008–8011.
80. Wang, Y., X. Wang, and M. Antonietti, *Polymeric graphitic carbon nitride as a heterogeneous organocatalyst: From photochemistry to multipurpose catalysis to sustainable chemistry.* Angewandte Chemie International Edition in English, 2012. **51**(1): pp. 68–89.
81. Niu, P., et al., *Graphene-like carbon nitride nanosheets for improved photocatalytic activities.* Advanced Functional Materials, 2012. **22**(22): pp. 4763–4770.
82. Hsu, H.-C., et al., *Graphene oxide as a promising photocatalyst for* CO_2 *to methanol conversion.* Nanoscale, 2013. **5**(1): pp. 262–268.

83. Wu, J., et al., *A metal-free electrocatalyst for carbon dioxide reduction to multi-carbon hydrocarbons and oxygenates.* Nature Communications, 2016. 7(1): p. 13869.

84. Tsuneoka, H., et al., *Adsorbed Species of CO_2 and H_2 on Ga_2O_3 for the Photocatalytic Reduction of CO_2.* Journal of Physical Chemistry C, 2010. 114(19): pp. 8892–8898.

85. Zhou, Y., et al., *High-yield synthesis of ultrathin and uniform Bi_2WO_6 square nanoplates benefitting from photocatalytic reduction of CO_2 into renewable hydrocarbon fuel under visible light.* CS Applied Materials & Interfaces, 2011. 3(9): pp. 3594–3601.

86. Cheng, H., et al., *An anion exchange approach to Bi_2WO_6 hollow microspheres with efficient visible light photocatalytic reduction of CO_2 to methanol.* Chemical Communications, 2012. 48(78): pp. 9729–9731.

87. Xi, G., et al., *Ultrathin W18O49 nanowires with diameters below 1 nm: Synthesis, near-infrared absorption, photoluminescence, and photochemical reduction of carbon dioxide.* Angewandte Chemie International Edition in English, 2012. 51(10): pp. 2395–2399.

88. Liu, M., et al., *Quantum-dot-derived catalysts for CO_2 reduction reaction.* Joule, 2019. 3(7): pp. 1703–1718.

89. Jianjian, T., et al., *Sn quantum dots for electrocatalytic reduction of CO_2 to HCOOH.* Journal of Inorganic Materials, 2021. 36(12).

90. Grigioni, I., et al., *CO_2 electroreduction to formate at a partial current density of 930 mA cm−2 with in P colloidal quantum dot derived catalysts.* ACS Energy Letters, 2020. 6(1): pp. 79–84.

91. Fu, J., et al., *Low overpotential for electrochemically reducing CO_2 to CO on nitrogen-doped graphene quantum dots-wrapped single-crystalline gold nanoparticles.* ACS Energy Letters, 2018. 3(4): pp. 946–951.

92. Yadav, R.M., et al., *Amine-functionalized carbon nanodot electrocatalysts converting carbon dioxide to methane.* Advanced Materials, 2022. 34(2): p. 2105690.

93. Zhu, Q., et al., *Carbon dioxide electroreduction to C_2 products over copper-cuprous oxide derived from electrosynthesized copper complex.* Nature Communications, 2019. 10(1): p. 3851.

94. Ye, R., et al., *Bandgap engineering of coal-derived graphene quantum dots.* ACS Applied Materials & Interfaces, 2015. 7(12): pp. 7041–7048.

95. Ye, R., et al., *Coal as an abundant source of graphene quantum dots.* Nature Communications, 2013. 4(1): p. 2943.

96. Marcano, D.C., et al., *Improved synthesis of graphene oxide.* ACS Nano, 2010. 4(8): pp. 4806–4814.

97. Zhang, T., et al., *Regulation of functional groups on graphene quantum dots directs selective CO_2 to CH_4 conversion.* ACS Nano, 2021. 12(1): p. 5265.

98. Wang, Q.H., et al., *Electronics and optoelectronics of two-dimensional transition metal dichalcogenides.* Nature Nanotechnology, 2012. 7(11): pp. 699–712.

99. Asadi, M., et al., *Robust carbon dioxide reduction on molybdenum disulphide edges.* Nature Communications, 2014. 5(1): p. 4470.

100. Bagger, A., et al., *Electrochemical CO_2 reduction: A classification problem.* Nano-Science Center, 2017. 18(22): pp. 3266–3273.

101. Han, N., et al., *Promises of main group metal-based nanostructured materials for electrochemical CO_2 reduction to formate*. Advanced Energy Materials, 2020. **10**(11): p. 1902338.
102. Zhao, S., et al., *Advances in Sn-based catalysts for electrochemical CO_2 reduction*. Nano-Micro Letters, 2019. **11**: pp. 1–19.
103. Mezzavilla, S., et al., *Structure sensitivity in the electrocatalytic reduction of CO_2 with gold catalysts*. Angewandte Chemie International Edition in English, 2019. **58**(12): pp. 3774–3778.
104. Trindell, J.A., J. Clausmeyer, and R.M. Crooks, *Size stability and H_2/CO selectivity for Au nanoparticles during electrocatalytic CO_2 reduction*. Journal of the American Chemical Society, 2017. **139**(45): pp. 16161–16167.
105. Li, S., et al., *Boosting CO_2 electrochemical reduction with atomically precise surface modification on gold nanoclusters*. Angewandte Chemie International Edition in English, 2021. **60**(12): pp. 6351–6356.
106. Reske, R., et al., *Particle size effects in the catalytic electroreduction of CO_2 on Cu nanoparticles*. Journal of the American Chemical Society, 2014. **136**(19): pp. 6978–6986.
107. Mistry, H., et al., *Exceptional size-dependent activity enhancement in the electroreduction of CO_2 over Au nanoparticles*. Journal of the American Chemical Society, 2014. **136**(47): pp. 16473–16476.
108. Dudley, B., *BP statistical review of world energy 2016*. In Bplc (ed.) *British Petroleum Statistical Review of World Energy*. Pureprint Group Limited, 2019.
109. Rao, B.M., and S. Roy, *Solvothermal processing of amorphous TiO_2 nanotube arrays: Achieving crystallinity at a lower thermal budget*. Journal of Physical Chemistry C, 2014. **118**(2): pp. 1198–1205.
110. Berardi, S., et al., *Molecular artificial photosynthesis*. Chemical Society Reviews, 2014. **43**(22): pp. 7501–7519.
111. Geary, W.J. *The use of conductivity measurements in organic solvents for the characterisation of coordination compounds*. Coordination Chemistry Reviews, 1971. **7**(81): pp. 81–122.
112. Li, H., et al., *Noble metal-free single-and dual-atom catalysts for artificial photosynthesis*. Advanced Materials, 2023: p. 2301307.
113. Yang, K.R., G.W. Kyro, and V.S. Batista, *The landscape of computational approaches for artificial photosynthesis*. Nature Computational Science, 2023. **3**(6): pp. 504–513.
114. Handoko, A.D., K. Li, and J. Tang, *Recent progress in artificial photosynthesis: CO_2 photoreduction to valuable chemicals in a heterogeneous system*. Current Opinion in Chemical Engineering, 2013. **2**(2): pp. 200–206.
115. Kim, D., et al., *Artificial photosynthesis for sustainable fuel and chemical production*. Angewandte Chemie International Edition in English, 2015. **54**(11): pp. 3259–3266.
116. Roy, N., et al., *Recent improvements in the production of solar fuels: From CO_2 reduction to water splitting and artificial photosynthesis*. Bulletin of the Chemical Society of Japan, 2019. **92**(1): pp. 178–192.
117. Sharma, A., et al., *Insight into ZnO/carbon hybrid materials for photocatalytic reduction of CO_2: An in-depth review*. Journal of CO_2 Utilization, 2022. **65**: p. 102205.

118. Wang, J., et al., *Reactor design for solar-driven photothermal catalytic CO_2 reduction into fuels*. Energy Conversion and Management, 2023. **281**: p. 116859.

119. Wan, L., et al., *Solar driven CO_2 reduction: From materials to devices*. Journal of Materials Chemistry A, 2023. **11**: p. 12499.

120. Zhang, Q., Z. Zuo, and D.J.C.C. Ma, *Plasmonic nanomaterials for solar-driven photocatalysis*. Chemical Communications, 2023. **59**(50): pp. 7704–7716.

Chapter 9

Photocatalytic Hydrogen Production

ABSTRACT

To achieve sustainable growth of nations, researchers are being compelled to investigate renewable energy resources due to the world's rapidly depleting energy resources, high energy consumption, and quick increase in demand for energy. The current requirements are being met by utilizing a large number of available resources. However, one major issue of the modern era is that these resources are also severely harming the environment. As a result, there is an urgent need to develop sustainable, environmentally friendly renewable energy sources. Since water is so abundant on earth, it is one of the best sources of hydrogen energy. This chapter goes into great detail about the different ways to split water and produce hydrogen, with a particular emphasis on the exciting process of photocatalysis. The properties of photocatalysts are influenced by their nature, composition, and other features; these properties in turn have an impact on the production of hydrogen and the splitting of water. Thus, a wide range of materials—including metal sulphides, metal oxides, and nanocomposites—used in photocatalytic hydrogen production are covered in this chapter. In addition, the pros and cons of the various material classes are analyzed to identify the best material class for water-splitting hydrogen production. Beginners in this sector will find this chapter useful in understanding the basic functioning of several hydrogen generation mechanisms, as well as the benefits and drawbacks of each. But it will also assist industry professionals and specialists in choosing the optimal class of materials for hydrogen evolution from water splitting.

9.1 INTRODUCTION

Many industries rely on energy, notably companies, manufacturing, educational institutions, the agricultural sector, and numerous other service sectors. Energy, as a need, has a huge impact on economic advancement and

DOI: 10.1201/9781003403357-11

has evolved into an aspect of social uplift. Energy has always been an essential component of a country's growth and development. Several third-world countries, including Pakistan, have experienced hardship as a consequence of energy crises. Such crises have hit Pakistan's industrial sector, cutting exports of goods, reverting foreign investment, decreasing agricultural output, boosting prices, creating unemployment, and pushing enterprises to close. The economy of Pakistan has made up for significant losses caused by the country's energy issues. Because electricity is the primary source of energy, an energy crisis or power outage is considered in this study. The term "energy crisis" means limited access to available resources or high prices. In both conditions, it is difficult for the average person to get enough energy resources to meet with their demands.[1] There are many factors behind such a crisis including the following key factors: disruptions in supply, geopolitical conflicts, not enough investment on infrastructure of energy,[2] and natural disasters as shown in the fishbone diagram that leads to energy wastage (Figure 9.1).

In order to speed up chemical reactions, photocatalysis uses photon energy, which is frequently in the form of ultraviolet (UV) or visible light. As the need for clean, sustainable energy sources rises, hydrogen production is a process of great scientific interest. This is where photocatalysis finds a crucial use. Basically, photocatalysis is the activation of a chemical reaction by light without the need for a catalyst. This occurs because a photocatalyst produces electron-hole pairs when it sufficiently absorbs

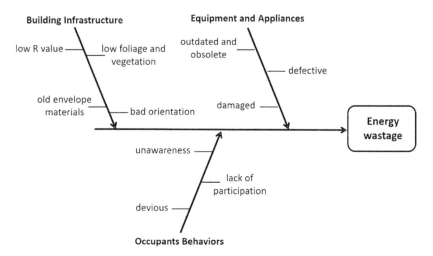

Figure 9.1 A fishbone diagram that leads to energy wastage.

Source: Adapted with permission under the terms of the Attribution Creative Commons (CC BY) AT (https://creativecommons.org/licenses/by/). Munguia, N., et al., *Energy efficiency in public buildings: A step toward the UN 2030 agenda for sustainable development*. 2020. *Sustainability*, 12(3): p. 1212.

light to initiate chemical reactions. Materials like TiO_2 and semiconductors like ZnO and H_2S are frequently utilized as photocatalysts for the creation of hydrogen.[3]

9.2 IMPORTANCE OF HYDROGEN AS AN ENERGY CARRIER

Hydrogen is very important as an energy carrier because it has the potential to greatly change how we use energy globally. People often consider hydrogen a clean and versatile fuel for a few important reasons. First, it is environmentally friendly, and when we use hydrogen as fuel, whether by burning it or in fuel cells, the only thing it produces is water vapor. This indicates that it emits no harmful emissions, which is essential in the fight against climate change and the reduction of greenhouse gas emissions.[4] Hydrogen holds significant importance in promoting cleaner and more sustainable energy sources by taking the place of fossil fuels in sectors such as transportation, industry, and power generation.[5]

Hydrogen is essential to photocatalysis as well because it facilitates sustainable and clean energy conversion. Hydrogen is essential as an energy carrier in photocatalysis for the key reasons discussed in the following sections.

9.2.1 Clean Energy Generation and Energy Storage

The hydrogen generated via photocatalysis functions as a sustainable energy source. Hydrogen and oxygen are the only by-products of photocatalytic water splitting, which uses sunlight to split water molecules. This results in a way of producing energy that emits less CO_2 and contributes to the reduction of greenhouse gases. One effective medium for storing energy obtained from photocatalytic reactions is hydrogen.[3]

Hydrogen serves as an effective medium for the storage of energy acquired through photocatalytic reactions. It is a useful way to store energy obtained from processes that use sunlight. It allows us to convert sunlight into a form of energy that can be kept for later use, like for making power or fueling vehicles.[6]

9.2.2 Transportation Fuel

By using sunlight to create hydrogen through a process called photocatalysis, one can generate a clean fuel suitable for fuel cell-powered vehicles. This advancement holds great importance in the effort to make transportation more environmentally friendly.[7] Hydrogen-powered vehicles produce only water vapor as a by-product, effectively addressing air pollution concerns and contributing to the reduction of carbon emissions in the transportation sector.

9.2.3 Grid Balancing and Load Shifting

Grid balancing and load shifting using photocatalytic hydrogen production mean using a special process to store extra solar energy when the sun is shining and then using that stored energy when people need electricity the most or when it is dark outside. This helps make sure the electricity grid stays steady and dependable, especially in places where a lot of the energy comes from renewable sources like the sun and wind.[8]

9.2.4 Carbon-Free Industrial Processes and Environmental Remediation

In the realm of industry, one can utilize hydrogen produced through photocatalysis to substitute for fossil fuels in applications like hydrogenation, ammonia synthesis, and metallurgical procedures. This substitution significantly lessens the release of carbon emissions stemming from these industrial activities.[8] Hydrogen generated through photocatalysis can be used to clean up the environment by removing contaminants from water, soil, and air. This helps make the environment cleaner and healthier.

9.2.5 Hydrogen Economy Advancement and Flexibility

The advancement of photocatalytic hydrogen generating systems is critical for the advancement of the hydrogen economy. These techniques aid in the shift from fossil fuels to a hydrogen-based energy system, which can enhance sustainability and increase energy security. Hydrogen is a versatile energy source that may be employed in a range of contexts, including households and companies. It is a versatile option for a range of energy requirements because it can be used for heating, producing electricity, and powering industrial processes.[4]

In conclusion, photocatalysis relies heavily on hydrogen as an energy carrier, which makes it possible to effectively transform sunlight into a clean, flexible energy source. Energy security, energy storage, and climate change are just a few of the major issues it may help with. It also makes it easier for solar radiation to be skillfully converted into a flexible, eco-friendly energy source. This in turn has numerous uses in industrial processes, energy storage, and transportation, among other areas, and is essential to the advancement of a sustainable and carbon-neutral future.[3]

9.3 ROLE OF PHOTOCATALYSIS IN HYDROGEN PRODUCTION

One of the most important steps in creating hydrogen, a clean and adaptable energy source, is photocatalysis. It splits water into hydrogen and oxygen using light. Materials with the ability to absorb light and produce charged particles, such as TiO_2 and more sophisticated ones like bismuth vanadate $(BiVO_4)$ and cadmium sulphide (CdS), are the reason for this. These charged particles aid in the chemical processes that convert water into hydrogen and oxygen on the catalyst's surface. Because photocatalytic hydrogen production is environmentally friendly and can use sunlight as an energy source, it is great because it does not release any harmful gases.[7] We could benefit from sustainable energy and a cleaner, greener future thanks to this technology. A schematic diagram of basic building blocks for hydrogen production is shown in Figure 9.2.

Beyond its benefits for the environment, photocatalysis is also highly versatile, as it can be applied in a range of settings, including room temperature and pressure. Because of its adaptability, it can be used for a wide range of purposes, from small, standalone systems in remote locations to large-scale industrial hydrogen production. Additionally, integrated energy systems that have the capacity to store excess energy as hydrogen can be created by combining photocatalytic hydrogen production with other renewable energy

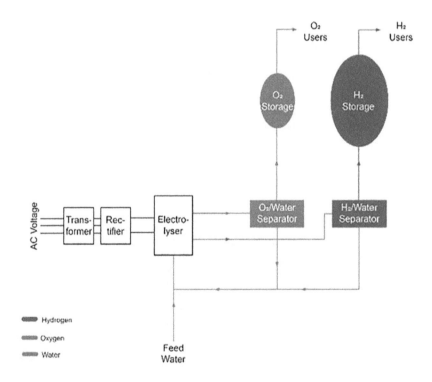

Figure 9.2 Basic building blocks for hydrogen production.

Source: Adapted with permission under the terms of the Creative Commons Attribution 4.0 International (CC BY 4.0) AT (https://creativecommons.org/licenses/by/4.0/at/mdpi). Jovan, D.J., and G.J.E. Dolanc, *Can green hydrogen production be economically viable under current market conditions*. 2020. *Energies*, 13(24): p. 6599.

technologies, such as solar panels and wind turbines. The hydrogen that has been stored can then be utilized as a clean energy source for a variety of applications, including fuel cells, electricity generation, and transportation. Its significance in sustainable energy is further increased by this factor.[6]

9.3.1 Challenges

Photocatalysis shows great potential, but there are still challenges to address. These include improving how efficient the process is, finding better catalyst materials, and dealing with issues related to the stability and scalability of the catalyst. Researchers are working on making photocatalytic systems work better and last longer so they can be used on a larger scale without being too expensive. As the world looks for cleaner and more sustainable ways to make energy and reduce our reliance on fossil fuels to fight climate change, photocatalysis for hydrogen production is becoming more important. It could help create a greener and more sustainable energy future.[9]

9.3.2 Fundamentals of Photocatalysis

Photocatalysis is a captivating area within chemistry that relies on the interaction between a photocatalyst, light energy, and chemical reactions. Essentially, it involves employing a photocatalyst, a substance capable of absorbing photons from a light source, typically UV or visible light. This absorbed energy is then used to hasten a chemical reaction without causing any lasting changes to the photocatalyst itself. Common examples of photocatalysts are materials like TiO_2 and ZnO.[10] This field finds diverse applications, ranging from cleaning the environment by breaking down organic pollutants into harmless by-products to generating clean energy, including photocatalytic hydrogen production and creating organic compounds.

9.3.3 Photocatalytic Reactions and Mechanisms

Photocatalytic reactions involve different processes, each with its own unique way of working, depending on what kind of reaction is happening. In environmental uses, photocatalysis is used to break down organic pollutants. This happens when light (photons) makes electrons in the catalyst material excited, leading to the creation of electron-hole pairs.[10] These electron-hole pairs then join in chemical reactions, producing reactive oxygen species (ROS), like hydroxyl radicals $(\cdot OH)$, which are really good at breaking down and getting rid of organic contaminants. On the other hand, when we want to use photocatalysis to make hydrogen gas (H_2) and oxygen gas (O_2) from water, it is a bit more complicated. It involves light being absorbed, creating electron-hole pairs, water sticking to the catalyst surface, and then chemical reactions happening there to produce hydrogen and oxygen. Photocatalysts can also help make various organic compounds through chemical reactions that need light to occur; without light, these reactions would not happen.[11]

9.3.3.1 Key Principles of Photocatalytic Hydrogen Production

The approach of producing hydrogen using photocatalysis involves splitting water into hydrogen and oxygen using photocatalysts when exposed to light, usually sunlight. The following are the primary principles of photocatalytic hydrogen production:

i.   *Photocatalysts and bandgap energy:* Photocatalyst materials are substances that absorb light energy and use it to break down water molecules into hydrogen and oxygen. Photocatalysts based on semiconductors, such as TiO_2, CdS, and other metal oxides, are frequently used. The energy difference between the valence band and the conduction band gives them a particular bandgap energy. Electrons become excited to form electron-hole pairs in the conduction band when the energy of light photons equals or exceeds the bandgap energy.[12] In photocatalysis, the generation of electron-hole pairs is a fundamental process. Holes in the valence band and electron in the conduction band can participate in redox chemical reactions because of their high reactivity.

ii. *Redox chemical reactions:* Holes in the valence band can oxidize water to produce oxygen gas (O_2), while electrons in the conduction band can reduce water (H_2O) to make hydrogen ions (H^+). Oxygen gas (O_2) and hydrogen gas (H_2) are produced concurrently by these redox chemical reactions.[12]

iii. *Light source:* Solar radiation is the most widely used light source for photocatalytic hydrogen production. Both the spectral distribution and light intensity affect the process's efficiency. The most sustainable and renewable method of producing hydrogen is through solar energy.

iv. *Reaction kinetics:* The kinetics of the photocatalytic process play a major role in how efficiently hydrogen is produced. The specific photocatalyst used, the pH, and the reaction temperature can all affect the rate of hydrogen generation.

v.   *Catalyst modification:* Researchers often alter the surface properties, composition, or structure of photocatalysts to improve their performance. This could enhance their ability to absorb light and separate charges.

vi. *Photocatalyst stability:* For practical uses in hydrogen production, photocatalyst long-term viability and stability are essential. In particular, photocatalysts must resist degradation over a long period of time when exposed to light.

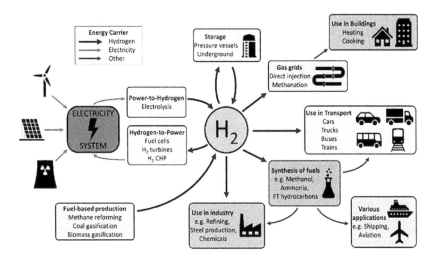

Figure 9.3 Flowchart of different uses of hydrogen.

Source: Adapted with permission under the terms of the Creative Commons Attribution 3.0 Unported (CC BY 3.0) AT (https://creativecommons.org/licenses/by/3.0/rcs.org). Quarton, C.J., et al., *The curious case of the conflicting roles of hydrogen in global energy scenarios.* 2020. *Sustainable energy & fuels,* 4(1): pp. 80–95.

vii. *Water source:* The photocatalytic process needs a source of water, usually in the form of liquid or vaporized water. The efficiency of producing hydrogen can be influenced by the purity and quality of the water source.

viii. *Integration with other systems:* For energy conversion and storage, photocatalytic production of hydrogen can be combined with energy storage devices like hydrogen fuel cells. It might therefore be a part of a cycle of sustainable energy.

ix. *Environmental impact*: It is crucial to assess and reduce the environmental impact of photocatalytic production of hydrogen. This involves interacting with concerns about non-toxic utilization of materials and waste management.

x. *Scale-up and practical applications:* Researchers must take into account the scalability of the process and its integration into practical applications (Figure 9.3), such as the production of hydrogen fuel for energy storage and transportation, in order to make photocatalytic hydrogen production cost-effective.

The development and optimization of photocatalytic hydrogen production processes are guided by these principles, thereby aiming to produce clean and renewable hydrogen fuel by using solar energy.[12]

9.4 PHOTOCATALYTIC MATERIALS FOR HYDROGEN PRODUCTION

Nanomaterials possess superior photocatalytic characteristics over their bulk equivalents, nanomaterials including CdS, SiC, CuInSe2, and TiO_2 can be used to make inexpensive and clean photocatalytic hydrogen. Numerous additional nanomaterials have been studied, including Nb_2O_5, Ta_2O_5, α-Fe_2O_3, ZnO, TaON, $BiVO_4$, and WO_3.[13-18] Bandgap limitation is a major problem in the majority of photocatalysts, contributing to limited H_2 generation, although certain photocatalysts have been modified using metal-ion implantation, ion doping, noble metal doping, and sensitization in order to address this problem. Pt has been proven to be the finest noble metal in noble metal doping; however, the cost is prohibitive. Thus, research has been done on alternative effective and less expensive metals, including Ni, Cu, Ag, Ru, Pd, and Ir.[19,20] Ion doping includes anions of nitrogen and sulphur, as well as ions of transition and rare-earth metals. Sensitization refers to the coupling of semiconductors and dye sensitization. It has been discovered that metal-ion implantation and dye sensitization are the most successful photoanode surface modification methods among those previously discussed. In addition, co-catalysts containing photocatalyst nanomaterials that can be employed for photocatalytic hydrogen production are of interest to researchers.[20]

9.4.1 Semiconductor Materials for Photocatalysis

Photocatalysis, a promising technique for producing hydrogen and producing clean, renewable energy, depends heavily on semiconductor materials. This method, often referred to as photoelectrochemical water splitting, splits water into its constituent parts, such as hydrogen and oxygen, by use of chemical processes triggered by light energy. Photocatalysts made of semiconductor materials are used to speed up this method of production.

Numerous semiconductors have been developed and evaluated as photocatalysts for hydrogen production in order to improve photocatalytic hydrogen production. Due to its low level of toxicity, physicochemical stability, and comparatively higher quantum efficiency, TiO_2 has been the semiconductor under investigation among these photocatalysts for the past 10 years, according to certain studies. Yet, it has also been discovered that several novel materials, including g-C_3N_4, MoS_2, CdS, ZnO, WO_3, Cu_2O, and compounds based on sulphides and bismuth, including ZnS, $ZnIn_2S_4$, and $BiVO_4$, show great potential as photocatalysts.[21]

In semiconductor photocatalysis, target molecules are reduced and oxidized by the photoexcited electrons and holes on a semiconductor surface during the photocatalytic reaction. The process is primarily limited by the quick recombination of electrons and holes produced by photogeneration. This explains why employing a photocatalyst of single-phase semiconductor results in low efficiency.[22] The improvement of photocatalytic activity through element doping, compositing, crystal facet engineering, and other methods has been the subject of numerous investigations. The most practical and efficient for enhancing photocatalytic activity among them are heterojunctions made of two or more semiconductors with complementary energy bands, as these enable the spatial separation of photo-generated electron-hole pairs at the heterojunction interface.

9.4.2 Bandgap Engineering for Efficient Hydrogen Generation

Bandgap engineering is a key tactic for increasing the effectiveness of photocatalytic hydrogen generation, which is a solar-powered process that splits water into hydrogen and oxygen using semiconductor materials. Researchers try to optimize solar radiation absorption, charge separation efficiency, and catalytic activity by precisely controlling the bandgap of semiconductor materials. It has been used to effectively develop semiconductor materials such as TiO_2, $BiVO_4$, and numerous metal sulphides and oxides for the efficient production of hydrogen.[23]

There are a lot of published practical studies on metal oxide nano-photocatalysts for producing H_2 through water splitting when exposed to UV and visible light; however, there are not many published theoretical studies based on density functional theory (DFT) techniques. Of them, nanostructured and modified TiO_2 photocatalysts have been the focus of a significant number of theoretical investigations. To produce H_2, for instance, Kaur et al. used DFT techniques to study amorphous TiO_2 as a photocatalyst.[24] Through a study of the electrical characteristics, the authors claimed that in comparison to crystalline TiO_2, amorphous TiO_2 might function as a more affordable, readily available, but relatively less effective photocatalyst. In order to enhance the photocatalytic activity of the TiO_2 nano-photocatalyst, TiO_2 was doped using various methods, such as reducing the bandgap of TiO_2 to increase its absorption of visible light in the solar spectrum and investigating the relationship between charge separation and recombination and the distance between the catalytic core and semiconductor surface.[25]

In a different study, Reynal et al.[25] discovered that photo-induced reduction of H_2 on a co-electrocatalyst immobilized on TiO_2 was 104 times faster than reverse charge recombination. The authors claim that both processes show exponential dependence with respect to the separation between the catalytic core and the semiconductor. When three related cobalt electrocatalysts were used to functionalize a semiconductor (TiO_2), their molecular structures changed the physical separation between the catalytic core and the semiconductor surface. The authors also computed the charge separation and recombination processes in these scenarios. This is illustrated in Figure 9.4a and b. In this hybrid system, the semiconductor functioned as a light-catching element while the attached molecular catalyst drove the evolution of H_2.[25]

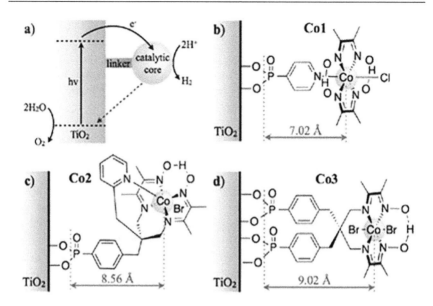

Figure 9.4 (a) The electron transport mechanisms in TiO₂ functionalized with a molecular catalyst for the generation of H₂ following stimulation by UV radiation. The charge separation and recombination are indicated, respectively, by the solid black and dashed arrows (b) for Co1, (c) for CO2, and (d) for CO3 display the molecular structures of the H₂ reduction catalysts; the charges have been removed for clarity. Based on the energy-minimized DFT calculations, the blue arrows show the distance (*r, A*) between the catalytic metal centre and the anchoring groups.

Second, metal cations and/or non-metal anions have been substituted in TiO₂ modifications aimed at improving visible-light absorption. The bandgap's decrease to the visible light portion of the solar spectrum may occasionally depend on the *DFT* approach from *DFT* simulations. In addition to bandgap reduction, doping strategy, and/or *DFT* methodology, critical factors including stability, solubility, and repeatability are also crucial for maximizing TiO₂'s photocatalytic activity because of minimum charge recombination. TiO₂ nanostructure engineering can also reduce charge recombination. In this context, the photocatalytic activity of TiO₂ can be enhanced by heterostructure approaches with other oxides, which can then be used to modify the structure to maximize visible light absorption. Last, in comparison to pure TiO₂, heterostructures as Ga₂O₃-TiO₂, MgO-TiO₂, and Bi₄Ti₃O₁₂-TiO₂ can decrease the bandgap, resulting in visible-light photoactivity, effective charge separation, and enhanced photocatalytic activity.[26]

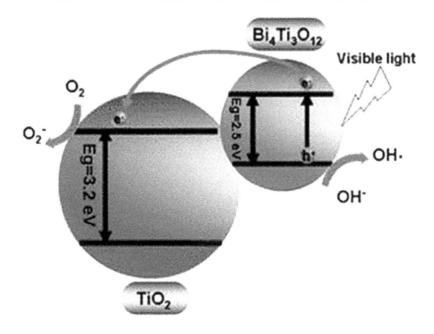

Figure 9.5 The $Bi_4Ti_3O_{12}$-TiO_2 heterostructure's energy band structure and electron-hole pair separation.

Source: Adapted with permission under the terms of the Creative Commons Attribution 3.0 Unported (CC BY 3.0) AT (https://creativecommons.org/licenses/by/3.0/rcs.org). Bhatt, M.D., and J.S. Lee, *Nanomaterials for photocatalytic hydrogen production: From theoretical perspectives*. RSC Advances, 2017. **7**(55): pp. 34875–34885.

The bandgap of $Bi_4Ti_3O_{12}$ (2.5 eV) is smaller than that of TiO_2 (3.2 eV) in the context of the $Bi_4Ti_3O_{12}$-TiO_2 heterostructure because *Bi 6s* mostly contributes to the valence band, and *Bi 6p* state primarily contributes to the conduction band. This results in a significant absorption of visible light. In Figure 9.5, an electron-hole pair separation diagram and energy band structure of the $Bi_4Ti_3O_{12}$-TiO_2 heterostructure are displayed.

WO_3 attracts the most attention of all the metal oxides because of its photosensitivity, high electron transport properties, stability against photocorrosion, and other qualities. Furthermore, it is appropriate for visible light absorption due to its narrower bandgap (around 2.8 eV) than other oxides like TiO_2 discussed earlier.[27,28]

9.4.3 Two-Dimensional Materials as Photocatalysts for H_2 Production

Promising photocatalysts are two-dimensional materials like graphene and stacked hexagonal (h-BN). On the other hand, (h-BN) shows a broad bandgap (about 5.5 eV) while graphene (monolayer) shows a zero bandgap.

By adjusting the bandgap and absolute energy levels through chemical modifications, these materials (let's say ternary B-C-N compounds)[29] can form the appropriate medium bandgap semiconductors. Similarly, charge separation and spontaneous redox reactions could be enhanced by graphene oxide using a carbon support. Due to their unusual sp^2 hybrid carbon networks, which show ultra-fast electron mobility at ambient temperature, conductivity, enormous theoretical surface area, high work function, etc., graphene-based heterogeneous photocatalytic nanomaterials have been receiving a lot of attention in recent years. Moreover, metal chalcogenides have been discovered to have the right bandgap width and band-edge location, making them attractive photocatalyst materials for photocatalytic H_2 production.

In conclusion, there are comparatively few theoretical studies based on DFT approaches on nano-photocatalysts (both metal oxide and non-metal oxide) and modified photocatalysts for photocatalytic H_2 production via water splitting. The few studies on pure TiO_2 and modified TiO_2 that were published focused on the electronic structures of modified TiO_2, which show gap reduction and effective charge separation as a result of doping TiO_2 to improve photocatalytic H_2 production. Owing to the advantageous properties of MoS_2 and graphene, comparatively little research has been published on the use of these materials and metal chalcogenides in multicomponent heterostructures (metal oxide/graphene/metal chalcogenides) and B-C-N ternary compounds for efficient photocatalytic hydrogen production.

Further DFT research should concentrate on either the experimental study of photocatalysts and modified photocatalysts or the exploration of novel photocatalytic materials. Theoretical researchers have a great opportunity to perform DFT calculations on metal oxide and non-metal oxide photocatalysts and their modified forms for extremely efficient photocatalytic H_2 production.

9.5 PHOTOCATALYTIC MECHANISMS FOR HYDROGEN PRODUCTION

As photocatalytic water splitting resembles the process of photosynthesis in green plants using sun energy, it is considered an artificial kind of photosynthesis. With the idea of photocatalytic reactions, it is possible to produce hydrogen from organic materials found in wastewater or water by taking use of the daily solar energy that reaches the earth's surface. Light energy is converted into chemical energy during this process, and the water-splitting reaction helps to accumulate Gibbs free energy. There are two ways to produce hydrogen: (i) the first is by photocatalytic water splitting in which water performs a redox reaction with electrons and holes, and (ii) the second is through photocatalytic organic reforming in which organic materials contribute electrons and oxidize to produce proton ions. These proton ions are ultimately transformed into H_2, when electrons are involved over the photocatalyst.

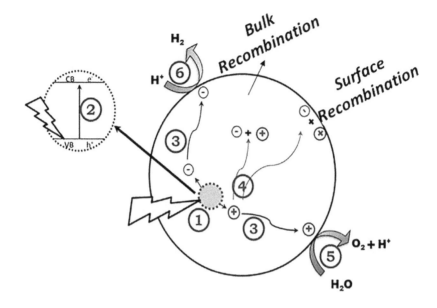

Figure 9.6 Showing photocatalytic water-splitting mechanism for the creation of H_2 when light irradiation produces e^-/h^+ pairs with energies greater than the semiconductor's bandgap.

In order to produce hydrogen, a photocatalytic system typically needs a reactant, a photocatalyst, a photo reactor, and a light source. Water either by itself or even a mixture containing a sacrificial reagent can act as the reactant. UV or visible light can be used to activate a photocatalyst; however, photocatalysts that absorb visible light can also ensure that an appropriate amount of solar energy is captured. An effective interaction between light, catalyst, and reactants would be necessary for the generation of hydrogen efficiently.[30]

In order to separate the filled valence band (VB) and vacant conduction band (CB), a semiconductor-based photocatalyst is first exposed to light with an energy higher than or equal to its bandgap. An electron (e^-) hole (h^+) pair is then separated as a result of an electron in VB being excited straight into the CB. The reduction process uses the electrons produced by photosynthesis, whereas the oxidation process uses up the holes. Figure 9.6 depicts the process of photocatalytic water splitting for the creation of hydrogen. Six main steps comprise photocatalysis: the first stage is light harvesting; the second is charge excitation; the third and fourth stages are charge separation and transfer; and the fifth and sixth stages are surface catalytic reactions. At the beginning of photocatalysis, light with an energy larger than or equal to the photocatalyst's bandgap is irradiated. VB and CB, which are divided from one another by a bandgap energy, make up the semiconductor of a photocatalyst in most cases.

When exposed to the proper photon excitation, the photocatalyst produces electronic transitions and e/h pairs (Equation 9.1). Second, the holes in the VB are created when the charges are separated and the electrons are excited from the VB to the CB. The reactions with water that involve reduction (stage 6) and oxidation (stage 5) require electrons and holes. Water undergoes oxidation, as demonstrated by Equation 9.3, and reduction occurs when H receives an electron to generate H_2, as demonstrated by Equation 9.4. Redox reactions take place on the photocatalyst surface when the oxidation and reduction potentials are, respectively, higher and lower than the CB and VB levels.

Equations 9.3 and 9.5 show that photoexcited holes are potent oxidants that have the ability to oxidize both water and organic materials like alcohols. Thermal dissociation of water at temperatures above 2070 K can also be used to carry out the reaction, although water splitting can also be done at room temperature using a photocatalyst under light irradiation with energy greater than the bandgap energy.[30-32]

When semiconductor materials are exposed to light in photocatalytic systems for hydrogen synthesis, a series of reactions occur that produce hydrogen gas (H_2) and oxygen gas (O_2) from water. The following fundamental reactions are involved in the process of photoreforming for hydrogen production[32]:

$$Catalyst \rightarrow Catalyst\left(e^- + h^+\right) \tag{9.1}$$

$$Catalyst\left(e^- + h^+\right) \rightarrow Catalyst \tag{9.2}$$

$$H_2O + h^+ \rightarrow \frac{1}{2}O_2 + 2H_2 \tag{9.3}$$

$$2e^- + 2H^+ \rightarrow H_2 \tag{9.4}$$

$$RCH_2OH + 2h^+ \rightarrow RCHO + 2H^+ \tag{9.5}$$

One of the main difficulties to photocatalytic water splitting is the recombination of charges (e/h). Equation 9.2 illustrates how the electron-hole pairs could recombine (stage 4) with the release waste that produces heat. Consequently, it reduces the H_2 generation's effectiveness. Also, using photocatalysts to produce water splitting for H_2 synthesis is difficult because of the rapid recombination of photo-generated charge carriers, especially in pure water. Therefore, when studying photocatalytic water splitting, sacrificial reagents (methanol, ethanol, and glycerol) and electrolytes (Na_2S and KI) are typically found. CB electrons and VB holes are not oxidizing or reducing the electrolytes. Electrolytes carry ions and electrons to nearby semiconductors. They will thereby enhance the photocatalytic water-splitting events. To improve the separation of charges, the sacrificial reagent or electron donors are interacting with V_B holes. The maximum performance for photocatalytic activity can be enhanced by studying thermodynamic analysis in terms of energy, bandgap, and redox potential, since the creation of H_2 from pure water has its constraints.[32,33]

9.6 FACTORS INFLUENCING PHOTOCATALYTIC HYDROGEN PRODUCTION

9.6.1 Bandgap Energy

The bandgap, or energy band, is commonly used to describe the electrical structure of semiconductor electrodes. Because of the energy difference between VB and CB, this electronic structure can be thought of as a continuum. Typically, the acceptor's relative potential level must be lower than the semiconductor's CB in order to meet thermodynamic requirements. Thermodynamically, H_2 is created by the first hydrogenation of intermediates; photoelectrons are used to decrease H^+ to H_2 at the active site. The bandgap of TiO_2, as shown in Figure 9.7a is suitable for water splitting because the bottom part of the CB (~0.5 V versus normal hydrogen electrode [NHE] at pH 7) is more negative than that of the H_2O/H_2 redox couple (~0/41 V versus NHE at pH 7), and the top of the VB (+2.7 V versus NHE at pH 7) is more positive in comparison to the O_2/H_2O redox couple (+1.23 V versus NHE). However, because of its 3.2 eV bandgap, TiO_2 is not appropriate for exposure to visible light. Therefore, TiO_2 needs to be doped with metals, non-metals, coupling semiconductors, and practical techniques to further minimize a bandgap in order to be able to absorb visible light irradiation. Bandgap of g-C_3N_4 is 2.7 eV with most negative CB level (~1.3 V versus NHE at pH 7), similar to $BiVO_4$ and WO_3, providing its widespread use in visible light photocatalysis.[33]

For this reason, co-doping g-C_3N_4 with extra photocatalysts is often suggested to improve semiconductors' ability to absorb visible light. Additionally, co-catalysts can contribute extra to semiconductor photostability and charge separation, which could lead to a potential increase in the rate at which hydrogen is produced. According to Figure 9.7b, non-metal doping can also reduce the bandgap of g-C_3N_4 to less than 2.7 eV and *Fe* metal-doped g-C_3N_4 narrows the bandgap by 2.5 eV. Still, combining TiO_2 with g-C_3N_4 and adjusting with metals or non-metals could serve as a potential approach to increase the photocatalytic activity for the production of hydrogen.[34]

(a)

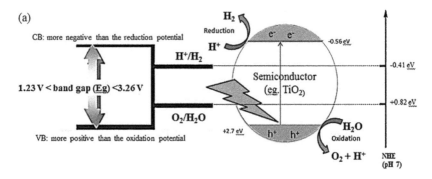

(b)

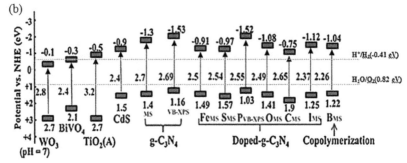

Figure 9.7 (a) Shows the basic idea behind the photocatalytic water-splitting reaction (b) and a schematic representation of the band structures of various photocatalyst types.

Source: Adapted from Wen, J., et al., *A review on g-C₃N₄-based photocatalysts.* Applied Surface Science, 2017. **391**: pp. 72–123. Copyright © 2017 with permission from Elsevier.

9.6.2 Structure/Surface Area

One surface modification technique to reduce the catalyst's bandgap is to alter the surface roughness. This has a substantial impact on the catalyst's crystalline structure and photocatalytic activity, which produces H_2. In the case of metal-modified TiO_2, the large size nanoparticles can be explained by a weaker metal-support interaction. More reactive sites will be available due to the bigger surface area, which will enhance photocatalytic activity. Reduced particle size of crystalline photocatalysts allows photo-generated electrons and holes to travel faster to the active sites of reaction on the surface of the catalyst, thereby reducing the possibility of recombination.

The synthesis method used for creating the catalyst significantly impacts the structure of the photocatalyst. When compared to untreated g-C₃N₄ under similar operating circumstances, the treated g-C₃N₄ with HNO_3 performed more effective photocatalytic activity. The formation of ultra-small pores and exfoliation on the structures are caused by the protonation of g-C₃N₄. This contributes to the photo-generated electrons and holes having a higher separation and transfer efficiency. Hydrothermal, mechanical mixing, solvothermal, sol-gel, chemical vapor deposition, and impregnation are the most often used synthesis techniques.

Different sizes, shapes, and structures of nanomaterials are also produced by varying the temperature at which the catalyst is prepared. From 6 nm at 250°C to 11 nm at 380°C, 30 to 45 nm at 550°C, over 45 nm at 650°C, and above 100 nm at 800°C, the size of crystallites increases with increasing temperature. Also, the catalyst's size may be affected by the pH that is utilized. Moreover, nanomaterial morphology, shape, and dimensionality are significantly influenced by the concentration of H^+ or OH^-.[35,36] According to Li et al., the hematite nanotubes' short charge collection distance and nanostructure consistently show enhanced water oxidation activity.[37] Additionally, in order to improve charge separation and light harvesting in photocatalytic H_2 production, Cai et al. proposed surface-disorder engineering of TiO_2 photonic crystal.[38]

9.6.3 Light Source and Intensity

By using light with energy higher than the activation threshold, photocatalytic water-splitting efficiency can be increased. In terms of the UV photon flux, the photocatalytic reaction falls into one of two phases: (i) Chemical processes consume electron-hole pairs more quickly than the recombination reaction, which is often found in fluxes from laboratory research projects that are 25 Wcm^2. (ii) If the intensity is higher, the half-order regime is used, where the rate of recombination is often dominating and has less of an impact on the rate of reaction. The catalyst's adsorption spectrum with a threshold equal to the band energy is followed by the change in reaction rate as a function of wavelength.

When light intensity was increased from 900 to 1000 Wm^2, photocatalytic hydrogen production utilizing ZnS showed a 20% improvement in photoactivity, according to Baniasadi et al.'s 2013 report. According to Tambago and Leon, the efficiency of $Cd_{0.4}Zn_{0.6}S$ for H_2 generation rises with increasing light intensity.[39]

9.6.4 Reaction Conditions

9.6.4.1 Temperature

From a thermodynamic perspective, temperature cannot stimulate photocatalysis because it does not contribute to the production of electrons and holes. In order to enhance the photocatalytic activity, temperature does, however, play a part in improving the desorption of products from the catalyst surface where the reaction rate is accelerated by temperature. Applying the same temperature to every catalyst is not the same. As a result, this parameter might be adjusted rapidly to improve photocatalytic activity.

Decreased temperature has a negative effect by reducing the rate at which H_2 is produced because the products' slower adsorption slows down the reaction. Higher electron transfers to higher energy levels in the valance band are made possible by high temperatures. Consequently, it assists the process to compete more successfully with charge-carrier recombination and promotes the production of electron holes, which may be used to start oxidation and reduction reactions, respectively. According to Boudjemaa et al., the evolution of H_2 increased with temperature, producing 59, 92, and 370 mol/g.s of H_2 at 30°C, 40°C, and 50°C, respectively. In a similar way, Pt/TiO_2 photocatalyst demonstrated that H_2 production was 4.71 mmolg^{-1} after 4 hours at 45°C, rising to 15.18 mmolg^{-1} at 55°C. The ideal temperature for photocatalytic research is said to be between 60°C and 80°C in another study from 2006.[40]

9.6.4.2 pH

As proton reduction by the photo-generated electron happens during water splitting, it can be considered that the formation of H_2 from water splitting depends upon the proton concentration, or pH of the solution. Because photoreformation needs the existence of a sacrificial organic species, this particular element is particularly significant. In weak basic pH solutions, H_2 can be produced more efficiently than in acidic or strong basic (>10) solutions, and the bandgap energy shift is affected by *pH* changes. CuO_x/TiO_2 produced the most H_2 in weak basic medium (pH 10), according to Wu et al.[41] However, pH 2 produced the least amount of H_2 due to the Cu (I) species' decreased stability on the TiO_2 surface in an acidic media. Similarly, because TiO_2 is less stable in strong acid or basic solutions, the $Si/CdS/TiO_2/Pt$ catalyst performs weakly in visible light.

According to Brahimi et al., 11 was the ideal pH for photocatalysis to produce H_2 over $CuAlO_2/TiO_2$. Catalyst bandgap changes may result from pH modification. After 4 hours, $Pt/r - TiO_2$ produced the most H_2 at pH 5.5 around 56.6 μmol, followed by pH 12 and pH 2.0.[42] Fujita's study on NiO/TiO_2 yielded a H_2 yield of 56.6 μmolg^{-1}h^{-1} and a pH value of roughly 6.6. Therefore, the mixture's pH affects the rate of H_2 evolution, with a pH near to zero-point charge being ideal.[43] Nada et al., however, reported that in an acidic environment, photosensitized TiO_2/RuO_2-MV^{2+} from a methanol water mixture produces more H_2. There is the probability that reduction of H^+ to H_2 by e$^-$ will be increased when more H^+ ions are adsorbed on the photocatalyst at an acidic pH.[44] According to a report, the H_2 evolution rate performed better in an acidic environment than in a basic one. In general, still, the photocatalytic reaction in the fundamental system offers more benefits to improve H_2 evolution, according to prior investigations.

9.6.5 Oxygen Vacancies

Oxygen vacancies can be affected by the addition of metal oxide by catalyst production, reduction, and doping.[45] Together with Ti^{3+} there are oxygen vacancies in TiO_2. More Ti^{3+} ions are created when the concentration of oxygen vacancies is greater, which results in the surface-disorder, Ti^{3+} defect state, and associated oxygen vacancies. The oxygen vacancy defect can trap and expand the lifespan of electrons because of the missing oxygen atom. On the other hand, Ti^{3+} and oxygen vacancies created the local state while electrons took over the oxygen atom's normal lattice. In the meantime, the electrons were excited to the CB of TiO_2 and the VB holes were created. At CB, a reduction process converts the H^+ to H_2. It is determined that this strategy works well to extend the lifespan of the charges without loading metals, which in turn prevents electron-hole recombination for increased H_2 generation.[46]

9.7 PHOTOCATALYTIC HYDROGEN PRODUCTION SYSTEMS

There are basically three types of photocatalytic hydrogen production systems. The first report of heterogeneous photocatalytic hydrogen production (HETPHP) systems by water splitting was made in 1972 by Fujishima and Honda. A semiconductor that serves as a light harvester and catalyst helps HETPHP systems. Five years later, another report on homogeneous photocatalytic hydrogen production (HOMPHP) systems from triethanolamine (TEOA) was published by Lehn et al. (1977). An organometallic complex is typically used as a catalyst in HOMPHP systems, while a second organic or organometallic chemical is used as a photosensitizer.[47,48] In more recent times, homogeneous catalysts and heterogeneous photosensitizer semiconductors—the system units of both HETPHP and HOMPHP systems—have been used to build hybrid photocatalytic hydrogen production (HYBPHP) systems.[49]

In recent years, analytical evaluations have examined each of the three system types independently; however, an integrated and comparative examination of the three systems has not been conducted yet. This comprehensive analysis offers an overview of the photocatalytic generation of H_2 and the effects of process factors and photocatalytic units on the rate and stability of H_2 production. This comparison was conducted using the following metrics: turnover frequency (TOF) for HOMPHP and HYBPHP systems, moles of H_2 per mole of catalyst, turnover number (TON), as described by Equations 9.6 and 9.7, and H_2 production per unit time per gram of catalyst ($mol_{H2}. g^{-1}_{cat}.time^{-1}$) for HETPHP systems. The "Quantum Yield" and "Apparent Quantum Yield" (Equations 9.8 and 9.9) serve as indicators that quantify the effectiveness of the energy capture and are used to determine the stability of the photocatalytic system and the process efficiency.

$$TOF = \frac{mol_{H2}}{mol_{cat} \cdot time} \tag{9.6}$$

$$TON = \frac{mol_{H2}}{mol_{cat}} \tag{9.7}$$

$$QY(\%) = \frac{Number\ of\ reacted\ electrons}{Number\ of\ absorbed\ photons} \times 100$$

$$= \frac{2 \cdot Number\ of\ hydrogen\ molecules}{Number\ of\ absorbed\ photons} \times 100 \tag{9.8}$$

$$AQY(\%) = \frac{Number\ of\ reacted\ electrons}{Number\ of\ incident\ photons} \times 100$$

$$= \frac{2 \cdot Number\ of\ hydrogen\ molecules}{Number\ of\ incident\ photons} \times 100 \tag{9.9}$$

The photocatalyst's stability is a crucial factor in the process design and scale-up of HETPHP systems. To support applications, aging techniques that could foresee photocatalyst lifetime are considered essential.[50]

9.8 INFLUENCE OF SYSTEM UNITS ON PHOTOCATALYTIC PERFORMANCE

HETPHP systems employ a semiconductor photocatalyst and a sacrificial agent, as was previously mentioned. In addition to serving as a photosensitizer or light harvester, the semiconductor catalyzes the production of H_2. The following processes are part of the H_2 production mechanism by HETPHP systems: (i) a pair of electrons is created when photons from a light source that have enough energy to cross the semiconductor photocatalyst's bandgap are absorbed, and the electrons are stimulated from the semiconductor's valence band to the conduction band, then (ii) the photocatalyst's surface can absorb electrons that can decrease H^+. The semiconductor's bandgap determines whether the holes oxidize the sacrificial agent or water. The mechanism of the recombination of non-reacting electrons and holes is shown in Figure 9.8.[50]

The photocatalyst conduction band potential (E_{CB}) must be more negative than the proton reduction potential ($E^+_{H/H2}$) in order to reduce protons

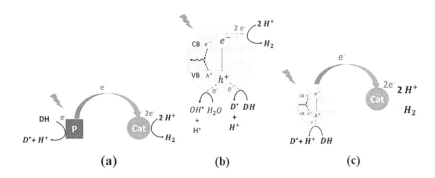

Figure 9.8 Photocatalytic H_2 production using three different systems: hybrid, homogeneous, and heterogeneous, where PS is a photosensitizer, D is an electron donor/sacrificial agent, and Cat is a catalyst.

Source: Reprinted from Corredor, J., et al., *Comprehensive review and future perspectives on the photocatalytic hydrogen production.* Journal of Chemical Technology & Biotechnology, 2019. **94**(10): pp. 3049–3063. Copyright © 2019 Society of Chemical Industry (wileyonlinelibrary.com/jctb).

to H_2. Furthermore, as shown in Figure 9.9a, the sacrificial agent oxidation potential ($E_{D/D+}$) must be less than the photocatalyst valence band potential (E_{VB}) in order for the sacrificial agent to be oxidized. If the E_{VB} is greater than the oxidation potential of water, then water can also undergo oxidation.[51]

A soluble molecular catalyst, a soluble molecular photosensitizer, and an electron donor contribute to a conventional HOMPHP system. The following steps are included in HOMPHP systems: (1) The photosensitizer absorbs photons from a light source to become excited. (2) An oxidative or reductive reaction pathway allows electron transfer to remove the excited state (Figure 9.10). It can also be thermally quenched by collisions with other constituents of the system. (2a) Reductive quenching of the stimulated photosensitizer produces the reduced state of the photosensitizer by transferring electrons from the electron donor to the excited photosensitizer.[51]

After then, the photosensitizer turns to its original state when an electron is moved to the catalyst. The first step in (2b) oxidative quenching is the transfer of electrons from the excited photosensitizer to the catalyst, which results in the oxidized form of the photosensitizer. The sacrificial agent is then oxidized before returning to its original form. Lastly, the catalyst lowers the energy barrier to proton reduction, increasing the rate at which H_2 is produced. Since two electrons and protons must couple to form H_2, there are a number of possible reaction pathways. These include two consecutive electron transfers to the catalyst, which subsequently react with two protons, or the disproportionation of two hybrid species that are momentarily formed by the catalyst's one-electron reduced form.

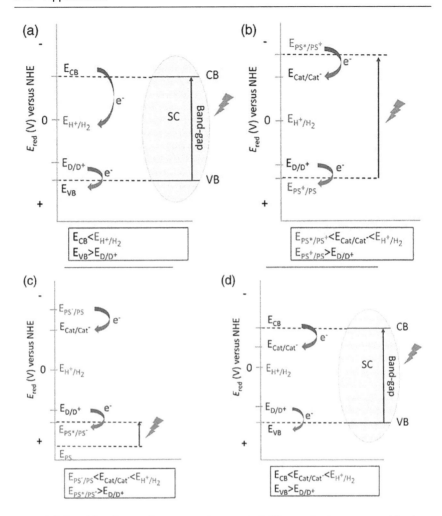

Figure 9.9 Possible schemes for various photocatalytic H_2 production systems: (a) a heterogeneous system, (b) a homogeneous system with an oxidative photosensitizer quench, (c) a homogeneous system with a reductive photosensitizer quench, and (d) a hybrid system photosensor (PS), electron donor (D), semiconductor (SC), and catalyst (Cat).

Source: Reprinted from Corredor, J., et al., *Comprehensive review and future perspectives on the photocatalytic hydrogen production.* Journal of Chemical Technology & Biotechnology, 2019. **94**(10): pp. 3049–3063. Copyright © 2019 Society of Chemical Industry (wileyonlinelibrary.com/jctb).

The reduction potential of the catalyst in its active form (E_{Cat/Cat^-}) must be greater than the oxidation potential of the excited photosensitizer (E_{PS^*/PS^+}) in the oxidative quenching case (Figure 9.9b) and the reduced photosensitizer ($E_{PS^-/PS}$) in the reductive quenching case (Figure 9.9c). Furthermore,

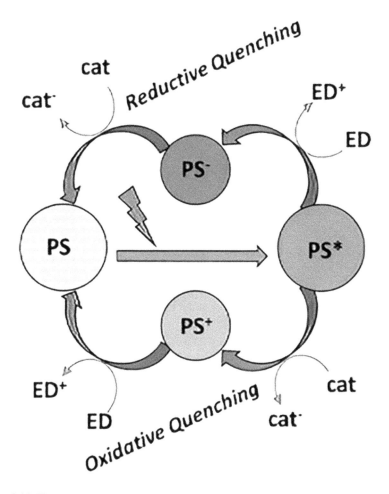

Figure 9.10 The oxidative and reductive quenching scheme. Photosensor (PS); electron donor (ED); catalyst (cat).

Source: Reprinted from Corredor, J., et al., *Comprehensive review and future perspectives on the photocatalytic hydrogen production*. Journal of Chemical Technology & Biotechnology, 2019. **94**(10): pp. 3049–3063. Copyright © 2019 Society of Chemical Industry (wileyonlinelibrary.com/jctb).

the oxidized photosensitizer's $(E_{PS^+/PS})$ reduction potential in the oxidative quenching situation and the excited photosensitizer's (E_{PS^+/PS^-}) reduction potential in the reductive case must both be greater than the electron donor oxidation potential (E_{D/D^+}). The identification of the redox potential at which the molecular catalysts are active is made more difficult by the potential for several proton reduction paths.

A sacrificial agent, a semiconductor acting as a photosensitizer, and a soluble molecular catalyst are all used in HYBPHP systems. The first and second phases of photocatalytic H_2 generation in this particular case are identical to those found in HETPHP systems. Third, the catalyst speeds up the process of proton reduction to H_2 through many reaction routes, similar to those found in homogeneous systems. This method is shown in Figure 9.8c.

Thermodynamically, E_{VB} must be more positive than the oxidation potential of the sacrificial agent ($E_{D/D}+$), and E_{CB} must be more negative than (E_{Cat/Cat^-}) as depicted in Figure 9.9d.

Consequently, for photocatalytic H_2 production to function as effectively as it can, the set of sacrificial agent, catalyst, photosensitizer, and solvent must be selected carefully.

9.8.1 Sacrificial Agent

An energy-intensive reaction occurs during the oxidation of H_2O, which supplies electrons to the photocatalytic system. Thus, sacrificial agents are used in the majority of artificial photocatalytic systems for H_2 generation. In addition to acting as hole scavengers to stop electron-hole recombination and supply electrons for proton reduction, they play the roles of electron donors to improve process efficiency. A single component serves as the sacrificial agent for HETPHP systems, while HOMPHP and HYBPHP systems can use the same compound, such as trimethylamine or ascorbic acid, for both purposes, or one compound as the electron donor and another as the proton donor. As a result, in heterogeneous systems, the phrase "sacrificial agent" is often employed, whereas in hybrid systems, the terms "electron donor" and "proton donor" are most frequently used.[50,52]

9.8.2 Catalyst

High-performing heterogeneous photocatalysts have the following functions: (i) prevent rapid electron-hole recombination; (ii) permit rapid electron and hole diffusion to the semiconductor's surface; (iii) have band potentials appropriate for the oxidation and reduction of sacrificial agents; (iv) exhibit photoactivity under visible light, which is correlated with the semiconductor's bandgap; (v) show good chemical stability; and (vi) be economically feasible.

Numerous semiconductor photocatalysts, including chalcogenides (ZnS, CdS, and CdSe),[53] metal oxides (TiO_2, Cu_2O, and ZrO_2),[54] carbonaceous materials (g-C_3N_4),[55] and solid solutions, have been studied. Nonetheless, due to its excellent stability across a broad pH range, resistance to photocorrosion, non-toxicity, and commercial availability, TiO_2 remains the most

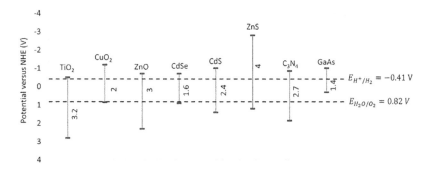

Figure 9.11 The pH 7 aqueous solution's water-splitting redox potentials and the band positions of a few semiconductor photocatalysts.

commonly used photocatalyst. The primary disadvantage of TiO_2 is the small fraction of the solar spectrum (c. 4%–8%) that it absorbs. A selection of materials used in photocatalytic H_2 production is shown in Figure 9.11. It is phenomenal that catalytic materials require a sacrificial agent because of their low valence band potential, which prevents them from oxidizing H_2O molecules.[55]

9.8.3 Photosensitizer

The most significant factors that could restrict the QY of the photocatalytic H_2 generation system are effective light harvesting, excited electron generation, and electron transfer. All of these depend on the photosensitizer's performance.

A high-performance photosensitizer must have the following characteristics: a long lifetime in excited states; high photostability to support long-term H_2 production, a very broad light absorption range, appropriate redox potentials for electron transfer from the electron donor to the photosensitizer, and good solubility in the reaction medium.[56]

In HETPHP systems, the semiconductor typically serves as both a catalyst and a photosensitizer at the same time. Yet, some methods have been suggested to increase the photoactivity in visible light, such as adding an organic dye and another semiconductor with a narrow bandgap, like CdS, but the photosensitizer and catalyst are typically two separate units in homogeneous configurations. The so-called supramolecular catalysts, which are catalysts that are chemically bonded to photosensitizers, have been used in a number of studies. The majority of the photosensitizers utilized in a HOMPHP system exhibit noticeable response.[57]

9.8.4 Solvent

The solubility of the catalyst is not a problem in HETPHP systems. Additionally, it is simple to recover the system units from the treated solution. In the process of producing H_2, H_2O is typically used as the solvent, which can also serve as a proton donor. Hydroxyl radicals produced by water splitting can also carry out the indirect oxidation of the sacrificial agent. Studies have indicated that H_2 production is not favoured when the reaction mixture has low H_2O ratios.[58]

The choice of solvent in HOMPHP and HYBPHP systems is determined by the characteristics of the photosensitizer and molecular catalyst. Certain organic solvents, including acetonitrile, acetone, dimethylformamide (DMF), and tetrahydrofuran (THF), can dissolve some of them.[59] The solubility issue can be solved by encasing the catalyst inside micelles or cyclodextrins; this tactic has been applied to [FeFe]-hygrogenase models with low solubility.[60] The concentration of the species that react with the photosensitizer readily can be greatly reduced by the weak dissociation of the sacrificial agent caused by the organic solvents. Consequently, it is necessary to screen for the ideal volume ratio to provide the best photocatalytic performance when using mixtures of organic solvents and H_2O.[61,62]

9.9 CONCLUSION AND FUTURE PERSPECTIVES

This chapter provides a systematic discussion of recent developments in the modification of TiO_2 and g-C_3N_4 for photocatalytic H_2 production. The following factors can be the focus of future efforts to improve photocatalytic water splitting:

i. Both TiO_2 and g-C_3N_4 modification techniques have benefits and drawbacks. However, innovative ideas and engineering based on TiO_2 and g-C_3N_4 can get around these restrictions. It is expected that combining TiO_2 and g-C_3N_4 with other semiconductors using Z-scheme construction will result in photocatalysts for water splitting that are more effective. Z-scheme performance is still low, but in the near future, it is expected to have great impact on catalyst innovation.

ii. It is commonly known that operational factors like pH and temperature can also increase the photocatalysts' efficiency. In addition, effective H_2 production is made possible by reducing the bandgap while controlling the morphology of catalyst. Thus, further studies on surface modification and catalyst synthesis are required to adjust the bandgap and maximize visible light absorption while reducing charge recombination.

iii. Because of their high semiconductor photocatalyst stability and capacity to be recovered from the treated solution, HETPHP systems are the most compatible with large-scale applications. They also typically offer longer operation times. Additionally, it can be said that HETPHP systems are appropriate for treating wastewater and simultaneously generating H_2.

iv. The only area of research on water splitting using the current sacrificial agent is its performance in producing H_2. The reason and process behind water splitting in the presence of a sacrificial agent, which produces H_2, need more attention. When sacrificial reagents are present, the gas phase water-splitting reaction in particular is essential for the production of H_2.

In conclusion, in order to enable their practical implementation, affordable photocatalytic H_2 producing systems should emphasize the use of visible light–sensitive and noble metal–free system components. Research on the prior treatment of waste effluents should be conducted with respect to the sacrificial agent. With this thorough understanding, a more efficient system and improved photocatalytic performance are possible in the future.

REFERENCES

1. Coyle, E.D., and R.A. Simmons, *Understanding the Global Energy Crisis*. Purdue University Press, 2014.
2. Newton, D.E., *World Energy Crisis: A Reference Handbook*. ABC-CLIO, 2013.
3. Preethi, V., and S. Kanmani, *Photocatalytic hydrogen production*. Materials Science in Semiconductor Processing, 2013. 16(3): pp. 561–575.
4. Gupta, N.M., *Factors affecting the efficiency of a water splitting photocatalyst: A perspective*. Renewable and Sustainable Energy Reviews, 2017. 71: pp. 585–601.
5. Dubey, P.K., et al., *Synthesis of reduced graphene oxide–TiO₂ nanoparticle composite systems and its application in hydrogen production*. International Journal of Hydrogen Energy, 2014. 39(29): pp. 16282–16292.
6. Tee, S.Y., et al., *Recent progress in energy-driven water splitting*. Advanced Science, 2017. 4(5): p. 1600337.
7. Muradov, N.Z., and T.N. Veziroğlu, *"Green" path from fossil-based to hydrogen economy: An overview of carbon-neutral technologies*. International Journal of Hydrogen Energy, 2008. 33(23): pp. 6804–6839.
8. Ahmad, H., et al., *Hydrogen from photo-catalytic water splitting process: A review*. Renewable and Sustainable Energy Reviews, 2015. 43: pp. 599–610.
9. Acar, C., and I. Dincer, *Impact assessment and efficiency evaluation of hydrogen production methods*. International Journal of Energy Research, 2015. 39(13): pp. 1757–1768.

10. Abe, R., *Recent progress on photocatalytic and photoelectrochemical water splitting under visible light irradiation.* Journal of Photochemistry and Photobiology C: Photochemistry Reviews, 2010. **11**(4): pp. 179–209.

11. Shi, N., et al., *Artificial chloroplast: Au/chloroplast-morph-TiO$_2$ with fast electron transfer and enhanced photocatalytic activity.* International Journal of Hydrogen Energy, 2014. **39**(11): pp. 5617–5624.

12. Johar, M.A., et al., *Photocatalysis and bandgap engineering using ZnO nanocomposites.* Advances in Materials Science and Engineering, 2015. **2015**(6):1–22.

13. Chen, X., et al., *Enhanced activity of mesoporous Nb$_2$O$_5$ for photocatalytic hydrogen production.* Applied Surface Science, 2007. **253**(20): pp. 8500–8506.

14. Takahara, Y., et al., *Mesoporous tantalum oxide. 1. Characterization and photocatalytic activity for the overall water decomposition.* Chemistry of Materials, 2001. **13**(4): pp. 1194–1199.

15. Kudo, A., et al., *Photocatalytic O$_2$ evolution under visible light irradiation on BiVO$_4$ in aqueous AgNO$_3$ solution.* Catalysis Letters, 1998. **53**(3–4): pp. 229–230.

16. Lin, Y., et al., *Semiconductor nanostructure-based photoelectrochemical water splitting: A brief review.* Chemical Physics Letters, 2011. **507**(4–6): pp. 209–215.

17. Hochbaum, A.I., and P. Yang, *Semiconductor nanowires for energy conversion.* Chemical Reviews, 2010. **110**(1): pp. 527–546.

18. Kim, J.Y., et al., *A stable and efficient hematite photoanode in a neutral electrolyte for solar water splitting: Towards stability engineering.* Advanced Energy Materials, 2014. **4**(13): p. 1400476.

19. Szabó-Bárdos, E., H. Czili, and A. Horváth, *Photocatalytic oxidation of oxalic acid enhanced by silver deposition on a TiO$_2$ surface.* Journal of Photochemistry and Photobiology A: Chemistry, 2003. **154**(2–3): pp. 195–201.

20. Kamat, P.V., M. Flumiani, and A. Dawson, *Metal-metal and metal-semiconductor composite nanoclusters.* Colloids and Surfaces A: Physicochemical and Engineering Aspects, 2002. **202**(2–3): pp. 269–279.

21. Yan, S., et al., *Organic-inorganic composite photocatalyst of gC$_3$N$_4$ and TaON with improved visible light photocatalytic activities.* Dalton Transactions, 2010. **39**(6): pp. 1488–1491.

22. Yi, Z., et al., *An orthophosphate semiconductor with photooxidation properties under visible-light irradiation.* Nature Materials, 2010. **9**(7): pp. 559–564.

23. Wang, C., et al., *Efficient hydrogen production by photocatalytic water splitting using N-doped TiO$_2$ film.* Applied Surface Science, 2013. **283**: pp. 188–192.

24. Johnson, K.A., and N. Ashcroft, *Corrections to density-functional theory bandgaps.* Physical Review B, 1998. **58**(23): p. 15548.

25. Reynal, A., et al., *Distance dependent charge separation and recombination in semiconductor/molecular catalyst systems for water splitting.* Chemical Communications, 2014. **50**(84): pp. 12768–12771.

26. Bhatt, M.D., and J.S. Lee, *Nanomaterials for photocatalytic hydrogen production: From theoretical perspectives.* RSC Advances, 2017. **7**(55): pp. 34875–34885.

27. Deb, S.K., *Opportunities and challenges in science and technology of WO₃ for electrochromic and related applications.* Solar Energy Materials and Solar Cells, 2008. **92**(2): pp. 245–258.

28. Deb, S., *Optical and photoelectric properties and colour centres in thin films of tungsten oxide.* Philosophical Magazine, 1973. **27**(4): pp. 801–822.

29. Novoselov, K.S., et al., *Electric field effect in atomically thin carbon films.* Science, 2004. **306**(5696): pp. 666–669.

30. Fajrina, N., and M. Tahir, *A critical review in strategies to improve photocatalytic water splitting towards hydrogen production.* International Journal of Hydrogen Energy, 2019. **44**(2): pp. 540–577.

31. Wen, J., et al., *A review on g-C₃N₄-based photocatalysts.* Applied Surface Science, 2017. **391**: pp. 72–123.

32. Etacheri, V., et al., *Visible-light activation of TiO₂ photocatalysts: Advances in theory and experiments.* Journal of Photochemistry and Photobiology C: Photochemistry Reviews, 2015. **25**: pp. 1–29.

33. Tahir, M., and N.S. Amin, *Advances in visible light responsive titanium oxide-based photocatalysts for CO₂ conversion to hydrocarbon fuels.* Energy Conversion and Management, 2013. **76**: pp. 194–214.

34. Dong, G., K. Zhao, and L. Zhang, *Carbon self-doping induced high electronic conductivity and photoreactivity of g-C₃N₄.* Chemical Communications, 2012. **48**(49): pp. 6178–6180.

35. Han, Q., et al., *One-step preparation of iodine-doped graphitic carbon nitride nanosheets as efficient photocatalysts for visible light water splitting.* Journal of Materials Chemistry A, 2015. **3**(8): pp. 4612–4619.

36. Wang, T., and J. Gong, *Single-crystal semiconductors with narrow bandgaps for solar water splitting.* Angewandte Chemie International Edition, 2015. **54**(37): pp. 10718–10732.

37. Li, C., et al., *Surviving high-temperature calcination: ZrO₂-induced hematite nanotubes for photoelectrochemical water oxidation.* Angewandte Chemie, 2017. **129**(15): pp. 4214–4219.

38. Cai, J., et al., *Synergetic enhancement of light harvesting and charge separation over surface-disorder-engineered TiO₂ photonic crystals.* Chem, 2017. **2**(6): pp. 877–892.

39. Tambago, H.M.G., and R.L. de Leon, *Intrinsic kinetic modeling of hydrogen production by photocatalytic water splitting using cadmium zinc sulfide catalyst.* International Journal of Chemical Engineering and Applications, 2015. **6**(4): pp. 220–227.

40. Boudjemaa, A., et al., *Fe₂O₃/carbon spheres for efficient photo-catalytic hydrogen production from water and under visible light irradiation.* Solar Energy Materials and Solar Cells, 2015. **140**: pp. 405–411.

41. Wu, Y., G. Lu, and S. Li, *The role of Cu (I) species for photocatalytic hydrogen generation over CuOₓ/TiO₂.* Catalysis Letters, 2009. **133**: pp. 97–105.

42. Brahimi, R., et al., *CuAlO₂/TiO₂ heterojunction applied to visible light H₂ production.* Journal of Photochemistry and Photobiology A: Chemistry, 2007. **186**(2–3): pp. 242–247.

43. Fujita, S.-I., et al., *Photocatalytic hydrogen production from aqueous glycerol solution using NiO/TiO₂ catalysts: Effects of preparation and reaction conditions.* Applied Catalysis B: Environmental, 2016. **181**: pp. 818–824.

44. Nada, A., et al., *Enhancement of photocatalytic hydrogen production rate using photosensitized TiO$_2$/RuO$_2$-MV^{2+}*. International Journal of Hydrogen Energy, 2008. **33**(13): pp. 3264–3269.
45. Lu, J., et al., *In situ synthesis of mesoporous C-doped TiO$_2$ single crystal with oxygen vacancy and its enhanced sunlight photocatalytic properties*. Dyes and Pigments, 2017. **144**: pp. 203–211.
46. Zhang, D., et al., *Enhanced photocatalytic hydrogen evolution activity of carbon and nitrogen self-doped TiO$_2$ hollow sphere with the creation of oxygen vacancy and Ti^{3+}*. Materials Today Energy, 2018. **10**: pp. 132–140.
47. Cha, G., et al., *Double-side co-catalytic activation of anodic TiO$_2$ nanotube membranes with sputter-coated pt for photocatalytic H$_2$ generation from water/methanol mixtures*. Chemistry–An Asian Journal, 2017. **12**(3): pp. 314–323.
48. Wrighton, M.S., et al., *Photo-assisted electrolysis of water by irradiation of a titanium dioxide electrode*. Proceedings of the National Academy of Sciences, 1975. **72**(4): pp. 1518–1522.
49. Wang, M., et al., *Integration of organometallic complexes with semiconductors and other nanomaterials for photocatalytic H$_2$ production*. Coordination Chemistry Reviews, 2015. **287**: pp. 1–14.
50. Corredor, J., et al., *Comprehensive review and future perspectives on the photocatalytic hydrogen production*. Journal of Chemical Technology & Biotechnology, 2019. **94**(10): pp. 3049–3063.
51. Sabur, M.A., et al., *Temperature-dependent infrared and calorimetric studies on arsenicals adsorption from solution to hematite nanoparticles*. Langmuir, 2015. **31**(9): pp. 2749–2760.
52. Pellegrino, F., et al., *The role of surface texture on the photocatalytic H$_2$ production on TiO$_2$*. Catalysts, 2019. **9**(1): p. 32.
53. Oros-Ruiz, S., et al., *Comparative activity of CdS nanofibers superficially modified by Au, Cu, and Ni nanoparticles as co-catalysts for photocatalytic hydrogen production under visible light*. Journal of Chemical Technology & Biotechnology, 2016. **91**(8): pp. 2205–2210.
54. García-Mendoza, C., et al., *Synthesis of Bi$_2$S$_3$ nanorods supported on ZrO$_2$ semiconductor as an efficient photocatalyst for hydrogen production under UV and visible light*. Journal of Chemical Technology & Biotechnology, 2017. **92**(7): pp. 1503–1510.
55. Cao, S., and J. Yu, *g-C$_3$N$_4$-based photocatalysts for hydrogen generation*. Journal of Physical Chemistry Letters, 2014. **5**(12): pp. 2101–2107.
56. Jin, Z., et al., *Improved quantum yield for photocatalytic hydrogen generation under visible light irradiation over eosin sensitized TiO$_2$—investigation of different noble metal loading*. Journal of Molecular Catalysis A: Chemical, 2006. **259**(1–2): pp. 275–280.
57. Yuan, Y.-J., et al., *Metal-complex chromophores for solar hydrogen generation*. Chemical Society Reviews, 2017. **46**(3): pp. 603–631.
58. Melián, E.P., et al., *Study of the photocatalytic activity of Pt-modified commercial TiO$_2$ for hydrogen production in the presence of common organic sacrificial agents*. Applied Catalysis A: General, 2016. **518**: pp. 189–197.
59. McLaughlin, M.P., et al., *A stable molecular nickel catalyst for the homogeneous photogeneration of hydrogen in aqueous solution*. Chemical Communications, 2011. **47**(28): pp. 7989–7991.

60. Berardi, S., et al., *Molecular artificial photosynthesis*. Chemical Society Reviews, 2014. **43**(22): pp. 7501–7519.

61. Na, Y., et al., *Visible light–driven electron transfer and hydrogen generation catalyzed by bioinspired [2Fe₂S] complexes*. Inorganic Chemistry, 2008. **47**(7): pp. 2805–2810.

62. Das, A., et al., *Photogeneration of hydrogen from water using CdSe nanocrystals demonstrating the importance of surface exchange*. Proceedings of the National Academy of Sciences, 2013. **110**(42): pp. 16716–16723.

Chapter 10

Future Prospects of Photocatalysis

ABSTRACT

This chapter describes the future prospects of photocatalysis wastewater treatment which is an eco-friendly and practical method for managing water pollution and promoting a cleaner, healthier environment by considering certain environmental sustainability factors. The goal of ongoing research and development should be to increase the sustainability, economic viability, and applicability of photocatalysis to a wider spectrum of pollutants and water sources. An innovative and environmentally friendly method of treating wastewater is photocatalysis. However, adherence to pertinent regulatory frameworks is crucial to ensuring the technology's safe and efficient application. Photocatalytic wastewater treatment may considerably enhance water quality and environmental protection while adhering to moral and ethical obligations provided it complies with the applicable regulations and rules. One of the most ecologically responsible ways to combat the issue of global warming brought on by the emission of extremely high amounts of CO_2 is through photocatalytic processes. Future studies and commercialization of solar-driven CO_2 reduction are anticipated to concentrate on the creation of integrated systems, such as photovoltaic-based photoelectrocatalysis and photovoltaic-based electrocatalysis CO_2 reduction reaction, in order to guarantee high solar-to-full efficiency and low operating costs.

10.1 COMMERCIALIZATION OF PHOTOCATALYSIS FOR WASTEWATER TREATMENT

In the environment, water pollution is a very serious issue to solve. It is crucial to establish clean and sustainable teleology in order to deal with significant environmental contaminants. The use of photocatalytic reactors

DOI: 10.1201/9781003403357-12

for organic pollutants degradation has been widely documented in the recent decade.[1] The fundamental technological difficulty for scientists when it comes to photocatalysis application in environmental remediation is always efficiency. The improved photocatalytic efficiency is always dependent upon material development. However, photoreactor design lags behind photocatalyst development, limiting the broad application of photocatalysis technology for environmental cleanup. Nanoparticle separation, restriction on mass transfer, and photonic efficiency have always been issues that limit photoreactor high photocatalytic efficiency in wastewater treatment.[2]

Photocatalytic reactors are an emerging and new way to purify and treat wastewater. The reactors use the photocatalyst for the degradation of polluted particles in air and water and remove it through a method called photocatalysis.[3] Different kinds of reactors are used for this particular purpose which degrades the pollutants from water by using photocatalysis application. The main thing used is photocatalysis materials including ZnO, TiO_2, and WO_3.[1] The working part of a photocatalytic reactor uses light energy i.e. ultraviolet (UV) light. When light is exposed through photocatalyst, an electron-hole pair is generated, which creates recombination and high species. It breaks down the water contamination into radicals of hydroxyl (OH). These radicals separate the contamination from affected organic compounds. It decreases the harmful effect in the environment and makes it simple to use. To partially reduce total organic carbon, macromolecular organic molecules will divide into pieces, boost biodegradability, and lower the toxicity of generated water, and photocatalysis is recommended to be integrated with other treatment procedures, such as biological therapies.[1,3]

At the commercial level, the first and foremost goal is always to maintain environmental sustainability. The manufacturing of a photocatalytic reactor at a commercial level has an average budget prospect that every middle industry can use it. People must inform about photocatalysis and photocatalytic reactors for wastewater treatment and tell about future prospects. Standard operating procedures and rules for using it must be developed that pass through government in a legal way. A flowchart of all points is shown in Figure 10.1. Approximately 300 to 400 million tonnes of organic pollutants that have not been cleaned are created annually, which causes water contamination issues, particularly close to industrial locations.[1] Most nations have implemented stringent restrictions to reduce environmental pollution in order to address this problem. Additionally, it draws scientists' focus to carefully examine the most advanced technologies in this field of study in the hopes of reducing environmental pollution and enhancing environmental wellness.[4]

10.1.1 Environmental Sustainability

Environmental sustainability is a very important topic that affects the life of human beings through the environment. It is a crucial part of wastewater treatment methods and emerging technology. Photocatalysis is the most

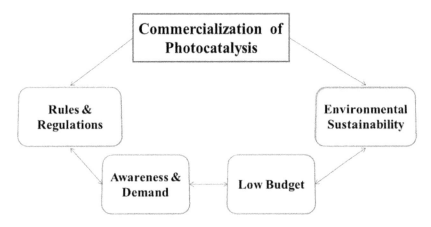

Figure 10.1 Commercialization of photocatalysis.

usable method that degrades contaminants from water and makes it useful. It increases its efficiency and validity for the long term. In resources of consumption, photocatalysis uses a catalyst which has a less hazardous effect on the environment; it should be nature-friendly and energy efficient. The form of energy used in photocatalysis is mostly solar energy. Solar energy is available and free of cost. It fulfils the requirements and has no toxic effects.[5] In a reaction, the by-products made during its process must be non-hazardous to both the environment and human health.[6] When these by-products react with water, most of them form acids that destroy the abyssal life. The other main concern is the method through which mostly effective by-products are produced that are used in other reactions.[7] Photocatalysis requires minimal water consumption and is easy to manage. Large-scale photocatalysis wastewater should not be included in sea or oceans. It has the potential to affect the wildlife. Understanding the photocatalysis process's total environmental effect from raw material extraction to the last stage of the treatment process requires doing a life cycle study of the method. Life cycle assessment aids in identifying opportunities for development and directs choices towards more sustainable methods. Photocatalysis is also a process that has long-term duration and performance.[8,9]

10.1.2 Rules and Regulations

At a commercial level, almost all countries have rules and regulations for cleaning wastewater and better use of scientific phenomenon. In the United States, the Environmental Protection Agency works for treatment of water to save human life and the environment. This agency introduced the Clean Water Act, and every company must follow its guidelines. It requires permits and has specific ranges and limits for specific pollutants.[10] In European countries, the European Union water framework directive sets limits for the use of chemicals, and an approval from government is needed to perform experiments on the commercial level. Many countries set standards that everyone must meet or their project could be cancelled. Some parameters that are highlighted are regarding wastewater treatment in a photocatalytic reactor.[11] The energy consumed in photocatalysis for water treatment changes when the light source is changed. This method should be easy to handle and with less risks.[12]

10.1.3 Awareness and Demand

People are aware of all crises related to water and must have knowledge about the treatment of wastewater. There are so many methods for purification of water, but photocatalysis is most authentic because of its use of renewable energy sources.[13] In aquatic habitats, algal blooms can pose a serious hazard to the ecology. The effects of excessive algal growth on the environment include a drop in the level of dissolved oxygen, a reduction in the nutrients available to other species due to algae's ability to store nutrients, and a decrease in light penetration for other species due to the blooming algal biomass's ability to absorb sunlight. So, everyone must have information that water that is full of chemicals will destroy the abyssal and wildlife. When affected water is thrown into forests, it kills plants and animals. Due to less plants, human life and the environment will be damaged.[14] That is why photocatalysis is used, and in order to boost demand for and acceptance of novel technologies for environmental solutions, public awareness is a critical factor. Photocatalytic reactors have the potential to totally transform how we deal with water pollution in the context of wastewater treatment.

Education to save water and provide positive points about photocatalytic reactors is necessary. New technology is crucial in our lives.[4] It is critical at this point in time to save our environment for the sake of life. By sharing successful stories about technology, we can enhance the interest of people towards science and explain the effectiveness of technology. Through exhibitions and different events, positive results and outcomes can be shared directly. Safety precautions and the non-hazardous nature of photocatalysis and reactors can be presented. By using social media, the information related to photocatalytic reactors can be spread. These methods are the easiest, and the significance of water to life can be explained.[14]

10.1.4 Low Budget

The selection of photocatalyst is very important and will be made on its impact on the budget and environmental effect. Researchers always try their best to choose the low-budget and most effective photocatalyst that is least harmful for the environment.[15] Another way to decrease cost is to decrease the quantity of the catalyst but make sure that it does not affect the efficiency and properties of the materials. For example, as a light source, the inexpensive light-emitting diodes can be used and get the better result.[16] Solar energy can be chosen over artificial energy sources in the long term due to its lower cost of energy. When a photocatalytic reactor is manufactured, the maximum amount of water is produced with no contaminants. Enhance the quantity of photocatalytic reactors while also improving their scalability and efficiency.[17]

10.1.5 Technological Challenges

Photocatalysis is an oxidation process, and it is used for the treatment of wastewater, to purify the air, and for conversion of solar energy into a usable form. But in the whole process, there are some technological issues that can be faced.[18] These challenges include solving problems with reactor design, catalyst recovery, and the degradation of complex contaminants, as well as improving photocatalyst efficiency and extending their light absorption spectrum.[19] We also discuss the significance of reducing the environmental impact, regulating environmental conditions, and guaranteeing the long-term stability and economic viability of photocatalytic processes.[20] Understanding and exploring reactors is one of the most difficult technological issues for scientists and academics to solve. For a green and healthy environment, the causes of pollution will be minimized by using new technology with new methods. Technology challenges play a very essential role in enhancing the efficiency of technologies.[12, 21]

Figure 10.2 shows some steps of technological problems that affect the future of photocatalysis and the photocatalytic reactor. The efficiency of the reactor can always be improved. It enhances its activity and production. The catalyst which is used is easily available at low cost, and it would be stable with maximum materials with low quantity. The engineering which is used is understandable, energy efficiency must increase, and the by-products or remaining can be used or in any other way broken down to minimize its toxicity to maintain environmental safety. The by-products produced during the photocatalytic process should be decomposed, otherwise human life as well as animal and bird life would be badly affected.

It ignores significant issues with technology transfer and contributes to academic research and practical use. In this point, we critically study the development of photocatalytic water treatment research, determining the most promising applications and evaluating the viability of proposed applications.[22] For scientists and engineers who want to assist research efforts to develop industrially applicable photocatalytic water treatment methods, many approaches are suggested. A negative evaluation in different areas does not limit the sharing of photocatalysis for water treatment with other specialized applications, since

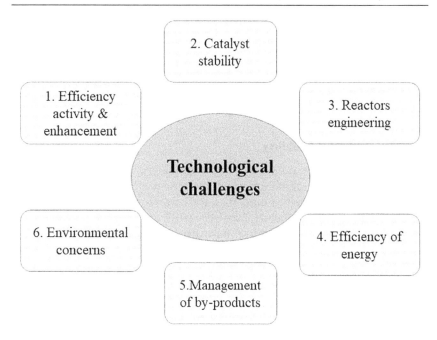

Figure 10.2 Technological challenges of photocatalysis and photocatalytic reactor.

the technique still offers significant and distinctive advantages, even though the potential may not live up to original academic expectations.[23]

10.1.5.1 Efficiency and Activity Enhancement

For the benefits of photocatalysis and the photocatalytic reactor, a lot of technological issues must be highlighted to increase its working and performance. The efficiency of the reactor completely relies on the design and working of the reactor for the purification of wastewater.[24] If titanium oxide is used as a catalyst, then as per studies it has some boundaries that it will not absorb the visible light but if light is absorbed then it will decrease the efficiency of the reactor and sample.[25] So, it is compulsory to use such a catalyst that absorbs much light with maximum efficiency. For usage of low cost and more sustainable energy, solar energy is a choice that does not have toxic by-products. To improve solar light harvesting and maximize the use of available sunlight for photocatalytic processes, researchers are actively investigating a variety of ways, including bandgap engineering and the use of plasmonic nanoparticles. The disintegration of complex and rigid contaminants that block full mineralization frequently presents a problem for photocatalysis.[26] Researchers are looking at developing multicomponent or composite photocatalysts that give beneficial interactions, resulting in more effective pollutant degradation, to increase activity to deal with harmful pollutants. By increasing the surface area of materials by doping and co-doping, it will also enhance the usage of photocatalysis.[13,26] Effective geometries and shapes of reactors have different ways to operate, such as slurry and immobilized photocatalytic reactors which were discussed in previous chapters in detail.

10.1.5.2 Catalyst Stability

Catalyst stability plays a vital role in photocatalysis. The choice of catalyst totally depends on research for which purpose it is being used. By analyzing the properties related to outcomes, the catalyst is chosen.[26] In photocatalysis, catalysts are frequently used repeatedly or continuously over long periods of time. The photocatalyst must sustain its activity and effectiveness over time without noticeably deteriorating or becoming inactive.[27] When operating for an extended period of time, a stable catalyst can maintain its effectiveness, eliminating the need for regular replacement and the resulting downtime. As long as the catalyst is stable, photocatalytic activity will be steady and predictable. Unstable catalysts may experience changes in activity or diminishing efficacy, which might provide variable therapy results. The photocatalytic method is more trustworthy for environmental cleanup and water treatment when a stable catalyst is used, since they produce consistent and reproducible outcomes.[28] The surface area, charge transport, and chemical reactivity of photocatalysts can be improved by altering their surface characteristics by surface modification or doping with certain elements. This strategy increases the activity overall by enhancing the interaction between the photocatalyst and the target pollutants. It might be difficult to go from small-scale laboratory research to extensive industrial applications.[29] In order for photocatalysis to be widely used in practical applications, it is essential to design photocatalytic reactors that are scalable, affordable, and reliable during continuous operation.[27-29]

10.1.5.3 Reactor Engineering

The following factors depend on the efficiency and working of the photocatalytic reactor:

i. *Catalyst:* The choice of catalyst is very important because any catalyst is not suitable for every reaction. It totally affects the reaction occurring in the reactor.[30] A catalyst's bandgap varies which changes their ability to absorb light and their reaction.[31]

ii. *Light source:* The light source (electromagnetic waves) and intensity work efficiently in a reactor. The intensity is directly related to the wavelength of lights.[16] For activation of the reaction, the range of absorption light must have the same frequency range with the light wavelength.[32,33]

iii. *Reactant concentration:* The rate of reaction can be increased or decreased by controlling the concentration of the reaction. Higher concentrations may result in more collisions between reactant molecules and photocatalyst surfaces, which may result in faster reaction rates up to a point.[30]

iv. *Design:* The reactor's design, size, shape, and flow pattern can all affect the interaction with photocatalyst and reactants, diminishing the reactor's efficiency.[34]

v. *Temperature:* Temperature has a footprint on the reaction rate and the adsorption of reactants on the catalyst surface. For peak performance efficiency, some photocatalytic processes may have an optimal temperature range.[33]

vi. *Solutions:* The material used in the photocatalytic reactor is in solution form. The pH of the solution can affect the surface charge of a photocatalyst as well as the degree of ionizing of the reactants, both of which influence the reaction kinetics.[35]

vii. *Mass transfer rate:* The photocatalytic process implies efficient mass transport of reactants to the catalyst surface. Improper mass transfer can result in fewer reaction rates and imperfect conversions.[26]

viii. *Recombination:* The photocatalytic method requires efficient mass transport of reactants to the catalyst surface. Improper mass transfer can result in decreased reaction rates and ineffective conversions.[30]

ix. *Degradation of contaminates:* Impurities or catalyst degradation can decrease the photocatalytic activity over time, requiring suitable reactor maintenance and catalyst regeneration actions.[27]

x. *Exposure time:* The radiation dosage and photocatalyst exposure period will have an effect on the whole reaction efficiency.[15]

xi. *Different operating conditions:* The photocatalytic reaction rates can also be affected by operating variables such as pressure, temperature, gas flow, and volume.[36]

10.1.5.4 Efficiency of Energy

Energy efficiency may be increased by minimizing side reactions and maximizing the intended photocatalytic process by proper optimization of reaction parameters, such as temperature, pH, and catalyst loading.[37] Depending on the size of the operation, the energy efficiency might change. Scaling up to industrial levels may provide new difficulties that affect total efficiency, even while laboratory-scale research may attain high energy efficiencies.[38] The photocatalyst can be made more stable and reusable by immobilizing it on a support material, which will increase its energy efficiency across a number of operation cycles.[39] The complexity and composition of the wastewater that has to be treated might have an impact on energy efficiency. Some wastewaters could have toxins that are harder to get rid of and need more energy to be successfully treated. Energy efficiency may be increased by understanding the makeup of the wastewater and designing the process accordingly.[40]

10.1.5.5 Management of By-products

It is crucial to recognize and describe the by-products created during the photocatalytic process. Analytical methods including chromatography, spectroscopy, and mass spectrometry can be used to accomplish this.[41] Understanding the effectiveness of the photocatalytic process and the type of by-products generated via it may be done with regular monitoring. There are by-products that might be poisonous or bad for the environment. To make sure that the procedure does not create new environmental concerns, a toxicology study of the treated wastewater, including the by-products, is required.[42] This procedure aids in assessing the general safety of the therapeutic process. The catalysis which is being used can be recycled and sometimes toxic by-products can be produced. So, further treatments are required to clean water and remove the toxicity of it. By-products may need to be further treated or eliminated from the treated wastewater depending on their toxicity and persistence.[43] To get rid of or to reduce the amount of toxic by-products, further treatment stages including adsorption, biological degradation, or sophisticated oxidation techniques may be used. If the by-products cannot be completely removed or degraded, their safe disposal must be ensured. Conducting an environmental impact assessment of the treated wastewater and by-products is important to understand their potential effects on the environment.[41,43]

10.1.5.6 Environmental Concerns

Photocatalysis efficiency is influenced by the availability and power of light sources. The photocatalysts may require large quantities of artificial light to function, which might result in a rise in greenhouse gas (GHG) emissions and participate in climate change. The amount of water needed for the treatment process might be large, depending on the design and manufacturing of the photocatalytic reactor. This can lead to questions regarding water supply and sustainability in areas where there is a water shortage.[44]

10.1.5.7 Research Problems and Solutions

In order to determine the viability and scalability of the technology, efforts are being made to show the practical use of photocatalytic reactors at pilot size or in actual wastewater treatment facilities.[45] In order to comprehend reaction kinetics and improve operating conditions for increased photocatalytic efficiency, researchers are using mathematical modelling and optimization methodologies. By combining photocatalysis with membrane filtration, pollutants may be removed and separated simultaneously, improving water quality and reducing reactor fouling.[46]

10.1.6 Economic and Market Challenges

10.1.6.1 Introduction

Researchers, businesses, and governments must work together to address these commercial and economic concerns. Unlocking the full potential of this promising technology in numerous industries requires ongoing research and development to boost productivity, save costs, and optimize photocatalytic systems.[47] Additionally, promoting photocatalysis and teaching prospective users about its advantages might increase its acceptance and market penetration. Despite the fact that basic photocatalytic research has received a lot of attention, there is still some distance between laboratory and commercial use.[48] Laboratory tests can only demonstrate the viability and mechanism of the photocatalytic system; they cannot account for costs, catalyst recycling, energy consumption, environmental protection, or other difficulties.[49] However, there are a number of unpredictable elements in the real manufacturing process in the case of increasing industrial use, and the catalyst preparation conditions will not be as stable and controllable as in the laboratory.

To realize the industrial use of photocatalytic systems, therefore, large-scale preparation procedures that are affordable, practical, and stable must be developed.[50] For production, the first thing to be observed is its economical demand and what kind of challenges must be faced for any material formation. Figure 10.3 illustrates the large initial investment required for any

Figure 10.3 Economic and market challenges of photocatalysis and photocatalytic reactor.

project that also considers the economics. It depends on how well it works in a team or on its own initiative. Furthermore, because a scientist has to verify the one-time cost or lifetime of photocatalysis, operating cost is also very important. To ensure more efficacy and better outcomes, the photocatalytic system has to be compared against other technologies. Market constraints force physicists to do experimental research, yet the needs of customers may vary, while people should have appropriate advice and understanding of the benefits and uses of reactors.

10.1.6.2 High Investment

Indeed, a major commercial obstacle for photocatalysis and photocatalytic reactors is high investment. The high initial expenses associated with using these technologies may discourage corporations and other industries from doing so. When it comes to photocatalysis and photocatalytic reactors, costly materials, complex reactor layouts, and specialized machinery are frequently used. A photocatalytic system's initial capital cost might be prohibitive, especially for small and medium-sized organizations or businesses with limited financial means. It may be necessary to make major adjustments and alterations in order to integrate photocatalytic reactors into currently used industrial processes or infrastructure.[47] It can be expensive and time-consuming to adapt present systems to accept photocatalysis technology. A photocatalytic technology often goes through substantial research and development before it can be considered commercially viable. Costs for laboratory testing, prototyping, and optimization are incurred at this period. These costs can be substantial, particularly for start-ups and smaller businesses. Competition for photocatalysis comes from advances in technology that could be easier to adopt. Photocatalytic systems may find it difficult to compete with these well-known alternatives. Research and development (R&D) activities can be supported by government grants, research partnerships, and financing from the public and private sectors, which can enhance photocatalytic technologies and lower their development costs. Companies that embrace environmentally friendly technology like photocatalysis may get incentives, subsidies, or tax breaks from governments and regulatory agencies. These steps can reduce the cost of initial investment.[51]

10.1.6.3 Operational Cost and Management

When using photocatalysis and photocatalytic reactors, operational cost and management are important factors to take into account. Although these technologies have many advantages for the environment and the economy, their effective implementation depends on their efficacy and affordability. To guarantee the stability and effectiveness of photocatalytic systems over the long term, routine maintenance is necessary.[52] Cleaning, replacing worn-out parts, and examining the integrity of the system are some examples of maintenance procedures. The entire cost of operations may be impacted by maintenance planning and budgeting. Personnel in charge of running and maintaining photocatalytic systems must have adequate training. Competent personnel can manage the machinery with ease, diagnose problems, and avoid expensive mistakes. Photocatalysis and photocatalytic reactors can become economically viable solutions for a variety of environmental and industrial applications by carefully regulating operating costs and putting in place effective maintenance and operational practices. For widespread acceptance and long-term success, ongoing research and development projects aimed at improving system performance and cutting operational costs are necessary.[53]

10.1.6.4 Scale-Up and Industry Integration

Scaling up and integrating photocatalysis and photocatalytic reactors into the industrial landscape are essential stages in transitioning these technologies from the lab to real-world uses in a variety of sectors. For photocatalytic systems to be scaled up and integrated successfully, careful planning, optimization, and cooperation between academics, business, and policymakers are necessary and studies and demonstration projects must be carried out before large-scale implementation.[54] These activities shed important light on the effectiveness, difficulties, and possibilities of photocatalysis in practical contexts.

Possible investors and business partners may become more confident after seeing examples of successful implementations. It might be challenging to incorporate photocatalysis into current business operations or societal structures.[55] To make room for photocatalytic devices, existing systems may need to be modified, adapted, or completely redesigned. For smooth integration, it is essential to comprehend industry-specific requirements. The process of scaling up may be sped up by facilitating technology transfer and knowledge sharing between academics, developers, and industries. Rapid commercialization and broader acceptance may result from open cooperation.[55]

10.1.6.5 Competition with Established Technologies

The broad adoption of photocatalysis and photocatalytic reactors faces a considerable hurdle from competition with current technologies. Numerous conventional and well-known technologies have been in use for a long time and have demonstrated their dependability, performance, and affordability. Photocatalysis has to show that it is superior to these well-established alternatives in order to achieve a competitive edge. Catalytic converters and electrostatic precipitators are some of the air pollution control technologies that are now in use.[56] These devices are accepted by regulators and have a successful track record of reducing air pollution. The market potential for photocatalysis may be increased by showcasing flexibility in a range of applications, both in terms of the pollutants treated and the sectors served. It is essential for maintaining competitiveness and developing the technology to conduct ongoing research and development to enhance photocatalytic materials, reactor designs, and processes. Alternative chemical cleaning agents and surface coatings have been utilized in sectors where self-cleaning surfaces are crucial. These goods could function satisfactorily and are easily accessible on the market.[57]

10.1.6.6 Lack of Standardization

The lack of standardization in photocatalysis and photocatalytic reactors presents serious financial obstacles to the broad use and commercialization of the technology. To assure uniformity, repeatability, and dependability across many applications and sectors, standardization refers to the creation and deployment of consistent principles, methods, and protocols. Diverse research teams or corporations could employ diverse experimental setups, photocatalytic components, and reactor designs in the absence of standardized methodologies. It might be challenging to evaluate and judge the genuine efficacy of photocatalysis due to the uneven performance and results that can emerge from this lack of consistency. Industries may find it challenging to assess various photocatalytic technologies and choose the best solution for their unique applications in the absence of standardized performance measurements and testing processes. Investment in photocatalytic systems may be reluctant as a result of this uncertainty. Regulatory uncertainties and delays in getting permissions for the use of photocatalytic systems in different applications may be brought on by the absence of standardized data and performance criteria. This may hinder product uptake and market entrance.[58]

10.1.6.7 Perception and Awareness

The perception and understanding of the financial difficulties that photocatalysis and photocatalytic reactors face can have a big influence on whether or not they are adopted and accepted by the market. Despite the fact that these technologies provide potential answers to pressing business and environmental issues, a number of financial obstacles may prevent their broad adoption. For stakeholders, including businesses, investors, legislators, and the general public, the degree of knowledge and comprehension of these issues is critical in determining how eager they are to support photocatalysis. There may be a lack of understanding or awareness of photocatalysis and its potential advantages among many sectors and people.[59] Decision-makers may not be familiar with photocatalysis since it is still a relatively new technology, which might prevent them from considering it as a potential solution to a particular problem. Compared to developing technologies like photocatalysis, established technologies with a track record of success may be perceived as more dependable and financially sensible. Industries may be discouraged from looking into alternate solutions by the idea of intense rivalry. Photocatalysis and its economic potential may be better understood and perceived by industry experts, decision-makers, and investors through increased awareness generated through educational initiatives such as workshops, seminars, and instructional programmers. Collaboration between researchers, businesses, and politicians can help to encourage the use of technology as well as knowledge exchange.[60]

10.1.6.8 Market Limitations

Whereas photocatalysis has demonstrated success in fields including water purification, reducing air pollution, and self-cleaning surfaces, it might not be appropriate for all environmental and industrial problems. Its adaptability is occasionally constrained since its efficacy is frequently reliant on the particular pollutants or toxins present and the desired application. A multifaceted strategy is needed to address these market constraints, including ongoing research and development to increase efficiency and cost-effectiveness, standardization efforts, education and awareness campaigns, and cooperation between regulatory agencies, businesses, and academic institutions. Photocatalysis and photocatalytic reactors can find their place and contribute significantly to the solution of several environmental and industrial issues by proactively tackling these difficulties.[12]

10.1.7 Future Prospects and Opportunities

The capability of photocatalysis for the treatment of wastewater and drinking water is great. Future developments in photocatalytic components and reactor layouts should result in more effective and affordable water treatment solutions. This could help to approach the rising need for clean, safe water throughout the world. Particularly in metropolitan areas, photocatalytic devices have the potential to be extremely important for lowering air pollution. Photocatalysis might be integrated into urban infrastructure, such as roadways, building facades, and air purifiers, to help clean the air as cities work to improve air quality and battle the consequences of climate change.[61] A sustainable and effective method to clean up polluted places is photocatalysis. We may anticipate seeing photocatalytic technology used more frequently in the future to clean up contaminated surroundings and support ecological restoration initiatives. Their effectiveness and dependability may be improved by the creation of intelligent and self-sufficient photocatalytic devices. Without human interaction, these systems may be able to alter their functioning based on the environment and the amount of pollution present.[61]

The synthesis of advanced nanomaterials with distinctive features has been made possible by nanotechnology, which has played a crucial role in photocatalysis. Metal oxides and carbon-based nanomaterials are examples of nanoscale materials that offer greater surface area, higher light absorption, and improved catalytic activity. Titanium dioxide (TiO_2) and other conventional photocatalytic materials typically react to ultraviolet (UV) light. Given that visible light is more prevalent in sunlight, the range of applications has been increased by the discovery of visible light-responsive photocatalysts including metal sulphides and carbon nitrides. Co-catalysts and sensitizers can boost the selectivity and efficiency of photocatalytic systems. These extra parts help with charge transfer, encourage certain reactions, and broaden the spectrum of the response to various light wavelengths. By focusing and amplifying local electromagnetic fields, plasmonic materials like gold and silver nanoparticles can increase photocatalysis, resulting in better catalytic activity and faster reaction rates. The discovery process is being greatly accelerated by the use of computational techniques like density functional theory (DFT) and machine learning to anticipate and develop novel photocatalytic materials with desired features. Positive effects and enhanced performance are possible with photocatalytic composites, which are made of many materials. For instance, increasing light absorption and charge separation can be accomplished through combining semiconductor photocatalysts with plasmonic metal nanoparticles.[25]

10.1.8 Emerging Trends and Applications

Advanced nanomaterials-equipped photocatalytic reactors are being investigated for indoor air filtration in structures and other confined environments. In order to enhance indoor air quality and occupant health, these systems can efficiently break down dangerous volitile organic compounds, odors, and airborne pathogens.

Due to their widespread use, glass-ceramics, which are made by the controlled crystallization of glasses, have become an indispensable component of daily life for humans. These glass-ceramics have been studied again recently for environmental applications. Photocatalytic active glass-ceramics may be used to resolve significant environmental problems including water purification, hydrogen production, and bacterial disinfection, among others. Compared to conventional photocatalysts, photocatalytic active glass-ceramics have several benefits, including flexibility in shape and size; great physical, chemical, and thermal stability; ease of manufacture; and reproducible performance. As a result, these materials may be able to offer immediate answers to environmental issues. The notion of photocatalytic active glass-ceramics is introduced in the current review, along with its advantages over conventional photocatalysts and the underlying process of photocatalysis using glass-ceramics.[62]

The methods used to maintain the elemental life in space habitats and ships serve as a microcosm for the life cycles of the elements N, C, and O. The primary requirements for the space microenvironment are the provision of potable water, an atmosphere with 20.95% oxygen, and the ability to recycle atmospheric carbon dioxide. The preservation of the atmospheric composition and the production of nutrient-rich food may both benefit from plant development. For plant development to be supported and sustained, fertilizers must be accessible. To protect against the severe space conditions, every element from packing materials to tanks, cupboard-sized containers, and huge housing needs to be enclosed.[63]

With reasonable conditions, a straightforward procedure, and green technology, it can overcome or oxidize inorganic pollutants into harmless chemicals while converting organic pollutants in polluted water to clean water, CO_2, or some small molecules.[9,10] However, the unsteadiness of deactivated photocatalysts inside catalysts makes it subject to self-etching in this phenomenon. In order to increase photocatalytic performance, it has been under concentration for heterojunction structure building, doping, defect manufacturing, and the reestablishment of photocatalytic performance by reduction of deactivated photocatalysts recycling.[64]

10.2 INTRODUCTION TO PHOTOCATALYSIS FOR HYDROGEN GENERATION

Our energy system helped develop human civilization and enable the creation of technologies that raise our level of living. In terms of productivity, energy is just as crucial as labour, capital, and raw resources. Fossil fuels will likely run out soon because of the fact that the rate of global energy use has considerably outpaced the energy storage, and fossil fuels are typically controlled by a small number of countries throughout the world. As a result, the transportation and distribution of these fuels will take a substantial amount of time and money. Hydropower is the term for water-based energy that is then transformed into electricity. The most popular way to harness water power is to build a hydroelectric dam, where water is used to turn turbines, which are then used to power generators. The main drawback of hydropower is that it takes a long time for dams to become economical since they are so expensive to build. Large dam construction can also frequently result in significant geological as well as ecological harm. Another kind of energy is solar that is a free, limitless reserve that comes from the sun and may be used to create heat or electricity by employing solar cells or concentrators. Solar energy offers benefits over wind in addition to hydropower since the latter two require turbines by running parts that are loud and expensive to maintain. Solar energy's intermittent nature is, however, one of its biggest drawbacks. That is, a site's exposure to sunlight is considerably influenced by its location, the time of day, the season, and even the presence or absence of clouds. Deep underground fluids or rocks can be heated to provide geothermal energy. Geothermal energy, in contrast to wind and solar energy, is not erratic, making it a long-term stable energy source. The rate at which heat is normally removed from the rocks is substantially higher than the rate at which it is returned to the environment.[65]

10.2.1 Hydrogen Production

Although the aforementioned benefits should be considered, applying hydrogen technology may be subject to some restrictions. Due to its poor energy density, hydrogen requires compression to reduce its storage space in command to function as a useful fuel intended for transportation. Low volumetric energy hydrogen is often kept as a compressed gas and liquid, necessitating the use of sophisticated compression techniques. However, these procedures call for expensive machinery and a lot of energy, which raises the price of using hydrogen. Another option to compression for hydrogen storage is in the form of metal hydrides. Metal hydrides are frequently pricey. The technique is expensive and less useful since metal hydrides are frequently pricey, hefty, and have a short lifespan. The current efforts are focused on two different approaches when it comes to the use of hydrogen in road transportation. The first is to create hydrogen-powered automobiles, and the second is to create hydrogen fuel cells. The benefits of hydrogen vehicles comprise a decrease in the release of nitrogen oxides (NOx), as well as a significant decrease in the production of CO_2 and spare smog-producing contaminants. Unfortunately, the commercialization of hydrogen-powered vehicles has been hindered by a lack of hydrogen fueling infrastructure and high production costs as compared to cars powered by petroleum. These issues are actually related because consumers will not buy hydrogen cars unless there is a sufficient supply of fuel, automakers will not make cars that people will not buy, and fuel companies will not build hydrogen stations for cars that do not exist.[66]

Currently, a procedure known as "steam reforming" is used to manufacture the majority of the hydrogen in the world.[67] Since methane has the maximum hydrogen-to-carbon ratio of all the hydrocarbons employed in this process, the amount of by-products produced is kept to a minimum. The steam-methane reforming procedure typically has dual phases. One is reformation, which produces hydrogen and carbon monoxide (CO) from methane and steam when they are passed through a catalyst surface at high temperatures (700°C to 900°C) and high pressures (1.5 to 3 MPa). The shift reaction, which occurs in the second phase, is where the previous step's CO interacts with more steam to produce CO_2 and more hydrogen.

$$CH_4 + H_2O \rightarrow CO + 3H_2 \tag{10.1}$$

Coal gasification is another method for producing hydrogen that uses fossil fuels.[68] With the aid of oxygen and steam, the coal is subjected to partial oxidation in this process, which results in the production of a mixture of natural elements and other chemicals at high temperatures and pressures (5 MPa). At pressures of 1 bar and temperatures exceeding 1000°C, primarily hydrogen and CO are left. The subsequent reactions (Equations 10.2 and 10.3) can serve as a representation of the process.

$$C + \frac{1}{2}O_2 \rightarrow CO \tag{10.2}$$

$$C + H_2O \rightarrow CO + H_2 \tag{10.3}$$

Hydrogen may also be produced from biomass, such as plant and animal waste, using thermochemical and biological methods. Achievable thermochemical methods for producing hydrogen include pyrolysis[69] and gasification, while potential biological methods include biophotolysis, biological gas shift reaction, and fermentation.[70] Depending on the kind of biomass, pyrolysis involves rapidly heating it to a high temperature in non-existence of oxygen to generate various chemicals, including carbon, hydrogen, methane, CO, and CO_2. Pyrolysis is performed at temperatures of 400°C to 600°C and pressures of 0.1 to 0.5 MPa.

10.2.1.1 Hydrogen Production by Solar Energy

The three main ways that solar energy may be used to produce hydrogen are thermochemical water splitting, photo biological water splitting, and photocatalytic water splitting. The basic idea behind thermochemical water splitting is to employ concentrators to capture sunlight's heat, which could normally reach temperatures of approximately 2000°C, and then use that heat to carry out the water-splitting process in the company of a catalyst like ZnO.[71] Equations 10.4 and 10.5 depict the reactions. Even while this method seems simple, finding the right materials for heat resistance and heat management has proven to be difficult. Furthermore, the high temperature requirement necessitates the use of expansive solar concentrator systems; as a result, such a method is frequently expensive.

$$ZnO + Heat \rightarrow Zn + O_2 \tag{10.4}$$

$$Zn + H_2O + Heat \rightarrow ZnO + H_2 \tag{10.5}$$

According to the microorganisms used, the end-products produced, and the reaction processes involved, photobiological water splitting[72] may essentially be separated into two types. Water biophotolysis is the term used to describe the production of hydrogen by photosynthetic oxygenic cyanobacteria or green algae under light irradiation and anaerobic conditions, while organic biophotolysis is used to describe the production of hydrogen by photosynthetic and oxygenic bacterium under light irradiation and anaerobic conditions:

$$2H^+ + 2e^- \rightarrow H_2 \tag{10.6}$$

10.2.2 Water Splitting

The production of "clean" hydrogen through photocatalytic water splitting is another promising approach. It has the following advantages over thermo-chemical and photobiological water-splitting methods:

- A respectable solar-to-hydrogen conversion efficiency
- A low cost of production
- The aptitude to achieve single hydrogen and oxygen evolution through-out reaction
- Minor reactor systems appropriately designed for household applica-tions, creating a giant market opportunity

The following is a summary of the photocatalytic water-splitting ability to produce hydrogen.

Photocatalysis is the term used to describe the chemical reaction that results from photoirradiation in the presence of a catalyst, specifically a photocata-lyst.[73] Such a substance will aid chemical processes without being ingested or changed. Chlorophyll acts as the photocatalyst in photosynthesis, which is a well-known example of photocatalysis. The fundamental workings of photocatalysis are straightforward. An electron in the full valence band (VB) is excited into the unoccupied conduction band (CB) by light radiation with energy larger than the photocatalyst's bandgap, causing an electron-hole (e^-/ h^+) pair to form. Unless they combine to produce no net chemical reaction, these electrons and holes decrease and oxidize the corresponding chemical species on the surface of the photocatalyst. If an equal amount of electrons and holes are used in chemical reactions and recombination, the original structure of the photocatalyst is not modified.[74]

The applied bias photon to current efficiency (ABPE) and quantum effi-ciency are two terms that have been used to demonstrate the efficiency of changing solar energy. The photoresponse efficiency of a photoelectrode material under an applied voltage is often assessed using ABPE. The photo-conversion efficiency is another name for ABPE.[75] Such words cannot accu-rately describe the real photoconversion efficiency for photocatalytic water splitting due to the voltage utilized. So, efficiency is defined as follows:

$$\eta_{ABPE}\% = \left(\frac{Total power output - Electrical power output}{Light power input} \right) 100\% \qquad (10.7)$$

The photocatalyst is a key component of a photocatalytic water-splitting event. Titania (TiO_2) has up to this point been a frequently utilized photocat-alyst for photocatalytic water splitting due to its stability, non-corrosiveness, abundance, and cost-effectiveness. The energy levels are suitable to start the water-splitting process, which is more significant.[74] In other arguments, the VB is more positive than the oxidation energy level of water, whereas the CB of TiO_2 is more negative to the reduction energy level of water. TiO_2 has several benefits, but its photocatalytic water-splitting efficiency when using solar energy remains rather poor. This is mostly because of the following factors:

i.   The electrons produced by photosynthesis in the TiO_2 CB may quickly recombine with holes to release extra energy as waste heat or photons.

ii. The reverse process happens readily because the chemical reaction that turns water into hydrogen and oxygen has a significant positive energy which is known as Gibbs free energy (G = 237 kJ/mol).

iii. Because TiO_2 has a bandgap of roughly 3.2 eV, the photocatalyst can only be activated by UV light.

TiO_2's effectiveness in solar photocatalytic hydrogen synthesis is constrained by its inability to use visible light, which makes up roughly 50% of solar energy compared to UV light's about 4% contribution.[59] Figure 10.4 represents the flow of electrons and creation of holes from VB to CB. Bandgap between VB and CB has an energy hʋ along potentials. This figure represents the physics behind the water splitting through the photocatalytic mechanism.

It is challenging to perform water splitting for hydrogen generation using TiO_2 photocatalyst in pure water because photo-generated CB electrons and VB holes recombine quickly. The electron-hole separation can be improved, leading to increased quantum efficiency, by adding electron donors or

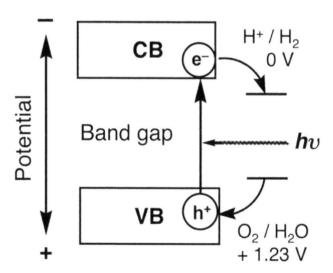

Figure 10.4 Photocatalytic water-splitting mechanism.

Source: Adapted with permission under the terms of the Creative Commons Attribution 3.0 International (CC BY 3.0) AT (https://creativecommons.org/licenses/by/3.0/at/mdpi. com). Liao, C.-H., C.-W. Huang, and J.C.J.C. Wu, *Hydrogen production from semiconductor-based photocatalysis via water splitting.* 2012. *Catalysts,* **2**(4): pp. 490–516.

sacrificial reagents to react with the photo-generated VB holes. According to reports, the organic contaminants were broken down in a way that generated hydrogen. In addition to the use of sacrificial agents, it was discovered that the inclusion of carbonate salts enhanced photocatalytic hydrogen synthesis by reducing its reverse reaction to create water.

10.2.2.1 High-Efficiency Photocatalytic Water-Splitting System

Due to TiO_2's inherent limitations, improved TiO_2 has enhanced photocatalytic activity concerning the water-splitting process, but its performance is still very low for what is needed in commercial use. As a result, work has begun to create additional possible photocatalysts to boost the effectiveness of the water-splitting reaction. The maximum of the high-efficiency photocatalysts created for photocatalytic water splitting to produce H_2 includes two or more complex components than TiO_2. The NiO-$SrTiO_3$ photocatalyst created by Domen and his whole team is one illustration. The surface of $SrTiO_3$ was initially coated with nickel oxide (NiO) as a co-catalyst, which was subsequently reduced and oxidized by hydrogen and oxygen, respectively, to produce a core of nickel shell (NiO) structure. It is thought that a co-catalyst having a core-shell structure makes it easier for electrons to go towards the photocatalysts surface, increasing photoactivity. However, the piling of NiO-Ni only enhanced the activity of $Sr_2Nb_2O_7$. A reason can be that the transfer of electrons from the CB of $Sr_2Ta_2O_7$ to that of NiO, the active site for hydrogen generation, is likely, whereas the transfer of electrons from the CB of $Sr_2Nb_2O_7$ to that of NiO is difficult because of their similar CB energy levels.[76]

LiTaO_3, NaTaO_3, KTaO_3, $MgTa_2O_6$, and $BaTa_2O_6$ show photocatalytic activity for water breakdown in the alkali and alkaline earth tantalites without the need for co-catalysts. On the other hand, $NiTa_2O_6$ did not require co-catalysts to create both H_2 and O_2 in the transition metal tantalites.[77]

10.2.2.2 Types of Water-Splitting Reactions

In general, two types of photochemical-cell reactions and photoelectrochemical (PEC)-cell reactions may be distinguished in the literature on hydrogen synthesis by photocatalytic water splitting. The water-splitting process occurs in a photochemical cell using powder photocatalyst as suspended particles in solution as shown in Figure 10.5. The majority of the photocatalytic water-splitting processes we have described up to this point are illustrations of photochemical-cell processes. In contrast, a PEC cell uses a tiny layer of photocatalyst to create a photoanode that conducts the water-splitting process in solution. The photo-generated electrons from the photoanode must be guided by an external circuit to the cathode, where hydrogen is developed. In a PEC cell, photocatalytic water splitting was originally shown by Fujishima and Honda in 1972.[78]

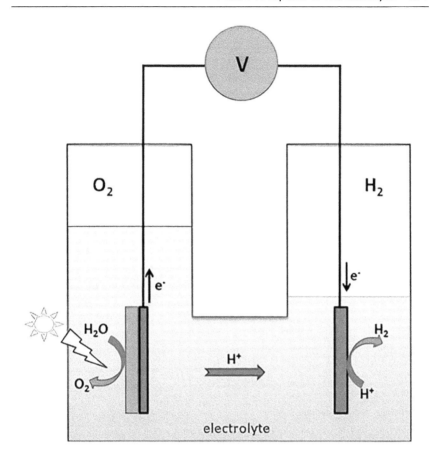

Figure 10.5 A photochemical cell.

Source: Adapted with permission under the terms of the Creative Commons Attribution 3.0 International (CC BY 3.0) AT (https://creativecommons.org/licenses/by/3.0/at/mdpi.com). Liao, C.-H., C.-W. Huang, and J.C.J.C. Wu, *Hydrogen production from semiconductor-based photocatalysis via water splitting*. 2012. *Catalysts*, 2(4): pp. 490–516.

Typically, photochemical cells benefit from a straightforward procedure because no extra film deposition or coating machinery is needed. Another benefit is that suspended photocatalyst often has more active sites for photocatalytic reaction due to its tendency to have a higher surface area per unit weight accessible for photocatalytic reaction. The benefit of a PEC cell is that the photoanode may readily produce an internal bias by combining several materials. The bias created will make it easier to separate electrons from holes and increase photocatalytic activity. Since water oxidation and reduction in PEC cell arises at different sites (electrodes), simultaneous separation of evolved O_2 and H_2 is possible, which is the biggest benefit of PEC cells.

Immediate separation of the produced O_2 and H_2 not only escapes the backward reaction of water splitting to form water again but also saves on the cost for further hydrogen separation before usage. The major benefit of PEC cells is the simultaneous separation of evolved O_2 and H_2, which is made feasible since water oxidation and reduction take place at distinct locations (electrodes) in the PEC cell. Instantaneous hydrogen separation before use not only eliminates the cost of extra hydrogen separation before use, but also prevents the reverse reaction of water splitting to generate water again. Furthermore, immediate separation makes the complete system non-toxic for scale-up and commercial operation because the combination of O_2 and H_2 is easily flammable. The H-type reactor system suggested by Anpo et al.[79] is one illustration that best illustrates the benefits of PEC cells. An H-type reactor, a photoelectrode, and a Nafion or proton-exchange membrane made up the reactor system.

Figure 10.6 illustrates how the photoelectrode and proton-exchange membrane divided the water solution within the reactor into two sections. The photoelectrode was composed of a Ti foil substrate sandwiched between a Pt cathode and a visible light-active TiO_2 photocatalyst anode that were both manufactured through sputtering. Improved charge separation and increased visible light absorption efficiency are two benefits of the dual-layer photoelectrode. By using photo voltammetry and a water-splitting reaction in an

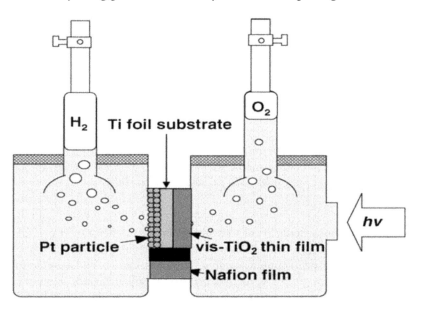

Figure 10.6 H-type reactor of photocatalytic water splitting.

Source: Adapted from Matsuoka, M., et al., *Photocatalysis for new energy production: Recent advances in photocatalytic water splitting reactions for hydrogen production.* Catalysis Today, 2007. 122(1–2): pp. 51–61. Copyright © 2007 with permission from Elsevier.

H-type reactor, the activity of the produced dual-layer photoelectrode was assessed under UV and visible light irradiation. When compared to a TiO_2-only photoelectrode, the dual-layer photoelectrode displayed increased photocurrent, which was later shown to be mostly due to the greater charge separation of the dual-layer structure. Additionally, the photocurrent findings, which showed dual-layer photoelectrodes with the maximum photoactivity, were compatible with the H_2 and O_2 yields produced from the water-splitting processes. A chemical deposition technique known as "evaporation-induced self-assembling" is used in addition to physical approaches like sputtering and electron beam–induced deposition.[80,81]

10.2.3 Technological Challenges

Molecular hydrogen may be produced from a variety of sources, including water, biomass, and fossil fuels. The energy required for recovering hydrogen from these sources must be present in abundance and be continuously available. Therefore, sustainable hydrogen production would be made possible by using the potential of renewable energy in hydrogen manufacturing methods. The primary hydrogen generation technology moving forward will be determined by the type of raw material used, whether it be fossil fuels or materials from nature.

10.2.3.1 Hydrogen Production from Fossil Fuels

Fossil fuels continue to dominate the world's hydrogen supply because of the close correlation between production costs and fuel prices, which are currently kept at reasonable levels. Hydrocarbon reforming and pyrolysis are now the most popular methods for producing hydrogen from fossil fuels. These methods virtually make it possible to produce all the hydrogen required. The method for producing hydrogen that has seen the most advancement is hydrocarbon reforming. Other reactants, such as steam or oxygen, also known as the partial oxidation reaction or steam reforming, are needed for this process in addition to hydrocarbons. Autothermic reforming is the term for the process that results from the combination of both processes and a net reaction enthalpy change of zero.[82]

10.2.4 Economic Challenges

For example, in Europe and the United States, where it contributed between 4% and 6.5% to employment and between 6% and 10% to output in 2002, the transport industry is of considerable economic importance. Production of vehicles accounts for almost 40% of these numbers. As a result, certain areas place a high political value on the transport sector's ability to compete internationally. When it comes to hydrogen, it is evident that the costs associated with hydrogen cars obviously dominate the structure of the necessary investments in hydrogen as an energy vector. If a hydrogen car is imported, it is extremely probable that the entire car, not just the hydrogen driving system, is involved. As a result, one of the major determinants of the growth of employment and gross domestic product (GDP) is the structure of the domestic automobile sector. According to the macroeconomic study for Europe, the adoption of hydrogen has favourable impacts on long-term employment, providing that the non-hydrogen technologies that will be replaced by the hydrogen technologies are competitive to a comparable extent as they are today.[83]

Recent technological development is an obstacle to offshoring production since it necessitates a greater degree of certification, more labours owing to a lower level of digitization, and more employees. The introduction of hydrogen into the energy system has a little overall effect on economic growth. The very little effects on GDP are further explained by the reality that hydrogen is only introduced in a portion of the energy system, in addition to the fact that net changes in spending patterns are minimal. It should be noted that the economic analysis is predicated on the idea that after receiving initial backing, hydrogen technologies would outperform traditional technologies in terms of cost. As a result, the major auto-producing nations face the following problem: On the one hand, there may be significant employment losses if these nations lose sales as a result of their late market entry. On the other hand, there are doubts about whether hydrogen vehicles will be successful on the market and about the possibility of losing several billion dollars because of early investments in hydrogen infrastructure and hydrogen car development. For nations without a sizable domestic automobile sector, the economic hazards of a hydrogen economy are significantly fewer than in these nations, and if the appropriate approach is adopted, these nations may experience huge increases in employment. For the plant and equipment branches, similar findings can be derived, albeit with less significant overall effects. Fuel cell cars will cause a sectoral job shift away from traditional automobile production and into, among other things, the fabricated metal, electrical, mechanical, and rubber/plastic industries. Early political action is crucial for planning for planned mass production since a trained work force and progressive increase in industrial capacity are both necessary.[84]

10.2.4.1 Global Scenario for Hydrogen

The different world energy forecasts greatly differ in how much hydrogen is anticipated to contribute to the future global energy system. The official reference scenarios do not mention hydrogen as a significant energy source. Only in situations with strong climate policies, high oil and gas costs, and also, a technological breakthrough in fuel cells and hydrogen storage, is hydrogen penetration expected. According to the most optimistic projections, hydrogen cars will account for 30% to 70% of worldwide vehicle sales, with the great majority distributed equitably across Europe, North America, and China. Oil usage would be reduced by 7 to 16 million barrels per day as a result. It is difficult to anticipate the worldwide hydrogen mix in 2050. In the case when all of the hydrogen needed up to 2050 is supplied by a single major energy source, this would have the greatest impact on available resources. There would be little effect on the depletion of fossil fuels (up to 4% of present natural gas reserves and up to 2% of hard coal reserves). In contrast, a large increase in the use of renewable energy sources specifically for the creation of hydrogen would be necessary. Up to six times as much biomass is being used globally, while the existing installed wind capacity might rise by up to 40 times. From a resource standpoint, it is impossible to draw any conclusions about whether hydrogen (from fossil fuels) is preferable to oil sands and oil shale because the primary energy required for their production while significantly higher than for the recovery of conventional oil yields more "mobility" than when it is used to produce hydrogen.[85]

10.2.5 Future Prospects

By bridging the gap between renewable energy and the transportation sector and converting biomass, wind, and solar energy into transportation fuel, hydrogen offers the opportunity to address all of the major energy policy objectives in the transport sector at once, including GHG emissions reduction, energy security, reduction of local air pollution, and noise reduction. Hydrogen may also be crucial as a way to store excess electricity generated by intermittent renewable energy sources. However, for hydrogen derived from fossil fuel, CO_2 carbon capture storage is a crucial requirement for fuels, notably coal, if total CO_2 reduction along the whole supply chain is to be achieved.

The development and improvement of the combustion engine took more than a century. Energy systems and technologies advance slowly. It will take several decades for a hydrogen infrastructure to develop and for hydrogen to significantly contribute to the fuel mix, and fuel cells and hydrogen are no different. Threats like global warming or the depletion of energy supplies might cause the market for hydrogen cars to grow quicker than was initially envisaged. When examining hydrogen's potential, it is important to compare it to its primary rivals rather than evaluating it alone because doing so might lead to inaccurate findings. The growth of the energy system as a whole should be taken into account while analyzing the introduction of hydrogen. In the long run, it appears that only electricity, combined with hydrogen, has the capacity to meet all the aforementioned needs for transport energy.[86]

Future markets for energy are uncertain to see the emergence of hydrogen and fuel cells without strong and advantageous regulatory support and incentives. Since significant industry investments are needed for vehicle manufacturing and infrastructure development well in advance of market forces, measures must be put in place and sustained long enough to raise the public's understanding of hydrogen, encourage consumers to approve of it, and provide investors with security. In the transportation industry, hydrogen will likely mostly replace oil-based fuels, with other energy sources like electricity continuing to be used. Consequently, using the phrase "hydrogen economy" might be deceptive. Hydrogen has the potential to address some of our energy issues via renewable energy, but increasing energy efficiency is equally essential for combating climate change and enhancing the security of our energy supply.

For splitting of water under UV as well as visible light irradiation, a number of semiconductor materials urbanized during the past few decades. In order to achieve high photoconversion efficiency, it has been found that photo-generated charge separation, inhibition backward reaction, and application of a significant portion of the incident energy are necessary. The activity of photocatalytic water splitting has been effectively enhanced using a variety of synthetic techniques, such as loading and/or doping metal or metal oxide particles on the photocatalyst and the creation of dye-sensitized or composite photocatalysts. These techniques work well for both limiting charge recombination and adjusting the bandgap of the material to capture more visible light. Numerous non-oxide semiconductor materials have also demonstrated increased photocatalytic water-splitting capability. The materials' stability continues to be a significant obstacle to their implementation.[87]

10.3 DEVELOPMENT OF PHOTOCATALYTIC CO_2 REACTORS

10.3.1 Introduction

The entire globe is currently experiencing an energy crisis, specifically since the start of the war between Russia and Ukraine. Despite the ongoing advancement and expanding usage of renewable energy sources, the extent of Europe's reliance on Russian gas imports has been made clear by this political chaos. Additionally, as the human population grows, so does the world's need for energy. Energy shortages and catastrophic global warming are being brought on by the overuse of fossil fuels in the energy sector. As a result, excessive CO_2 emissions from the burning of fossil fuels are worldwide problems that have a significant impact on climate change. Between 2020 and 2021, the amount of CO_2 produced globally by industrial operations, the burning of coal, and other fossil fuels grew by 6% to 36.3 Gt.[88] In 2020, the energy sector will account for 30% of all CO_2 emissions in the European Union (EU-27), followed by the transport sector (27%),

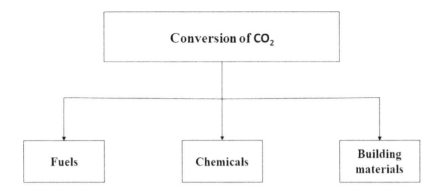

Figure 10.7 Conversion of CO_2 into beneficial compounds that include carbon.

and industry (24%), with other sectors contributing to a lesser amount. The projected total national GHG emissions in the United States in 2020 were 5.981 million metric tons of CO_2.[88]

CO_2 is a kind of GHG that traps heat and raises global temperatures. Over 60% of global warming is caused by it. To minimize CO_2 emissions to the environment, three primary strategies have been developed:

i.   Lowering CO_2 discharges at the source
ii. CO_2 capture and storage
iii. Salvaging CO_2 via its transformation into useful products that include carbon

For the synthesis of acids, alcohols, and other high-value compounds, CO_2 is a readily available, inexpensive, and non-toxic C_1 feedstock.[89] In the future, the "CO_2 economy" is predicted to be dependent on the transformation of CO_2 into fuels including methane and methanol. Additionally, CO_2 may be used as a feedstock to create construction materials including aggregates, cement, and concrete, salicylic acid for use in pharmaceuticals, urea for use in fertilizers and plastics, and polycarbonates (Figure 10.7).

10.3.2 Foundations of CO_2 Photoconversion

10.3.2.1 General Considerations

Over the past 10 years, interest in research on photocatalytic CO_2 conversion has gradually grown, although the amount of published scientific and review publications suggests that the area has not yet been sufficiently investigated. Owing to its eco-friendly and sustainable characteristics in energy and environmental challenges, photocatalysis has recently attracted more and more attention. In particular, converting CO_2 by photocatalysis into compounds with value-added properties is a promising method for addressing the energy crisis and lowering atmospheric CO_2 emissions. An essential requirement for renewable energy sources is the synthesis of fuels and chemicals from the photochemical conversion of CO_2 and earth-abundant materials without the need of chemical or exterior power inputs. Additionally, photocatalytic CO_2 conversion can be done at lower pressures and temperatures, which uses less energy. The benefit of solar-powered remediation technology is that it can be suited for independent, minimal schemes that can run even in places where there is no electrical power grid. In the process of reducing CO_2, a wide range of carbon compounds, including methanol and formic acid, as well as higher hydrocarbons like $CO_{(g)}$ and $CH_{3(g)}$ are produced.[90,91]

By utilizing clean, renewable solar energy to create value-added goods, the photocatalytic method has significant practical benefit for reducing CO_2. The majority of sunlight's wavelengths fall in the range of 250 to 2500 nm, with UV light making up just 5% of the solar spectrum, visible light 43%, and near-infrared light 52%. Adsorption, activation, as well as reaction at the catalyst surface are the standard steps for a catalytic reaction, but further processes like the absorption of pointed photons and complete process of e^- to h^+ pairs, electron-hole, are also necessary for the photocatalytic CO_2 conversion into the products of interest. The photo-generated electron-hole couples split and move to the active sites on the surface after the light is absorbed, which typically takes a few hundred picoseconds. Charge recombination can happen concurrently with charge transfer in a time window of picoseconds to tens of nanoseconds, which is on par with or quicker than the charge transfer process. Therefore, to address the issues with the bandgap energy, charge transfer, and suppression of photocatalytic activity owing to charge recombination, a new photocatalyst design is necessary.[92,93]

The effectiveness of light utilization, charge transference, and surface reaction are the major determinants of photocatalytic outputs of photocatalysts. Semiconductor materials have drawn a lot of interest and demonstrated strong CO_2 reduction capabilities. Differently doped materials, such as TiO_2, WO_3, g-C_3N_4, and perovskites, are the most often utilized materials for the photocatalytic reduction of CO_2. Efficiency is always the most crucial technological concern for scientists. The bandgap energy and charge transfer of the materials indicated earlier mostly dictate their effectiveness. These factors can be increased by doping, adding defects and co-catalysts. In accumulation to the materials mentioned earlier, ZnO, CdS, GaP, SiC, $BiVO_4$, indium oxide (In_2O_3), cerium oxide (CeO_2), and others are also being studied. TiO_2 has been extensively investigated as a photocatalyst for CO_2 reduction among all the known semiconductor materials because of its appropriate electronic and optical characteristics, low cost, availability, thermal stability, low toxicity, as well as strong photoactivity. When the energy CB is more negative as compared to the reduction potential of CO_2, which enables the flow of electrons from the photocatalyst to the CO_2, the photocatalytic reduction of CO_2 can take place. This is possible from a thermodynamic perspective. The VB energy is required to be higher than the reduction potential of the reactants in order to allow for the oxidation of water or hydrogen molecules (reducing agents), which permits the passing of holes through the catalyst to the reagents. The reduction potential of the photo-generated electrons is represented by the energy level at the bottom of the conduction band, whereas the oxidation potential of the photo-generated holes is represented by the energy level at the top of the VB.[94]

10.3.2.2 Importance of CO_2 Conversion for Addressing Climate Challenge

To stop the different negative impacts of climate change, CO_2 emissions into the atmosphere must be dramatically decreased. The use of renewable energy sources instead of fossil fuel–burning power plants, such as solar, wind, and water, has the added benefit of reducing our reliance on the world's finite supply of fossil fuels. However, the amount of energy that may be provided by renewable sources will be capped at 30% due to its sporadic nature unless methods for massive energy storage become accessible. As an alternative, CO_2 might be extracted from point sources like power plants and transformed into compounds of commercial value.[95] Potential end-products comprise methanol, CO, and formic acid. Our reliance on fossil fuels for chemical synthesis will be reduced through CO_2 conversion methods in addition to a reduction in GHG emissions. The benefit of electrochemical CO_2 reduction is that it could be a way to use extra energy from intermittent renewable sources instead of large-scale energy storage.[96]

CO_2 is reduced on the cathode of an electrolyze, while the oxygen evolution reaction occurs on the anode. The following is a list of the cathode's half-reactions during electrochemical CO_2 reduction into main products such CO, format, methane, and ethylene:

$$CO_2 + 2H^+ + 2e^- \rightarrow CO + H_2O \qquad (10.8)$$

$$CO_2 + H^+ + 2e^- \rightarrow HCOO^- \qquad (10.9)$$

$$CO_2 + 8H^+ + 8e^- \rightarrow CH_4 + 2H_2O \qquad (10.10)$$

$$2CO_2 + 12H^+ + 12e^- \rightarrow C_2H_4 + 4H_2O \qquad (10.11)$$

Since CO_2 electrolysis is essentially a fuel cell operating in reverse, there have been several lessons learnt in the development of catalysts over the past five or so centuries.[97] The establishment of effective CO_2 electrolysis techniques is applicable to the electrodes and cell designs of fuel cells. However, certain elements will be significantly different and call for distinct optimization techniques. For instance, since the cathode performance of both CO_2 electrolysis cells and low-temperature fuel cells is frequently a limiting factor, both aim to enhance sluggish cathode kinetics by creating more active catalysts. The catalyst for CO_2 reduction must, however, demonstrate great product selectivity alongside its activity to strongly favour the creation of desired products whereas suppressing undesirable reactions. Additionally, both CO_2 electrolysis cells and fuel cells require efficient product removal from the catalyst layer to prevent clogging active sites. The nature of the various goods, however, might lead to quite varied technological approaches to do this. In particular, water is produced by the oxygen reduction process in acidic fuel cells, which frequently results in problems with water management. Since the CO_2 reduction process in CO_2 electrolysis cells frequently results in the creation of both gaseous and liquid products, good gas/liquid phase separation is essential.[98]

10.3.2.3 Mechanisms of CO₂ Reduction

Since CO_2 has two double bonds between the carbon and oxygen atoms and a linear structure, it is a stable molecule ($\Delta G = -400$ kJ mol) that cannot be converted into value-added molecules without the use of a catalyst and external energy.[99] The enormous energy difference between the lowest unoccupied molecular orbital (LUMO) and the highest unoccupied molecular orbital, which is 13.7 eV, as well as the substantial electron affinity of CO_2 (-0.6 ± 0.2 eV), render it an inert molecule in addition to thermodynamic and kinetic constraints.[100] It has been demonstrated that CO_2 may be easily activated and transformed into valuable compounds via photocatalysis. The greatest technique to decrease a photocatalysts energy barrier is to adsorb CO_2, which exhibits high reactivity due to the transformation of CO_2's linear structure into a curved form. This is because when the CO_2 molecule bends, its degree of LUMO lowers. Consecutive chemical reactions are started by a one-electron transfer to create surface-bound CO_2 on a photocatalyst. The majority of these reactions include the transfer of an electron or proton, the breaking of CO bonds, and the formation of new CH bonds. The distribution of the final products might vary depending on the quantity and potential of the charge carriers involved in the chemical reaction, operating circumstances, and the type of reductant. A reducing agent for providing hydrogen is required in order to produce photocatalytic products like hydrocarbons. H_2O, H_2, CH_4, and CH_3OH are the primary reductants utilized to successfully reduce CO_2 emissions. Depending on the reductant reagent, photocatalytic CO_2 reduction follows distinct reaction routes, according to the various literatures. While the carbon-free reductants (H_2O and H_2) are thought to result in the development of C_1 products, carbon-containing species (CH_4 and CH_3OH) can also result in the formation of C_2 and C_3 products. Although little is known about how CO_2 photocatalysis really works, it is thought that the initial step is the creation of a CO_2 anion radical by removing an electron located in a photocatalysts CB. An acidic molecule with a high affinity for basic surfaces is CO_2. The interaction between CO_2 and the photocatalyst surface, which occurs via physisorbed processes, results in the creation of the anionic radical CO_2 on acidic metal oxides like TiO_2. Titania has more basic sites due to the abundance of Ti^{3+} sites on its surface, which results in the formation of a low-coordinated oxygen species.[101]

10.3.2.4 Photocatalytic CO_2 Reduction by Water

H_2O continues to be the most naturally abundant supply of hydrogen among all reductants and hole scavengers for the photocatalytic reduction of CO_2 because it is cheap and easily accessible. Hydrogen must be created simultaneously with the reduction of CO_2 by water in order to produce valuable carbon-based chemical compounds. H_2O/H_2 has a standard reduction potential of zero (when pH = 0.0) according to thermodynamics,[102] which is greater than the standard reduction potential of CO_2/. CO_2-(-1.9eV). Water reduction is a two-electron transfer process from a kinetics perspective, whereas CO_2 reduction to the most hydrogenated products requires four to eight electrons. These indicate that the synthesis of H_2 is preferable to the formation of other CO_2 reduction products. Because of this, it is preferable to monitor the volume of hydrogen and other products to determine whether water splitting competes with CO_2 reduction.[103]

$$H_2O + h_{VB}^+ \rightarrow H^+ + OH^- \qquad (10.12)$$

$$H^\wedge + e_{CB}^- \rightarrow H^\cdot \qquad (10.13)$$

$$H^\cdot + H^\cdot \rightarrow H_2 \qquad (10.14)$$

Additionally, a comparison of the polarities of H_2O and CO_2 shows that H_2O has a higher propensity to adsorb on photocatalyst surfaces like TiO_2 than CO_2, which has a lower dipole moment (0 D), owing to its polar nature (1.85 D). By placing one oxygen atom of CO_2 at the bridging oxygen vacancy (in the metal oxide photocatalysts), for example, the manufacture of photocatalysts with oxygen vacancy sites may improve the CO_2 adsorption.[104]

10.3.3 Future Prospects and Opportunities

The photocatalytic conversion of CO_2 faces a number of difficulties, including separation of charge and the formation of electron-hole pairs, activation of the semiconductor through visible light, increasing selectivity in choice of more recommended products, and enhancing photocatalysts efficiency in terms of product yields. Since there are several factors that impact the process itself, it takes tremendous work for scientists to review and evaluate all of the prior studies and offer recommendations to enhance the outcomes. For instance, various photoreactor types, photocatalyst synthesis techniques, and reaction conditions, as well as light sources, are all employed. The growing need for green energy is also a significant photochemical problem. Many viable photocatalysts have been developed in this context; however, the conversion of CO_2 is an energy-intensive and intricate process since there are numerous potential reaction pathways that might result in various products.[105] The production of high-value fuels and chemicals by the industrial-scale photocatalytic conversion of CO_2 is the next significant hurdle. As it uses naturally occurring solar energy to drive the reaction, sun-driven CO_2 reduction reaction (CO_2RR) with H_2O or H_2 is proving to be a viable strategy for the creation of numerous mono-carbon C_1 and multicarbon C_{2+} goods. As was previously said, a great deal of progress has been made thanks to the logical design of photocatalysts, the employment of various photoreactor designs, and the use of process-engineering techniques. Yet the regional sun intensity, which varies geographically, severely restricts the potential for CO_2 photoreduction on an industrial scale. Photocatalysis (PC), PEC, and photovoltaic-integrated systems are three of the most explored recent technical advancements in the field of photo-assisted CO_2 reduction.[106] Activation and reduction of CO_2 by photo-generated electrons in the conductive band and oxidation of H_2O by photo-generated holes in the VB should occur simultaneously in the photocatalytic process. This restricts the use of greater-bandgap semiconductors to offer the necessary redox possible for CO_2 reduction and H_2O corrosion. Additionally, they are ineffective at absorbing light in the visible spectrum, which severely reduces the efficiency of converting sunlight into fuel. When compared to the photocatalytic method, the photoelectrode-based PEC reduction of CO_2 can facilitate charge-carrier leave-taking and permit the use of less-bandgap photocathodes to boost light captivation, increasing the solar-to-fuel efficiency to 1% to 5%. Researchers have been concentrating their attention on integrated systems in recent years, such as photovoltaic-biased photoelectrocatalysis or electrocatalysis for CO_2RR, which straight couple a photovoltaic cell to electrochemical.[97]

10.4 SUMMARY

The demand for and implementation of photocatalytic reactors for the treatment of wastewater are sparked by public awareness. We can inspire support, trust, and industry investment in photocatalysis as a long-term solution for water pollution by educating the public about environmental issues and the advantages of this groundbreaking technology. Growing awareness may help create a cleaner, healthier environment that will benefit both the present and coming generations. It is impossible to overestimate the technical importance of catalyst stability in photocatalysis and photocatalytic reactors. Stable catalysts improve safety and environmental sustainability, save operational costs, make catalyst recovery and recycling easier, and contribute to constant and dependable photocatalytic performance. To promote the efficiency and viability of photocatalytic processes for many applications, ranging from wastewater treatment to air purification and renewable energy production, research and development funding to increase catalyst stability is crucial.

Standardization in photocatalysis can help the technique become more widely used and better able to address urgent economic and environmental problems while also enhancing its legitimacy, facilitating knowledge transfer, and encouraging larger investments. To realize their full potential and tackle urgent environmental concerns, further research, technical development, and cooperation between academics, companies, and governments will be necessary.

As a cross-disciplinary technique, photocatalytic water splitting necessitates the participation of specialists from several disciplines. Exploring possible semiconductor components and reactor designs that will produce the maximum solar-to-hydrogen efficiency requires collaboration. In order to create a low-cost and ecologically friendly water-splitting method for hydrogen generation, the development of new technologies necessitates collaboration with a solid theoretical basis for a better understanding of the hydrogen production mechanism. The usage of the majority of agents that are abundantly found in nature gives photocatalytic conversion of CO_2 to sustainable fuels an edge over other processes. The reactions happen at room temperature, which is more advantageous energetically than hot processes. The creation of novel photocatalytic materials, the efficient design of photoreactors, and the prediction of the reaction process may enhance photocatalytic activity with regards to mass transfer and light absorption by the photocatalyst.

REFERENCES

1. Alpert, D.J., et al., *Sandia National Laboratories' work in solar detoxification of hazardous wastes*. Solar Energy Materials, 1991. **24**(1–4): pp. 594–607.
2. Bekbölet, M., et al., *Photocatalytic detoxification with the thin-film fixed-bed reactor (TFFBR): Clean-up of highly polluted landfill effluents using a novel TiO_2-photocatalyst*. Solar Energy, 1996. **56**(5): pp. 455–469.
3. Butler, E.C., and A.P. Davis, *Photocatalytic oxidation in aqueous titanium dioxide suspensions: The influence of dissolved transition metals*. Journal of Photochemistry and Photobiology A: Chemistry, 1993. **70**(3): pp. 273–283.
4. Freudenhammer, H., et al., *Detoxification and recycling of wastewater by solar-catalytic treatment*. Water Science and Technology, 1997. **35**(4): pp. 149–156.
5. Guan, X., et al., *Application of titanium dioxide in arsenic removal from water: A review*. Journal of Hazardous Materials, 2012. **215**: pp. 1–16.

6. Enesca, A., *The influence of photocatalytic reactors design and operating parameters on the wastewater organic pollutants removal—A mini-review.* Catalysts, 2021. **11**(5): p. 556.

7. Ahmad, R., et al., *Photocatalytic systems as an advanced environmental remediation: Recent developments, limitations and new avenues for applications.* Journal of Environmental Chemical Engineering, 2016. **4**(4): pp. 4143–4164.

8. Guan, X., et al., *Application of titanium dioxide in arsenic removal from water: A review (vol 215, pg 1, 2012).* Journal of Hazardous Materials, 2012. **221**: pp. 303–303.

9. Haenel, A., et al., *Photocatalytic activity of TiO_2 immobilized on glass beads.* Physicochemical Problems of Mineral Processing, 2010. **45**: pp. 49–56.

10. Silva, L.I.D., et al., *Phosphorus-solubilizing microorganisms: A key to sustainable agriculture.* Agriculture, 2023. **13**(2): p. 462.

11. Bano, K., S. Kaushal, and P.P. Singh, *A review on photocatalytic degradation of hazardous pesticides using heterojunctions.* Polyhedron, 2021. **209**: p. 115465.

12. Koe, W.S., et al., *An overview of photocatalytic degradation: Photocatalysts, mechanisms, and development of photocatalytic membrane.* Environmental Science and Pollution Research, 2020. **27**: pp. 2522–2565.

13. Kümmerer, K., *Drugs in the environment: Emission of drugs, diagnostic aids and disinfectants into wastewater by hospitals in relation to other sources–a review.* Chemosphere, 2001. **45**(6–7): pp. 957–969.

14. Oller, I., S. Malato, and J. Sánchez-Pérez, *Combination of advanced oxidation processes and biological treatments for wastewater decontamination—A review.* Science of the Total Environment, 2011. **409**(20): pp. 4141–4166.

15. Qi, N., et al., *CFD modelling of hydrodynamics and degradation kinetics in an annular slurry photocatalytic reactor for wastewater treatment.* Chemical Engineering Journal, 2011. **172**(1): pp. 84–95.

16. Wang, D., et al., *Engineering and modeling perspectives on photocatalytic reactors for water treatment.* Water Research, 2021. **202**: p. 117421.

17. Ong, C., et al., *Investigation of submerged membrane photocatalytic reactor (sMPR) operating parameters during oily wastewater treatment process.* Desalination, 2014. **353**: pp. 48–56.

18. Andreozzi, R., et al., *Advanced oxidation processes (AOP) for water purification and recovery.* Catalysis Today, 1999. **53**(1): pp. 51–59.

19. Augugliaro, V., et al., *The combination of heterogeneous photocatalysis with chemical and physical operations: A tool for improving the photoprocess performance.* Journal of Photochemistry and Photobiology C: Photochemistry Reviews, 2006. **7**(4): pp. 127–144.

20. Bahnemann, D., *Photocatalytic water treatment: Solar energy applications.* Solar Energy, 2004. **77**(5): pp. 445–459.

21. Loeb, S.K., et al., *The Technology Horizon for Photocatalytic Water Treatment: Sunrise or Sunset?* ACS Publications, 2018.

22. Legrini, O., E. Oliveros, and A. Braun, *Photochemical processes for water treatment.* Chemical Reviews, 1993. **93**(2): pp. 671–698.

23. Mukherjee, P.S., and A.K. Ray, *Major challenges in the design of a large-scale photocatalytic reactor for water treatment.* Chemical Engineering &

Technology: Industrial Chemistry-Plant Equipment-Process Engineering-Biotechnology, 1999. **22**(3): pp. 253–260.

24. Mills, A., and S. Le Hunte, *An overview of semiconductor photocatalysis.* Journal of Photochemistry and Photobiology A: Chemistry, 1997. **108**(1): p. 1–35.

25. Do, Y., et al., *The effect of WO$_3$ on the photocatalytic activity of TiO$_2$.* Journal of Solid State Chemistry, 1994. **108**(1): pp. 198–201.

26. Abrahams, J., R.S. Davidson, and C.L. Morrison, *Optimization of the photocatalytic properties of titanium dioxide.* Journal of Photochemistry, 1985. **29**(3–4): pp. 353–361.

27. Sinar Mashuri, S.I., et al., *Photocatalysis for organic wastewater treatment: From the basis to current challenges for society.* Catalysts, 2020. **10**(11): p. 1260.

28. Carbonaro, S., M.N. Sugihara, and T.J. Strathmann, *Continuous-flow photocatalytic treatment of pharmaceutical micropollutants: Activity, inhibition, and deactivation of TiO$_2$ photocatalysts in wastewater effluent.* Applied Catalysis B: Environmental, 2013. **129**: pp. 1–12.

29. Klamerth, N., et al., *Degradation of emerging contaminants at low concentrations in MWTPs effluents with mild solar photo-Fenton and TiO$_2$.* Catalysis Today, 2009. **144**(1–2): pp. 124–130.

30. Cassano, A.E., and O.M. Alfano, *Reaction engineering of suspended solid heterogeneous photocatalytic reactors.* Catalysis Today, 2000. **58**(2–3): pp. 167–197.

31. Motegh, M., et al., *Photocatalytic-reactor efficiencies and simplified expressions to assess their relevance in kinetic experiments.* Chemical Engineering Journal, 2012. **207**: pp. 607–615.

32. Casado, C., et al., *Critical role of the light spectrum on the simulation of solar photocatalytic reactors.* Applied Catalysis B: Environmental, 2019. **252**: pp. 1–9.

33. Martín-Sómer, M., et al., *Influence of light distribution on the performance of photocatalytic reactors: LED vs mercury lamps.* Applied Catalysis B: Environmental, 2017. **215**: pp. 1–7.

34. Abhang, R., D. Kumar, and S. Taralkar, *Design of photocatalytic reactor for degradation of phenol in wastewater.* International Journal of Chemical Engineering and Applications, 2011. **2**(5): p. 337.

35. Khodadadian, F., et al., *Model-based optimization of a photocatalytic reactor with light-emitting diodes.* Chemical Engineering & Technology, 2016. **39**(10): pp. 1946–1954.

36. Ong, C., et al., *The impacts of various operating conditions on submerged membrane photocatalytic reactors (SMPR) for organic pollutant separation and degradation: A review.* RSC Advances, 2015. **5**(118): pp. 97335–97348.

37. Jamali, A., et al., *A batch LED reactor for the photocatalytic degradation of phenol.* Chemical Engineering and Processing: Process Intensification, 2013. **71**: pp. 43–50.

38. Gogoi, A., et al., *Occurrence and fate of emerging contaminants in water environment: A review.* Groundwater for Sustainable Development, 2018. **6**: pp. 169–180.

39. Sundar, K.P., and S. Kanmani, *Progression of photocatalytic reactors and it's comparison: A review.* Chemical Engineering Research and Design, 2020. **154**: pp. 135–150.

40. Natarajan, T.S., et al., *Energy efficient UV-LED source and TiO$_2$ nanotube array-based reactor for photocatalytic application.* Industrial & Engineering Chemistry Research, 2011. **50**(13): pp. 7753–7762.

41. Abdelraheem, W.H., et al., *Hydrothermal synthesis of photoactive nitrogen-and boron-codoped TiO$_2$ nanoparticles for the treatment of bisphenol A in wastewater: Synthesis, photocatalytic activity, degradation byproducts and reaction pathways.* Applied Catalysis B: Environmental, 2019. **241**: pp. 598–611.

42. Vincent, G., P.-M. Marquaire, and O. Zahraa, *Abatement of volatile organic compounds using an annular photocatalytic reactor: Study of gaseous acetone.* Journal of Photochemistry and Photobiology A: Chemistry, 2008. **197**(2–3): pp. 177–189.

43. Mozia, S., *Photocatalytic membrane reactors (PMRs) in water and wastewater treatment. A review.* Separation and Purification Technology, 2010. **73**(2): pp. 71–91.

44. Khan, S.H., and B. Pathak, *Zinc oxide based photocatalytic degradation of persistent pesticides: A comprehensive review.* Environmental Nanotechnology, Monitoring & Management, 2020. **13**: p. 100290.

45. Sacco, O., V. Vaiano, and D. Sannino, *Main parameters influencing the design of photocatalytic reactors for wastewater treatment: A mini review.* Journal of Chemical Technology & Biotechnology, 2020. **95**(10): pp. 2608–2618.

46. Ray, A.K., and A.A. Beenackers, *Development of a new photocatalytic reactor for water purification.* Catalysis Today, 1998. **40**(1): pp. 73–83.

47. Melchionna, M., and P. Fornasiero, *Updates on the roadmap for photocatalysis.* ACS Catalysis, 2020. **10**(10): pp. 5493–5501.

48. Wang, Y., E. Chen, and J. Tang, *Insight on reaction pathways of photocatalytic CO$_2$ conversion.* ACS Catalysis, 2022. **12**(12): pp. 7300–7316.

49. Paul Guin, J., J.A. Sullivan, and K.R. Thampi, *Challenges facing sustainable visible light induced degradation of poly-and perfluoroalkyls (PFA) in water: A critical review.* ACS Engineering Au, 2022. **2**(3): pp. 134–150.

50. Liu, Z., et al., *Photocatalytic conversion of methane: Current state of the art, challenges, and future perspectives.* ACS Environmental Au, 2023. **3**(5): pp. 252–276.

51. Peighambardoust, N.S., E. Sadeghi, and U. Aydemir, *Lead Halide Perovskite Quantum Dots for Photovoltaics and Photocatalysis: A Review.* ACS Applied Nano Materials, 2022. **5**(10): pp. 14092–14132.

52. Samadi, M., and A.Z. Moshfegh, *Recent developments of electrospinning-based photocatalysts in degradation of organic pollutants: Principles and strategies.* ACS Omega, 2022. **7**(50): pp. 45867–45881.

53. Faisal, M., et al., *A novel Ag/PANI/ZnTiO$_3$ ternary nanocomposite as a highly efficient visible-light-driven photocatalyst.* Separation and Purification Technology, 2021. **256**: p. 117847.

54. Chaudhuri, A., et al., *Scale-up of a heterogeneous photocatalytic degradation using a photochemical rotor–stator spinning disk reactor.* Organic Process Research & Development, 2022. **26**(4): pp. 1279–1288.

55. Kuckhoff, T., et al., *Photocatalytic hydrogels with a high transmission polymer network for pollutant remediation.* Chemistry of Materials, 2021. **33**(23): pp. 9131–9138.

56. Mazzanti, S., et al., *Carbon nitride thin films as all-in-one technology for photocatalysis.* ACS Catalysis, 2021. **11**(17): pp. 11109–11116.

57. Song, H., et al., *Solar-driven hydrogen production: Recent advances, challenges, and future perspectives.* ACS Energy Letters, 2022. **7**(3): pp. 1043–1065.

58. Coleman, H., et al., *Rapid loss of estrogenicity of steroid estrogens by UVA photolysis and photocatalysis over an immobilised titanium dioxide catalyst.* Water Research, 2004. **38**(14–15): pp. 3233–3240.

59. Zhong, L., and F. Haghighat, *Photocatalytic air cleaners and materials technologies—Abilities and limitations.* Building and Environment, 2015. **91**: pp. 191–203.

60. Schneider, J., et al., *Understanding TiO$_2$ photocatalysis: Mechanisms and materials.* Chemical Reviews, 2014. **114**(19): pp. 9919–9986.

61. Escobedo, S., and H. de Lasa, *Synthesis and performance of photocatalysts for photocatalytic hydrogen production: Future perspectives.* Catalysts, 2021. **11**(12): p. 1505.

62. Singh, G., M. Sharma, and R. Vaish, *Emerging trends in glass-ceramic photocatalysts.* Chemical Engineering Journal, 2021. **407**: p. 126971.

63. Do, T.-O., and S. Mohan, *Special issue on "emerging trends in TiO$_2$ photocatalysis and applications".* Catalysts, 2020. **10**(6): p. 670.

64. Singh, P., et al., *Emerging trends in photodegradation of petrochemical wastes: A review.* Environmental Science and Pollution Research, 2016. **23**: pp. 22340–22364.

65. Acar, C., I. Dincer, and C. Zamfirescu, *A review on selected heterogeneous photocatalysts for hydrogen production.* International Journal of Energy Research, 2014. **38**(15): pp. 1903–1920.

66. Dawood, F., M. Anda, and G. Shafiullah, *Hydrogen production for energy: An overview.* International Journal of Hydrogen Energy, 2020. **45**(7): pp. 3847–3869.

67. Palo, D.R., R.A. Dagle, and J.D. Holladay, *Methanol steam reforming for hydrogen production.* Chemical reviews, 2007. **107**(10): pp. 3992–4021.

68. Nowotny, J., et al., *Solar-hydrogen: Environmentally safe fuel for the future.* International Journal of Hydrogen Energy, 2005. **30**(5): pp. 521–544.

69. Czernik, S., R. Evans, and R. French, *Hydrogen from biomass-production by steam reforming of biomass pyrolysis oil.* Catalysis Today, 2007. **129**(3–4): pp. 265–268.

70. Ni, M., et al., *An overview of hydrogen production from biomass.* Fuel Processing Technology, 2006. **87**(5): pp. 461–472.

71. Steinfeld, A., *Solar hydrogen production via a two-step water-splitting thermochemical cycle based on Zn/ZnO redox reactions.* International Journal of Hydrogen Energy, 2002. **27**(6): pp. 611–619.

72. Akkerman, I., et al., *Photobiological hydrogen production: Photochemical efficiency and bioreactor design.* International Journal of Hydrogen Energy, 2002. **27**(11–12): pp. 1195–1208.

73. Das, D., and T.N. Veziroglu, *Advances in biological hydrogen production processes.* International Journal of Hydrogen Energy, 2008. **33**(21): pp. 6046–6057.
74. Guan, Y., et al., *Two-stage photo-biological production of hydrogen by marine green alga* Platymonas subcordiformis. Biochemical Engineering Journal, 2004. **19**(1): pp. 69–73.
75. Khan, S.U., M. Al-Shahry, and W.B. Ingler, Jr., *Efficient photochemical water splitting by a chemically modified n-TiO₂.* Science, 2002. **297**(5590): pp. 2243–2245.
76. Domen, K., et al., *Photocatalytic decomposition of water into hydrogen and oxygen over nickel (II) oxide-strontium titanate (SrTiO₃) powder. 1. Structure of the catalysts.* Journal of Physical Chemistry, 1986. **90**(2): pp. 292–295.
77. Midilli, A., et al., *On hydrogen and hydrogen energy strategies: I: Current status and needs.* Renewable and Sustainable Energy Reviews, 2005. **9**(3): pp. 255–271.
78. Fujishima, A., and K. Honda, *Electrochemical photolysis of water at a semiconductor electrode.* Nature, 1972. **238**(5358): pp. 37–38.
79. Matsuoka, M., et al., *Photocatalysis for new energy production: Recent advances in photocatalytic water splitting reactions for hydrogen production.* Catalysis Today, 2007. **122**(1–2): pp. 51–61.
80. Liao, C.-H., C.-W. Huang, and J.C. Wu, *Novel dual-layer photoelectrode prepared by RF magnetron sputtering for photocatalytic water splitting.* International Journal of Hydrogen Energy, 2012. **37**(16): pp. 11632–11639.
81. Huang, C.-W., et al., *Hydrogen generation from photocatalytic water splitting over TiO₂ thin film prepared by electron beam-induced deposition.* International Journal of Hydrogen Energy, 2010. **35**(21): pp. 12005–12010.
82. Iqbal, W., et al., *Assessment of wind energy potential for the production of renewable hydrogen in Sindh province of Pakistan.* Processes, 2019. **7**(4): p. 196.
83. Ishaq, H., I. Dincer, and C. Crawford, *A review on hydrogen production and utilization: Challenges and opportunities.* International Journal of Hydrogen Energy, 2022. **47**(62): pp. 26238–26264.
84. Yu, S.-C., et al., *A novel membrane reactor for separating hydrogen and oxygen in photocatalytic water splitting.* Journal of Membrane Science, 2011. **382**(1–2): pp. 291–299.
85. Hanley, E.S., J. Deane, and B.Ó. Gallachóir, *The role of hydrogen in low carbon energy futures—A review of existing perspectives.* Renewable and Sustainable Energy Reviews, 2018. **82**: pp. 3027–3045.
86. Liao, C.-H., C.-W. Huang, and J.C. Wu, *Hydrogen production from semiconductor-based photocatalysis via water splitting.* Catalysts, 2012. **2**(4): pp. 490–516.
87. Zhao, X., G. Zhang, and Z. Zhang, *TiO₂-based catalysts for photocatalytic reduction of aqueous oxyanions: State-of-the-art and future prospects.* Environment international, 2020. **136**: p. 105453.
88. Li, D., et al., *Photocatalytic CO₂ reduction over metal-organic framework-based materials.* Coordination Chemistry Reviews, 2020. **412**: p. 213262.
89. He, M., Y. Sun, and B. Han, *Green carbon science: scientific basis for integrating carbon resource processing, utilization, and recycling.* Angewandte Chemie International Edition, 2013. **52**(37): pp. 9620–9633.

90. Choi, W., et al., *Solar Energy Utilization and Photo (Electro) Catalysis for Sustainable Environment* (pp. 940–941). ACS Publications, 2022.

91. Hong, J., et al., *Photocatalytic reduction of CO_2: A brief review on product analysis and systematic methods.* Analytical Methods, 2013. 5(5): pp. 1086–1097.

92. Levinson, R., P. Berdahl, and H. Akbari, *Solar spectral optical properties of pigments—Part I: Model for deriving scattering and absorption coefficients from transmittance and reflectance measurements.* Solar Energy Materials and Solar Cells, 2005. 89(4): pp. 319–349.

93. Lu, H., Z. Wang, and L. Wang, *Photocatalytic and photoelectrochemical carbon dioxide reductions toward value-added multicarbon products.* ACS ES&T Engineering, 2021. 2(6): pp. 975–988.

94. Bratovčić, A., and V. Tomašić, *Design and development of photocatalytic systems for reduction of CO_2 into valuable chemicals and fuels.* Processes, 2023. 11(5): p. 1433.

95. Centi, G., and S. Perathoner, *Opportunities and prospects in the chemical recycling of carbon dioxide to fuels.* Catalysis Today, 2009. 148(3–4): pp. 191–205.

96. Tahir, M., and N.S. Amin, *Advances in visible light responsive titanium oxide-based photocatalysts for CO_2 conversion to hydrocarbon fuels.* Energy Conversion and Management, 2013. 76: pp. 194–214.

97. Ma, S., and P.J. Kenis, *Electrochemical conversion of CO_2 to useful chemicals: Current status, remaining challenges, and future opportunities.* Current Opinion in Chemical Engineering, 2013. 2(2): pp. 191–199.

98. Yano, H., et al., *Electrochemical reduction of CO_2 at three-phase (gas| liquid| solid) and two-phase (liquid| solid) interfaces on Ag electrodes.* Journal of Electroanalytical Chemistry, 2002. 533(1–2): pp. 113–118.

99. Ganesh, I., *Conversion of carbon dioxide into methanol—A potential liquid fuel: Fundamental challenges and opportunities (a review).* Renewable and Sustainable Energy Reviews, 2014. 31: pp. 221–257.

100. Jean, Y., F. Volatron, and J.K. Burdett, *An Introduction to Molecular Orbitals.* Oxford University Press, 1993.

101. Yuan, L., and Y.-J. Xu, *Photocatalytic conversion of CO_2 into value-added and renewable fuels.* Applied Surface Science, 2015. 342: pp. 154–167.

102. Maeda, K., *Photocatalytic water splitting using semiconductor particles: History and recent developments.* Journal of Photochemistry and Photobiology C: Photochemistry Reviews, 2011. 12(4): pp. 237–268.

103. Indrakanti, V.P., J.D. Kubicki, and H.H. Schobert, *Photoinduced activation of CO_2 on Ti-based heterogeneous catalysts: Current state, chemical physics-based insights and outlook.* Energy & Environmental Science, 2009. 2(7): pp. 745–758.

104. Kočí, K., et al., *Influence of reactor geometry on the yield of CO_2 photocatalytic reduction.* Catalysis Today, 2011. 176(1): pp. 212–214.

105. Francis, A., et al., *A review on recent developments in solar photoreactors for carbon dioxide conversion to fuels.* Journal of CO_2 Utilization, 2021. 47: p. 101515.

106. Li, X., J. Yu, and M. Jaroniec, *Hierarchical photocatalysts.* Chemical Society Reviews, 2016. 45(9): pp. 2603–2636.

Index

Note: Page numbers in *italics* indicate a figure on the corresponding page.

For Product Safety Concerns and Information please contact our EU
representative GPSR@taylorandfrancis.com
Taylor & Francis Verlag GmbH, Kaufingerstraße 24, 80331 München, Germany

www.ingramcontent.com/pod-product-compliance
Ingram Content Group UK Ltd.
Pitfield, Milton Keynes, MK11 3LW, UK
UKHW021117180425

457613UK00005B/123